P9-EMK-147

NAPOLEON'S BUTTONS

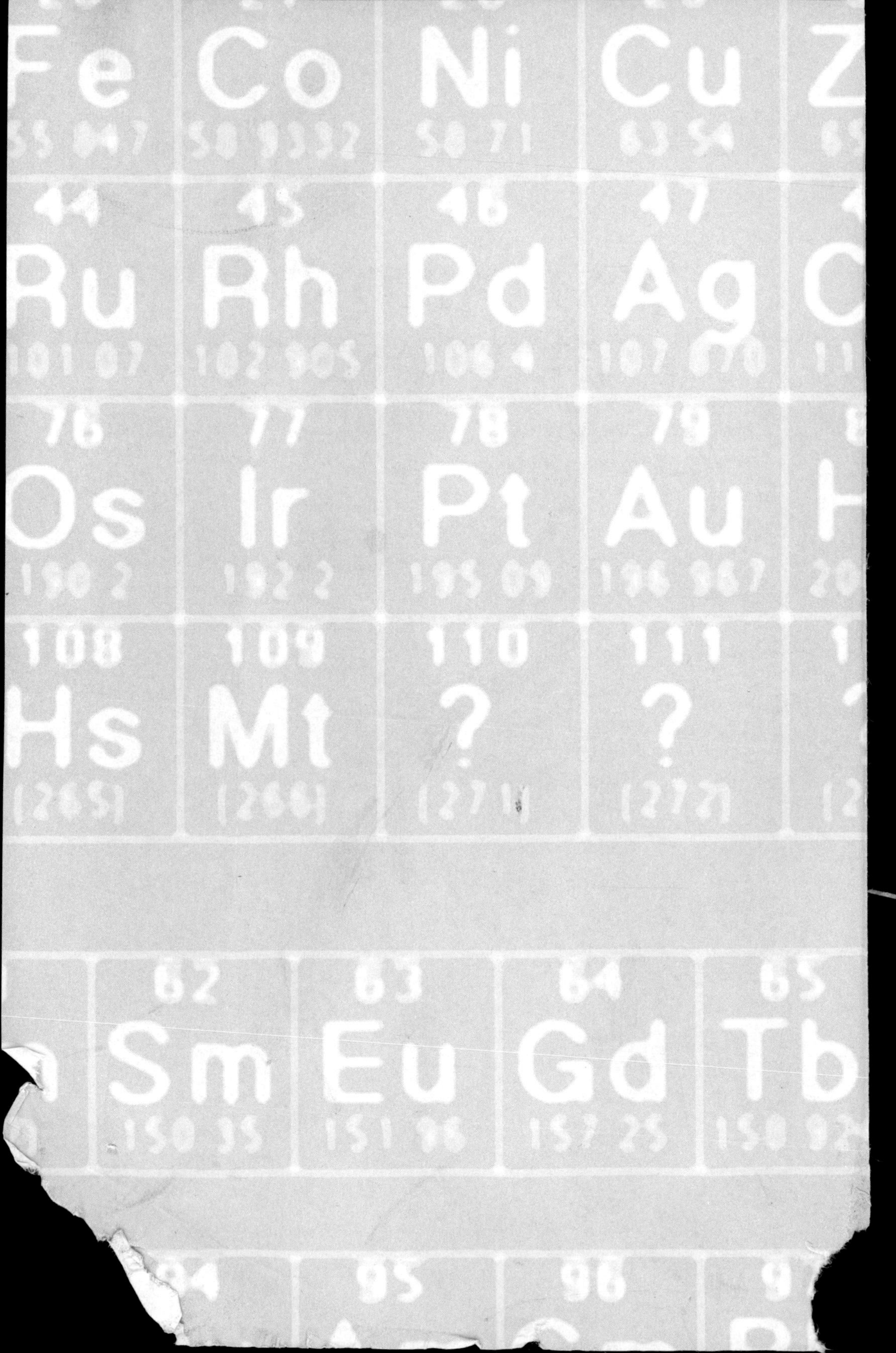

Fe
Co
Ni
Cu
55 847
58 9332
58 71
63 54
44
45
46
47
Ru
Rh
Pd
Ag
101 07
102 905
106 4
107 870
76
77
78
79
Os
Ir
Pt
Au
190 2
192 2
195 09
196 967
108
109
110
111
Hs
Mt
?
?
(265)
(266)
(271)
(272)
62
63
64
65
Sm
Eu
Gd
Tb
150 35
151 96
157 25
94
95
96

NAPOLEON'S BUTTONS

HOW 17 MOLECULES CHANGED HISTORY

PENNY LE COUTEUR
JAY BURRESON

Jeremy P. Tarcher/Putnam
a member of Penguin Putnam Inc.
New York

Most Tarcher/Putnam books are available at special quantity discounts for bulk purchases for sales promotions, premiums, fund-raising, and educational needs. Special books or book excerpts also can be created to fit specific needs. For details, write Putnam Special Markets, 375 Hudson Street, New York, NY 10014.

Jeremy P. Tarcher/Putnam
a member of
Penguin Putnam Inc.
375 Hudson Street
New York, NY 10014
www.penguinputnam.com

Published simultaneously in Canada

Library of Congress Cataloging-in-Publication Data

Le Couteur, Penny, date.
Napoleon's buttons : how 17 molecules changed history/
by Penny Le Couteur and Jay Burreson.
p. cm.
Includes bibliographical references and index.
ISBN 1-58542-220-7 (acid-free paper)
1. Chemistry—Popular works. 2. Chemistry—History.
I. Burreson, Jay, date. II. Title.
QD37.L34 2003 2002032247
540—dc21

Printed in the United States of America
3 5 7 9 10 8 6 4 2

This book is printed on acid-free paper. ♾

Book design by Carol Malcolm Russo/Signet M Design, Inc.

For our families

CONTENTS

INTRODUCTION

For the want of a nail the shoe was lost.
For the want of a shoe the horse was lost.
For the want of a horse the rider was lost.
For the want of a rider the battle was lost.
For the want of a battle the kingdom was lost.
And all for the want of a horse-shoe nail.

OLD ENGLISH NURSERY RHYME

IN JUNE 1812, Napoleon's army was 600,000 strong. By early December, however, the once proud Grande Armée numbered fewer than 10,000. The tattered remnants of Napoleon's forces had crossed the Berezina River, near Borisov in western Russia, on the long road of retreat from Moscow. The remaining soldiers faced starvation, disease, and numbing cold—the same enemies that had defeated their comrades as surely as had the Russian army. More of them were to perish, ill clad and ill equipped to survive the bitter cold of a Russian winter.

Napoleon's retreat from Moscow had far-reaching consequences on the map of Europe. In 1812, 90 percent of the Russian population consisted of serfs, the outright property of a landowner, bought, sold, or traded at whim, a situation closer to slavery than serfdom ever was in western Europe. The principles and ideals of the French Revolution of 1789–1799 had followed Napoleon's conquering army, breaking down the medieval order of society, changing political boundaries, and fomenting the concept of nationalism. His legacy was also practical. Common civil administration and legal codes replaced the widely varying and confusing system of regional laws and regulations, and new concepts of individual, family, and property rights were introduced. The decimal system of weights and measures became the standard instead of the chaos of hundreds of different local scales.

What caused the downfall of the greatest army Napoleon had led? Why did Napoleon's soldiers, victorious in previous battles, falter in the Russian campaign? One of the strangest theories to be advanced can be captured by paraphrasing an old nursery rhyme: "all for the want of a button." Surprising as it may seem, the disintegration of Napoleon's army may be traceable to something as small as the disintegration of a button—a tin button, to be exact, the kind that fastened everything from the greatcoats of Napoleon's officers to the trousers and jackets of his foot soldiers. When temperatures drop, shiny metallic tin starts to change into a crumbly nonmetallic gray powder—still tin, but with a different structural form. Is this what happened to the tin buttons of Napoleon's army? At Borisov one observer described Napoleon's army as "a mob of ghosts draped in women's cloaks, odd pieces of carpet or greatcoats burned full of holes." Were Napoleon's men, as the buttons on their uniforms fell apart, so weakened by the chilling cold they could no longer function as soldiers? Did the lack of buttons mean that hands were used to hold garments together rather than carry weapons?

There are numerous problems in determining the veracity of this theory. "Tin disease," as the problem was called, had been known in northern Europe for centuries. Why would Napoleon, a great believer in keeping his troops fit for battle, have permitted its use in their garments? And the disintegration of tin is a reasonably slow process, even at the very low temperatures of the 1812 Russian winter. It makes a good story, though, and chemists enjoy quoting it as a chemical reason for Napoleon's defeat. And if there is some truth to the tin theory, then one has to wonder whether, if tin did not deteriorate in the cold, the French might have continued their eastward expansion. Would the yoke of serfdom have been lifted from the Russian people half a century earlier than it was? Would the distinction between western and eastern Europe, which roughly parallels the extent of Napoleon's empire—a testament to his lasting influence—still be apparent today?

Throughout history metals have been pivotal in shaping human events. Apart from its possibly apocryphal role in Napoleon's buttons, tin from the Cornish mines in southern England was highly sought after by the Romans and was one reason for the extension of the Roman Empire into Britain. By 1650 an estimated sixteen thousand tons of silver from the mines of the New World had enriched the coffers of Spain and Portugal, much of it to be used supporting wars in Europe. The search for gold and silver had an immense impact on exploration, settlement, and the environment of many regions; for example, the gold rushes of the nineteenth century in California, Australia, South Africa, New Zealand, and the Canadian Klondike did much to open up those countries. As well, our language contains many words or phrases invoking this metal—*goldbrick, gold standard, good as gold, golden years.* Whole epochs have been named in tribute to the importance of metals. The Bronze Age, when bronze—an alloy or mixture of tin and copper—was used for weapons and tools was followed by the Iron Age, characterized by smelting of iron and the use of iron implements.

But is it only metals like tin and gold and iron that have shaped history? Metals are elements—substances that cannot be decomposed into simpler materials by chemical reactions. There are only ninety naturally occurring elements, and tiny amounts of another nineteen or so have been made by man. But there are about seven million compounds, substances formed from two or more elements, *chemically* combined in fixed proportions. Surely there must be compounds that have also been pivotal in history, compounds without which the development of human civilization would have been very different, compounds that changed the course of world events. It's an intriguing idea, and it is the principal unifying theme underlying each chapter of this book.

In looking at some common and not-so-common compounds from

this different perspective, fascinating stories emerge. In the Treaty of Breda of 1667 the Dutch ceded their only North American possession in exchange for the small island of Run, an atoll in the Banda Islands, a tiny group in the Moluccas (or Spice Islands), east of Java in present-day Indonesia. The other signatory nation to this treaty, England, gave up its legitimate claim to Run—whose only asset was its groves of nutmeg trees—to gain the rights to another small piece of land halfway around the world, the island of Manhattan.

The Dutch had staked their claim to Manhattan shortly after Henry Hudson, seeking a Northwest Passage to the East Indies and the fabled Spice Islands, visited the area. In 1664 the Dutch governor of New Amsterdam, Peter Stuyvesant, was forced to surrender the colony to the English. Protests by the Dutch over this seizure and other territorial claims kept the two nations at war for nearly three years. English sovereignty over Run had angered the Dutch, whose monopoly of the nutmeg trade needed only the island of Run to be complete. The Dutch, with a long history of brutal colonization, massacres, and enslavement in the region, were not about to allow the English to keep a toehold in this lucrative spice trade. After a four-year siege and much bloody fighting, the Dutch invaded Run. The English retaliated by attacking the richly laden ships of the Dutch East India Company.

The Dutch wanted compensation for English piracy and the return of New Amsterdam; the English demanded payment for the Dutch outrages in the East Indies and the return of Run. With neither side about to back down nor able to claim victory in the sea battles, the Treaty of Breda offered a face-saving opportunity for both sides. The English would keep Manhattan in return for giving up their claims to Run. The Dutch would retain Run and forgo further demands for Manhattan. As the English flag was raised over New Amsterdam (renamed New York), it seemed that the Dutch had got the better part of the deal. Few could see the worth of a small New World settlement of about a thousand people compared to the immense value of the nutmeg trade.

Why was nutmeg so valued? Like other spices, such as cloves, pepper, and cinnamon, nutmeg was used extensively in Europe in the preservation of food, for flavoring, and as medicine. But it had another, more important role as well. Nutmeg was thought to protect against plague, the Black Death that sporadically swept across Europe between the fourteenth and eighteenth centuries.

Of course, we now know that the Black Death was a bacterial disease transmitted from infected rats through the bites of fleas. So wearing a nutmeg in a small bag around the neck to ward off the plague may seem just another medieval superstition—until we consider the chemistry of nutmeg. The characteristic smell of nutmeg is due to *isoeugenol.* Plants develop compounds like isoeugenol as natural pesticides, as defenses against grazing predators, against insects, and fungi. It's entirely possible that the isoeugenol in nutmeg acted as a natural insecticide to repel fleas. (Then again, if you were wealthy enough to afford nutmeg, you probably lived in less crowded conditions with fewer rats and fewer fleas, thus limiting your exposure to the plague.)

Whether nutmeg was effective against the plague or not, the volatile and aromatic molecules it contained were undoubtedly responsible for its esteem and value. The exploration and exploitation that accompanied the spice trade, the Treaty of Breda, and the fact that New Yorkers are not New Amsterdamers can be attributed to the compound isoeugenol.

Considering the story of isoeugenol has led to contemplating many other compounds that have changed the world, some of them well known and still vitally important to world economy or to human health, and others that have faded into obscurity. All of these chemicals have been responsible for either a key event in history or for a series of events that altered society.

We decided to write this book to tell the stories of the fascinating connections between chemical structures and historical episodes, to uncover how seemingly unrelated events have depended on similar chemical structures, and to understand the extent to which the development of society has depended on the chemistry of certain compounds. The idea that momentous events may depend on something as small as a molecule—a group of two or more atoms held together in a definite arrangement—offers a novel approach to understanding the growth of human civilization. A change as small as the position of a bond—the link between atoms in a molecule—can lead to enormous differences in properties of a substance and in turn influence the course of history. So this book is not about the history of chemistry; rather it is about chemistry in history.

The choice of which compounds to include in this book was a personal one, and the final selection is by no means exhaustive. We have chosen those compounds we found the most interesting for both their stories and their chemistry. Whether the molecules we selected are definitely the most important in world history is arguable; our colleagues in the chemical profession would no doubt add other molecules to the list or remove some of the ones we discuss. We will explain why we believe certain molecules were the impetus for geographic exploration, while others made possible the ensuing voyages of discovery. We will describe molecules that were critical to the development of trade and commerce, that were responsible for human migrations and colonization, and that led to slavery and forced labor. We will discuss how the chemical structure of some molecules has changed what we eat, what we drink, and what we wear. We will look at molecules that spurred advances in medicine, in public sanitation, and in health. We will consider molecules that have resulted in great feats of engineering, and molecules of war and peace—some responsible for millions of deaths while others saving millions of lives. We will explore how many changes in gender roles, in human cultures and society, in law, and in the environment can be attributed to the chemical structures of a small number of crucial molecules. (The seventeen molecules we have chosen to focus on in these chapters—the seventeen molecules referred to in the title—are not always individual molecules. Often they will be groups of molecules with very similar structures, properties, and roles in history.)

The events discussed in this book are not arranged in chronological historical order. Instead, we have based our chapters on connections—the links between similar molecules, between sets of similar molecules, and even between molecules that are quite different chemically but have properties that are similar or can be connected to similar events. For example, the Industrial Revolution owes its start to the profits reaped from a slave-grown compound (sugar) on plantations in the Americas, but it was another compound (cotton) that fueled major economic and social changes in England—and chemically the latter compound is a big brother, or maybe a cousin, of the former compound. The late-nineteenth-century growth of the German chemical industry was due, in part, to the development of new dyes that came from coal tar (a waste material arising from the production of gas from coal). These same German chemical companies were the first to develop man-made antibiotics, composed of molecules with similar chemical structures to the new dyes. Coal tar also provided the first antiseptic, phenol, a molecule that was later used in the first artificial plastic and is chemically related to isoeugenol, the aromatic molecule from nutmeg. Such chemical connections are abundant in history.

We were also intrigued by the role serendipity has been accorded in numerous chemical discoveries. Luck has often been cited as crucial to many important findings, but it seems to us that the ability of the discoverers to realize that something unusual has happened—and to question why it occurred and how it could be useful—is of greater importance. In many instances in the course of chemical experimentation an odd but potentially important result was ignored and an opportunity lost. The ability to recognize the possibilities in an unexpected result deserves to be lauded rather than dismissed as a fortuitous fluke. Some of the inventors and discoverers of the compounds we discuss were chemists, but others had no scientific training at all. Many of them could be described as characters—unusual, driven, or compulsive. Their stories are fascinating.

ORGANIC—ISN'T THAT GARDENING?

To help you understand the chemical connections in the following pages, we'll first provide a brief overview of chemical terms. Many of the compounds discussed in this book are classified as *organic* compounds. During the last twenty or thirty years the word organic has taken on a meaning quite different from its original definition. Nowadays the term *organic,* usually in reference to gardening or food, is taken to mean agriculture conducted without artificial pesticides or herbicides and with no synthetic fertilizers. But *organic* was originally a chemical term dating back nearly two hundred years to Jöns Jakob Berzelius, a Swedish chemist who in 1807 applied the word organic to compounds that were derived from living organisms. In contrast, he used the word *inorganic* to mean compounds that did not come from living things.

The idea that chemical compounds obtained from nature were somehow special, that they contained an essence of life even though it could not be detected or measured, had been around since the eighteenth century. This special essence was known as *vital energy.* The belief that there was something mystical about compounds derived from plants or animals was called *vitalism.* Making an organic compound in the laboratory was thought to be impossible by definition, but ironically one of Berzelius's own students did just that. In 1828, Friedrich Wöhler, later professor of chemistry at the University of Göttingen in Germany, heated the inorganic compound ammonia with cyanic acid to produce crystals of urea that were exactly the same as the organic compound urea isolated from animal urine.

Although vitalists argued that cyanic acid was organic because it was obtained from dried blood, the theory of vitalism began to crack. Over the next few decades it shattered completely as other chemists were able to produce organic compounds from totally inorganic sources. Though some scientists were reluctant to believe what seemed to be heresy, eventually the death of vitalism was commonly acknowledged. A new chemical definition of the word *organic* was needed.

Organic compounds are now defined as compounds that contain the element carbon. Organic chemistry, therefore, is the study of the compounds of carbon. This is not a perfect definition, however, as there are a number of carbon-containing compounds that chemists have never considered organic. The reason for this is mainly traditional. Carbonates, compounds with carbon and oxygen, were known to come from mineral sources and not necessarily from living things well before Wöhler's defining experiment. So marble (or calcium carbonate) and baking soda (sodium bicarbonate) have never been labeled organic. Similarly, the element carbon itself, either in the form of diamond or graphite—both originally mined from deposits in the ground although now also made synthetically—has always been thought of as inorganic. Carbon dioxide, containing one carbon atom joined to two oxygen atoms, has been known for centuries but has never been classified as an organic compound. Thus the definition of *organic* is not completely consistent. But in general an organic compound is a compound that contains carbon, and an inorganic compound is one that consists of elements other than carbon.

More than any other element, carbon has tremendous variability in the ways it forms bonds and also in the number of other elements to which it is able to bond. Thus there are many, many more compounds of carbon, both naturally occurring and man-made, than there are compounds of all the other elements combined. This may account for the fact that we will be dealing with many more organic than inorganic molecules in this book; or perhaps it is because both the authors are organic chemists.

Chemical Structures: Do We Have To?

In writing this book, our biggest problem was determining how much chemistry to include in its pages. Some people advised us to minimize the chemistry, to leave it out and just tell the stories. Especially, we've been told, do not draw any chemical structures. But it is the connection between chemical structures and what they do, between how and why a compound has the chemical properties it has, and how and why that affected certain events in history, that we find the most fascinating. While you can certainly read this book without looking at the structures, we think understanding the chemical structures makes the interwoven relationship between chemistry and history come alive.

Organic compounds are mainly composed of only a few types of atoms: carbon (with chemical symbol C), hydrogen (H), oxygen (O), and nitrogen (N). Other elements may be present as well; for example, bromine (Br), chlorine (Cl), fluorine (F), iodine (I), phosphorus (P), and sulfur (S) are also found in organic compounds. The structures in this book are generally drawn to illustrate differences or similarities between compounds; mostly all that is required is to look at the drawing. The variation will often be arrowed, circled, or indicated in some other way. For example, the only difference between the two structures shown below is in the position where OH is attached to a C; it's pointed out by an arrow in each case. For the first molecule the OH is on the second C from the left; for the second molecule the OH is attached to the first C from the left.

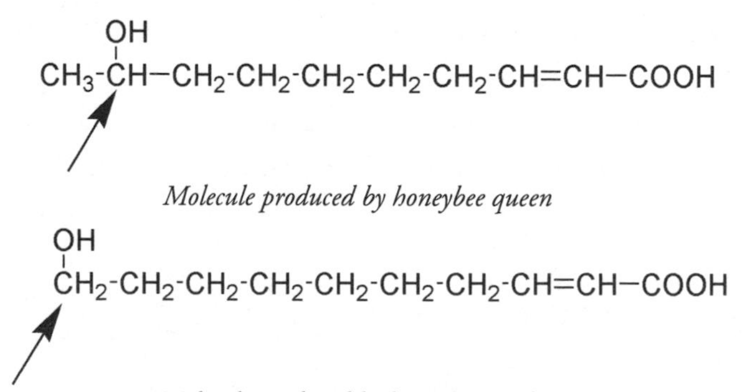

Molecule produced by honeybee queen

Molecule produced by honeybee worker

This is a very small difference, but is hugely important if you happen to be a honeybee. Queen honeybees produce the first molecule. Bees are able to recognize the difference between it and the second molecule, which is produced by honeybee workers. We can tell the difference between workers and queens by looking at the bees.

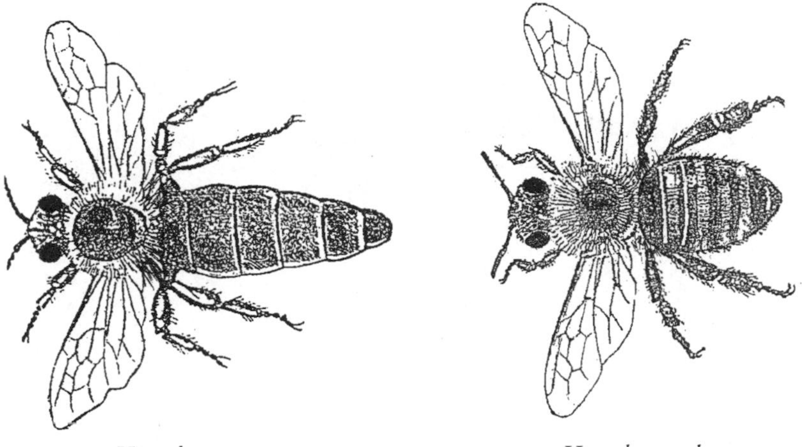

Honeybee queen *Honeybee worker*

(Courtesy of Raymond and Sylvia Chamberlin)

Bees use chemical signaling to tell the difference. We could say they see through chemistry.

Chemists draw such structures to depict the way atoms are joined to each other through chemical bonds. Chemical symbols represent atoms, and bonds are drawn as straight lines. Sometimes there is more than one bond between the same two atoms; if there are two it is a double bond and shown as =. When three chemical links exist between the same two atoms, it is a triple bond and drawn as ≡.

In one of the simplest organic molecules, methane (or marsh gas), carbon is surrounded by four single bonds, one to each of four hydrogen atoms. The chemical formula is given as CH_4, and the structure is drawn as:

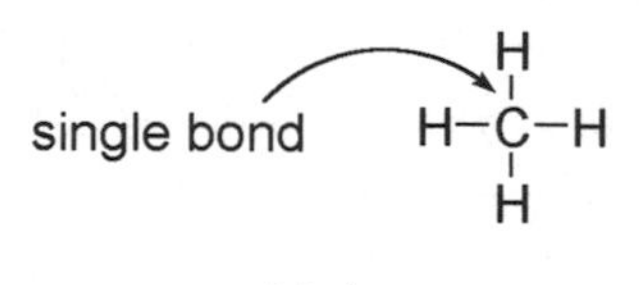

Methane

The simplest organic compound that has a double bond is ethene (also called ethylene) with a formula of C_2H_4 and the structure:

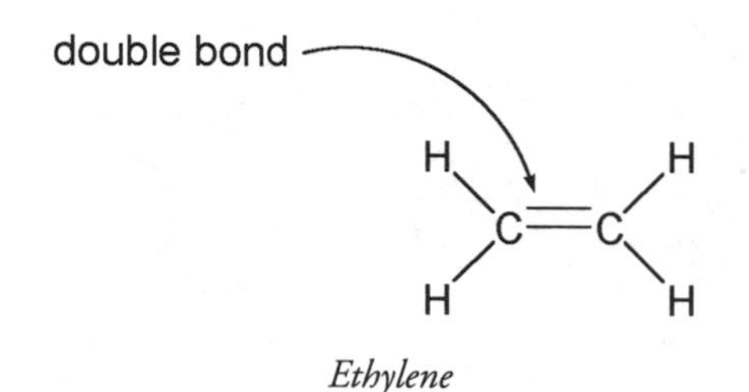

Ethylene

Here carbon still has four bonds—the double bond counts as two. Despite being a simple compound, ethylene is very important. It is a plant hormone that is responsible for promoting the ripening of fruit. If apples, for example, aren't stored with appropriate ventilation, the ethylene gas they produce will build up and cause them to overripen. This is why you can hasten the ripening of a hard avocado or kiwi fruit by putting it in a bag with an already ripe apple. Ethylene produced by the mature apple increases the rate of ripening of the other fruit.

The organic compound methanol, also known as methyl alcohol or wood alcohol, has the formula CH_4O. This molecule contains an oxygen atom and the structure is drawn as:

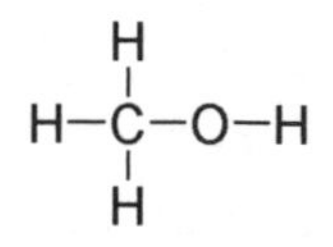

Methanol

Here the oxygen atom, O, has two single bonds, one connected to the carbon atom and the other to a hydrogen atom. As always,carbon has a total of four bonds.

In compounds where there is a double bond between a carbon atom and an oxygen atom, as in acetic acid (the acid of vinegar), the formula, written as $C_2H_4O_2$, does not directly indicate where the double bond is. This is the reason we draw chemical structures—to show exactly which atom is attached to which other atom and where the double or triple bonds are.

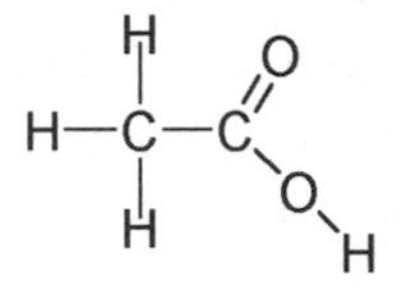

Acetic acid

We can draw these structures in an abbreviated or more condensed form. Acetic acid could also be drawn as:

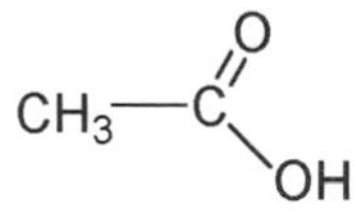

or even CH_3—COOH

where not all the bonds are shown. They are, of course, still there, but these shortened forms are faster to draw and show just as clearly the relationships among atoms.

This system of drawing structures works well for smaller examples, but when the molecules get bigger, it becomes time consuming and difficult to follow. For example, if we return to the queen honeybee recognition molecule:

OH

CH_3-CH—CH_2-CH_2-CH_2-CH_2-CH_2-CH=CH—COOH

and compare it to a fully drawn-out version showing all the bonds, the structure would look like:

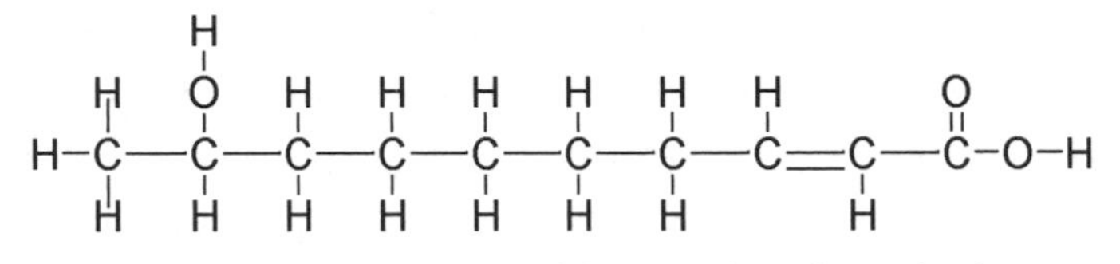

Fully drawn-out structure of the queen honeybee molecule

This full structure is cumbersome to draw and looks very cluttered. For this reason, we often draw compounds using a number of shortcuts, the most common of which is to leave out many of the H atoms. This does not mean that they are not there; we just do not show them. A carbon atom always has four bonds, so if it does not look as if C has four bonds, be assured it does—the ones that are not shown bond to hydrogen atoms.

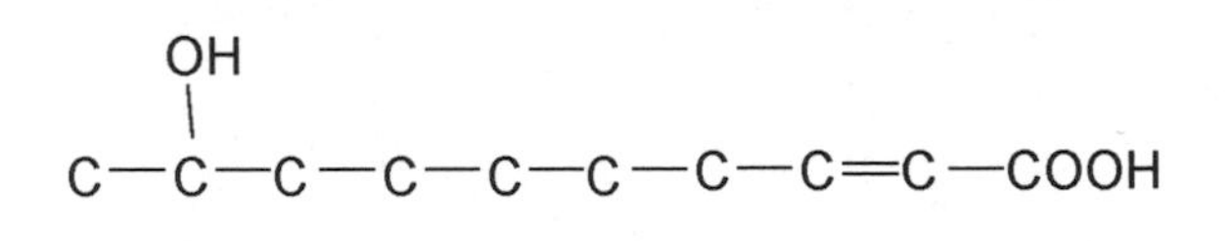

Queen honeybee recognition molecule

As well, the carbon atoms are often shown joined at an angle instead of in a straight line; this is more indicative of the true shape of the molecule. In this format the queen honeybee molecule looks like this:

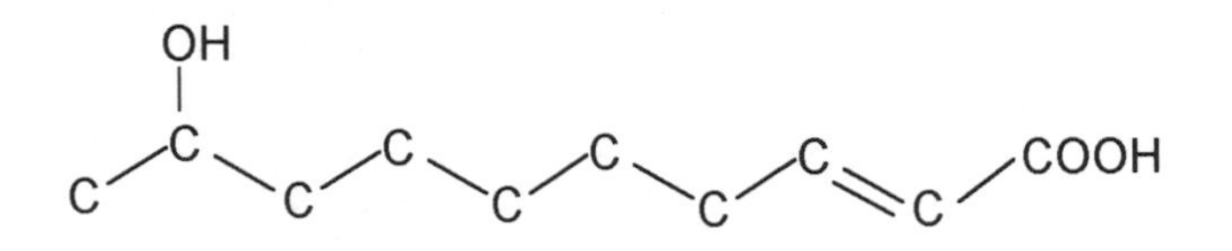

An even more simplified version leaves out most of the carbon atoms:

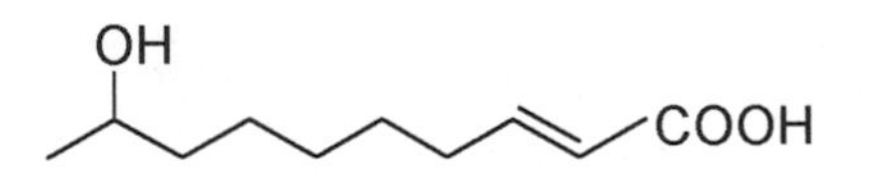

Here the end of a line and any intersection represent a carbon atom. All other atoms (except most of the carbons and hydrogens) are still shown. By simplifying in this manner, it is easier to see the difference between the queen molecule and the worker molecule.

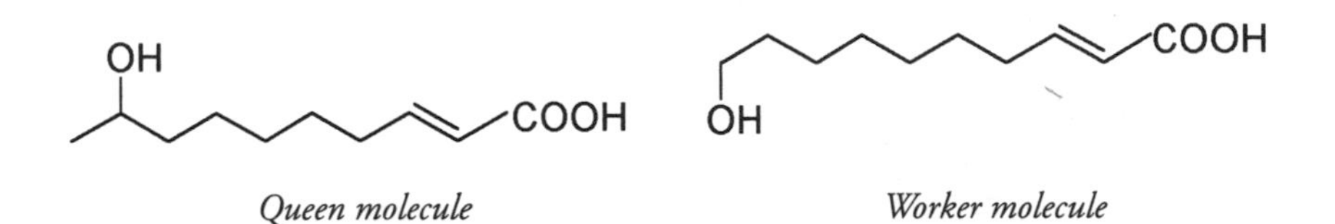

Queen molecule *Worker molecule*

It is now also easier to compare these compounds to those emitted by other insects. For example, *bombykol,* the pheromone or sex attractant molecule produced by the male silkworm moth, has sixteen carbon atoms (as opposed to the ten atoms in the honeybee queen molecule, also a pheromone), has two double bonds instead of one, and lacks the COOH arrangement.

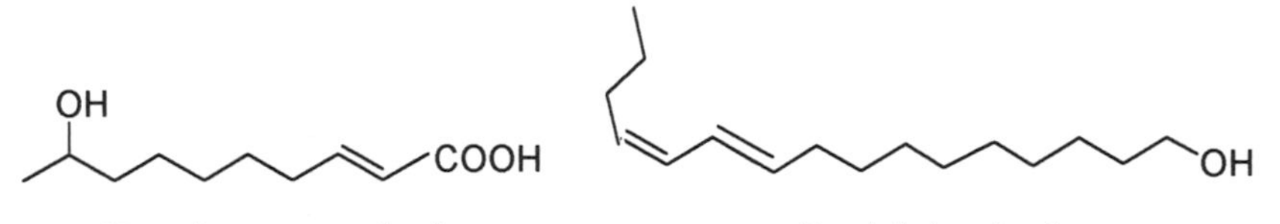

Honeybee queen molecule *Bombykol molecule*

It is particularly useful to leave out many of the carbon and hydrogen atoms when dealing with what are called cyclic compounds—a fairly common structure in which the carbon atoms form a ring. The following structure represents the molecule cyclohexane, C_6H_{12}:

Abbreviated or condensed version of the chemical structure of cyclohexane. Every intersection represents a carbon atom; hydrogen atoms are not shown.

If drawn out in full, cyclohexane would appear as:

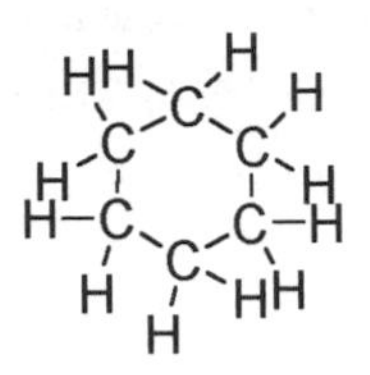

The fully drawn-out chemical structure of cyclohexane showing all atoms and all bonds

As you can see, when we put in all the bonds and write in all the atoms, the resulting diagram can be confusing. When you get to more complicated structures such as the antidepressant drug Prozac, the fully drawn-out version (below) makes it really hard to see the structure.

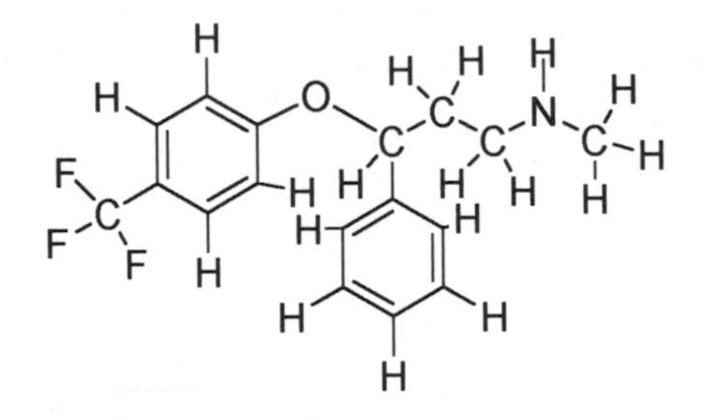

Fully drawn-out structure of Prozac

But the simplified version is much clearer:

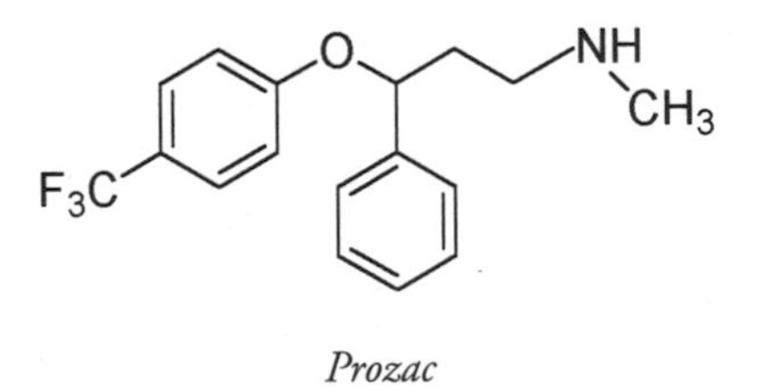

Prozac

Another term frequently used to describe aspects of a chemical structure is *aromatic.* The dictionary says that *aromatic* means "having a fragrant, spicy, pungent, or heady smell, implying a pleasant odor."

Chemically speaking, an aromatic compound often does have a smell, although not necessarily a pleasant one. The word *aromatic,* when applied to a chemical, means that the compound contains the ring structure of benzene (shown below), which is most commonly drawn as a condensed structure.

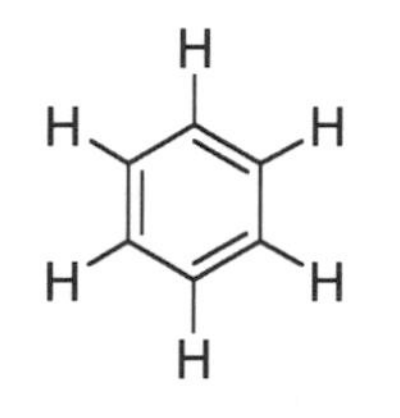

Structure of benzene

Condensed structure of benzene

Looking at the drawing of Prozac, you can see that it contains two of these aromatic rings. Prozac is therefore defined as an aromatic compound.

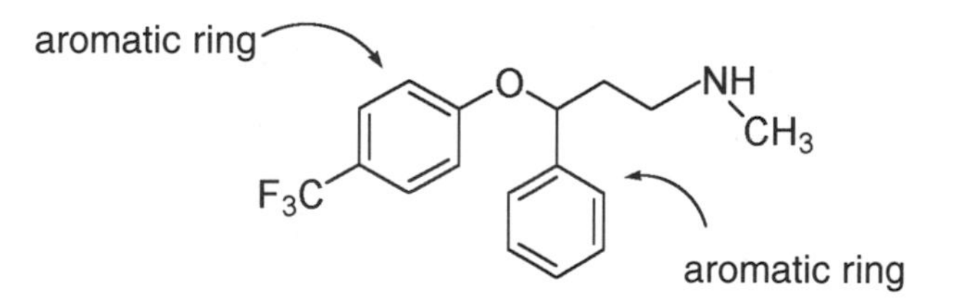

The two aromatic rings in Prozac

This is only a short introduction to organic chemical structures, but it is actually all that you need to understand what we describe in this book. We will compare structures to show how they differ and how they are the same, and we will show how extremely small changes to a molecule sometimes produce profound effects. Following the connections among the particular shapes and related properties of various molecules reveals the influence of chemical structures on the development of civilization.

1. PEPPERS, NUTMEG, AND CLOVES

CHRISTOS E ESPICIARIAS!—for Christ and spices—was the jubilant cry from Vasco da Gama's sailors as, in May 1498, they approached India and the goal of gaining untold wealth from spices that for centuries had been the monopoly of the merchants of Venice. In medieval Europe one spice, pepper, was so valuable that a pound of this dried berry was enough to buy the freedom of a feudal laborer bound to the estate of a nobleman. Although pepper now appears on dinner tables all over the world, the demand for it and for the fragrant molecules of cinnamon, cloves, nutmeg, and ginger fueled a global search that ushered in the Age of Discovery.

A Brief History of Pepper

Pepper, from the tropical vine *Piper nigrum,* originating in India, is still the most commonly used of all spices. Today its major producers are the equatorial regions of India, Brazil, Indonesia, and Malaysia. The vine is a strong, woody climber that can grow up to twenty feet or more. The plants begin to bear a red globular fruit within two to five years and under the right conditions continue to produce for forty years. One vine can produce ten kilograms of the spice each season.

About three-quarters of all pepper is sold as black pepper, produced by a fungal fermentation of unripe pepper berries. White pepper, obtained from the dried ripe fruit after removal of the berry skin and pulp, makes up most of the remainder. A very small percentage of pepper is sold as green pepper; the green berries, harvested just as they are beginning to ripen, are pickled in brine. Other colors of peppercorn, such as are sometimes found in specialty stores, are artificially dyed or are really other types of berries.

It is assumed that Arab traders introduced pepper to Europe, initially by the ancient spice routes that led through Damascus and across the Red Sea. Pepper was known in Greece by the fifth century B.C. At that time its use was medicinal rather than culinary, frequently as an antidote to poison. The Romans, however, made extensive use of pepper and other spices in their food.

By the first century A.D., over half the imports to the Mediterranean from Asia and the east coast of Africa were spices, with pepper from India accounting for much of this. Spices were used in food for two reasons: as a preservative and as a flavor enhancer. The city of Rome was large, transportation was slow, refrigeration was not yet invented, and the problem of obtaining fresh food and keeping it fresh must have been enormous. Consumers had only their noses to help them detect food that was off; "best before" labels were centuries in the future. Pepper and other spices disguised the taste of rotten or rancid fare and probably helped slow further decay. The taste of dried, smoked, and salted food could also be made more palatable by a heavy use of these seasonings.

By medieval times much European trade with the East was conducted through Baghdad (in modern Iraq) and then to Constantinople (now Istanbul) via the southern shores of the Black Sea. From Constantinople spices were shipped to the port city of Venice, which had almost complete dominance of the trade for the last four centuries of the Middle Ages.

From the sixth century A.D., Venice had grown substantially by marketing the salt produced from its lagoons. It had prospered over the centuries as a result of canny political decisions that let the city maintain its independence while trading with all nations. Almost two hundred years of holy Crusades, starting in the late eleventh century, allowed the merchants of Venice to consolidate their position as the world's spice kings. Supplying transport, warships, arms, and money to Crusaders from western Europe was a profitable investment that directly benefited the Republic of Venice. The Crusaders, returning from the warm countries of the Middle East to their cooler northern homelands, wanted to take with them the exotic spices they had come to appreciate on their journeys. Pepper may have initially been a novelty item, a luxury few could afford, but its ability to disguise rancidity, give character to tasteless dried food, and seemingly reduce the salty taste of salted food very soon made it indispensable. The merchants of Venice had gained a vast new market, and traders from all over Europe came to buy spices, especially pepper.

By the fifteenth century the Venetian monopoly of the spice trade was so complete and the profit margins so great that other nations started to look seriously at the possibility of finding competing routes to India—in particular a sea route around Africa. Prince Henry the Navigator, son of King John I of Portugal, commissioned a comprehensive shipbuilding program that produced a fleet of sturdy merchant ships able to withstand the extreme weather conditions found in the open ocean. The Age of Discovery was about to begin, driven in large part by a demand for peppercorns.

During the mid–fifteenth century Portuguese explorers ventured as far south as Cape Verde, on the northwestern coast of Africa. By 1483 the Portuguese navigator Diago Cão had explored farther south to the mouth of the Congo River. Only four years later, another Portuguese seaman, Bartholomeu Dias, rounded the Cape of Good Hope, establishing a feasible route for his fellow countryman, Vasco da Gama, to reach India in 1498.

The Indian rulers in Calicut, a principality on India's southwest coast, wanted gold in return for their peppercorns, which was not what the Portuguese had in mind if they were to take over world dominance in pepper. So five years later, da Gama, returning with guns and soldiers, defeated Calicut and brought the pepper trade under Portuguese control. This was the start of a Portuguese empire that eventually extended eastward from Africa through India and Indonesia and westward to Brazil.

Spain had also set its sights on the spice trade, pepper in particular. In 1492 Christopher Columbus, a Genoese navigator convinced that an alternative, and possibly shorter, route to the eastern edge of India could be found by sailing westward, persuaded King Ferdinand V and Queen Isabella of Spain to finance a voyage of discovery. Columbus was right in some but not in all of his convictions. You can get to India by going westward from Europe, but it is not a shorter route. The then unknown continents of North and South America as well as the vast Pacific Ocean are considerable obstacles.

What is it in pepper that built up the great city of Venice, that ushered in the Age of Discovery, and that sent Columbus off to find the New World? The active ingredient of both black and white pepper is *piperine,* a compound with the chemical formula $C_{17}H_{19}O_3N$ and this structure:

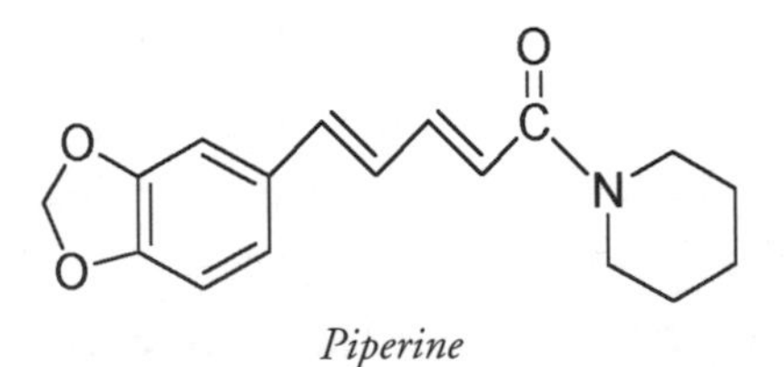

Piperine

The hot sensation we experience when ingesting piperine is not really a taste but a response by our pain nerves to a chemical stimulus.

How this works is not fully known, but it is thought to be due to the shape of the piperine molecule, which is able to fit onto a protein on the pain nerve endings in our mouths and other parts of the body. This causes the protein to change shape and sends a signal along the nerve to the brain, saying something like "Ow, that's hot."

The story of the hot molecule piperine and of Columbus does not end with his failure to find a western trade route to India. When he hit land in October 1492, Columbus assumed—or maybe hoped—that he had reached part of India. Despite the lack of grand cities and wealthy kingdoms that he had expected to find in the Indies, he called the land he discovered the West Indies and the people living there Indians. On his second voyage to the West Indies, Columbus found, in Haiti, another hot spice. Though it was totally different from the pepper he knew, he nevertheless took the chili pepper back to Spain.

The new spice traveled eastward with the Portuguese around Africa to India and beyond. Within fifty years the chili pepper had spread around the world and was quickly incorporated into local cuisines, especially those of Africa and of eastern and southern Asia. For the many millions of us who love its fiery heat, the chili pepper is, without a doubt, one of the most important and lasting benefits of Columbus's voyages.

HOT CHEMISTRY

Unlike the single species of peppercorn, chili peppers grow on a number of species of the *Capsicum* genus. Native to tropical America and probably originating in Mexico, they have been used by humans for at least nine thousand years. Within any one species of chili pepper, there is tremendous variation. *Capsicum annuum,* for example, is an annual that includes bell peppers, sweet peppers, pimentos, banana peppers, paprika, cayenne peppers, and many others. Tabasco peppers grow on a woody perennial, *Capsicum frutescens.*

Chili peppers come in many colors, sizes, and shapes, but in all of them the chemical compound responsible for their pungent flavor and often intense heat is *capsaicin,* with the chemical formula $C_{18}H_{27}O_3N$ and a structure that has similarities to that of piperine:

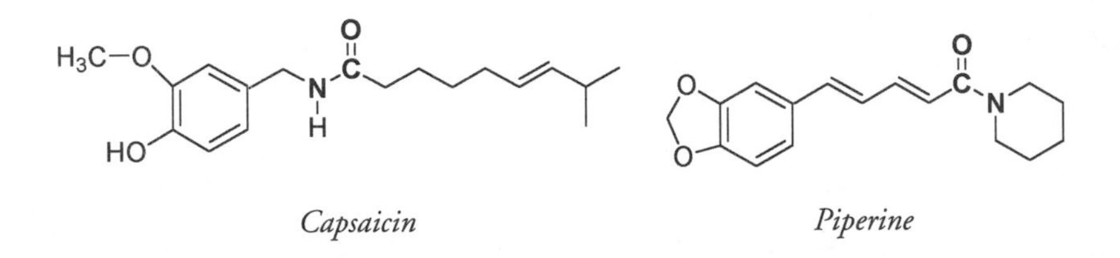

Capsaicin *Piperine*

Both structures have a nitrogen atom (N) next to a carbon atom (C) doubly bonded to oxygen (O), and both have a single aromatic ring with a chain of carbon atoms. That both molecules are "hot" is perhaps not surprising if the hot sensation results from the shape of the molecule.

A third "hot" molecule that also fits this theory of molecular shape is *zingerone* ($C_{11}H_{14}O$), found in the underground stem of the ginger plant, *Zingiber officinale.* Although smaller than either piperine or capsaicin (and, most people would argue, not as hot), zingerone also has an aromatic ring with the same HO and H_3C–O groups attached as in capsaicin, but with no nitrogen atom.

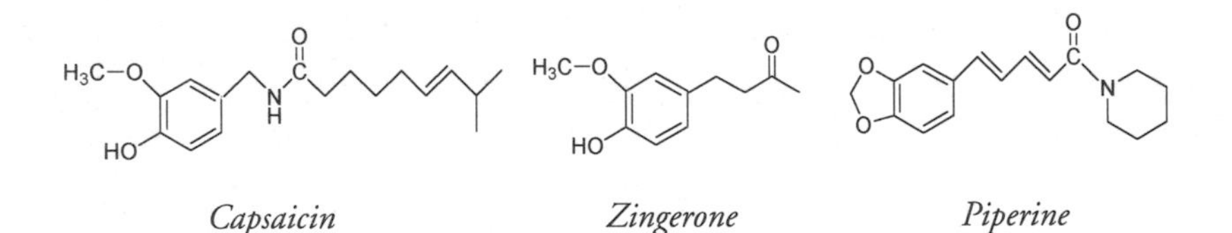

Capsaicin *Zingerone* *Piperine*

Why do we eat such pain-causing molecules? Perhaps for some good chemical reasons. Capsaicin, piperine, and zingerone increase the secretion of saliva in our mouths, aiding digestion. They are also thought to stimulate the movement of food through the bowel. Unlike taste buds that in mammals are mainly on the tongue, pain nerves, able to detect the chemical messages from these molecules, occur in other parts of the human body. Have you ever inadvertently rubbed your eyes while chopping up a chili pepper? Workers who harvest hot peppers need to wear rubber gloves and eye protection against the chili oil containing capsaicin molecules.

The heat we feel from peppercorns appears to be directly proportional to the amount of pepper in the food. Heat from a chili pepper, on the other hand, can be deceptive. Color, size, and region of origin all affect the "hotness" of a chili pepper. None of these guides are reliable; while small peppers are often associated with heat, large peppers are not always the mildest. Geography does not necessarily supply a clue, although the world's hottest chili peppers are said to grow in parts of East Africa. Heat generally increases as a chili is dried.

We often experience a feeling of satisfaction or contentment after eating a fiery meal, and this feeling may be due to endorphins, opiate-like compounds that are produced in the brain as the body's natural response to pain. This phenomenon may account for some people's seeming addiction to hot spicy food. The hotter the chili, the more the pain, so the greater the trace amounts of endorphins produced and ultimately the greater the pleasure.

Apart from paprika, which became well established in Hungarian food like goulash, the chili pepper did not invade the food of Europe the way it did African and Asian cuisine. For Europeans, piperine from the peppercorn remained the hot molecule of choice. Portuguese domination of Calicut and thus control of the pepper trade continued for about 150 years, but by the early seventeenth century the Dutch and the English were taking over. Amsterdam and London became the major pepper trading ports in Europe.

The East India Company—or to give the formal name by which it was incorporated in 1600, the Governor and Company of Merchants of London Trading into the East Indies—was formed to gain a more active role for England in the East Indian spice trade. The risks associated with financing a voyage to India that would return with a shipload of pepper were high, so merchants initially bid for "shares" of a voyage, thus limiting the amount of potential loss for any one individual. Eventually this practice turned into buying shares of the company itself and thus could be considered responsible for the beginning of capitalism. It may be only a bit of a stretch to say that piperine, which surely nowadays must be considered a relatively insignificant chemical compound, was responsible for the beginnings of today's complex economic structure of the world stock markets.

THE LURE OF SPICES

Historically, pepper was not the only spice of great value. Nutmeg and cloves were also precious and were a lot rarer than pepper. Both originated in the fabled Spice Islands or Moluccas, now the Indonesian province of Maluku. The nutmeg tree, *Myristica fragrans,* grew only on the Banda Islands, an isolated cluster of seven islands in the Banda Sea, about sixteen hundred miles east of Jakarta. These islands are tiny—the largest is less than ten kilometers long and the smallest barely a few kilometers. In the north of the Moluccas are the equally small neighboring islands of Ternate and Tidore, the only places in the world where *Eugenia aromatica,* the clove tree, could be found.

For centuries the people of both these island groups had harvested the fragrant product of their trees, selling spices to visiting Arab, Malay, and Chinese traders to be shipped to Asia and to Europe. Trade routes were well established, and whether they were transported via India, Arabia, Persia, or Egypt, spices would pass through as many as twelve hands before reaching consumers in western Europe. As every transaction could double the price, it was no wonder that the governor of Portuguese India, Afonso de Albuquerque, set his sights farther afield, landing first at Ceylon and later capturing Malacca, on the Malay Peninsula, then the center of the East Indian spice trade. By 1512 he reached the sources of nutmeg and cloves, established a Portuguese monopoly trading directly with the Moluccas, and soon surpassed the Venetian merchants.

Spain, too, coveted the spice trade. In 1518 the Portuguese mariner Ferdinand Magellan, whose plans for an expedition had been rejected by his own country, convinced the Spanish crown that it would not only be possible to approach the Spice Islands by traveling westward but that the route would be shorter. Spain had good reasons for supporting such an expedition. A new route to the East Indies would allow their ships to avoid Portuguese ports and shipping on the eastern passage via Africa and India. As well, a previous decree by Pope Alexander VI had awarded Portugal all non-Christian lands east of an imaginary north-south line one hundred leagues (about three hundred miles) west of the Cape Verde Islands. Spain was allowed all non-Christian lands to the west of this line. That the world was round—a fact accepted by many scholars and mariners of the time—had been overlooked or ignored by the Vatican. So approaching by traveling west could give Spain a legitimate claim to the Spice Islands.

Magellan convinced the Spanish crown that he had knowledge of a pass through the American continent, and he had also convinced himself. He left Spain in September 1519, sailing southwest to cross the Atlantic and then down the coasts of what are now Brazil, Uruguay, and Argentina. When the 140-mile-wide mouth of the estuary of the Río de la Plata, leading to the present-day city of Buenos Aires, turned out to be just that—an estuary—his disbelief and disappointment must have been enormous. But he continued southward, confident that a passage from the Atlantic Ocean to the Pacific was always just around the next headland. The journey for his five small ships with 265 crewmen was only to get worse. The farther south Magellan sailed, the shorter the days became and the more constant the gales. A dangerous coastline with surging tides, deteriorating weather, huge waves, steady hail, sleet and ice, and the very real threat of a slip from frozen rigging added to the misery of the voyage. At 50 degrees south with no obvious passageway in sight and having already subdued one mutiny, Magellan decided to wait out the remainder of the southern winter before sailing on to eventually discover and navigate the treacherous waters that now bear his name.

By October 1520, four of his ships had made it through the Strait of Magellan. With supplies running low, Magellan's officers argued that they should turn back. But the lure of cloves and nutmeg, and the glory and wealth that would result from wresting the East Indies spice trade from the Portuguese, kept Magellan sailing west with three ships. The nearly thirteen-thousand-mile journey across the vast Pacific, a far wider ocean than anyone had imagined, with no maps, only rudimentary navigational instruments, little food, and almost depleted water stores, was worse than the passage around the tip of South America. The expedition's landfall on March 6, 1521, at Guam in the Marianas, offered the crew a reprieve from certain death by starvation or scurvy.

Ten days later Magellan made his last landfall, on the small Philippine island of Mactan. Killed in a skirmish with the natives, he never did reach the Moluccas, although his ships and remaining crew sailed on to Ternate, the home of cloves. Three years after leaving Spain, a depleted crew of eighteen survivors sailed upriver to Seville with twenty-six tons of spices in the battered hull of the *Victoria,* the last remaining ship of Magellan's small armada.

The Aromatic Molecules of Cloves and Nutmeg

Although cloves and nutmeg come from different plant families and from remote island groups separated by hundreds of miles of mainly open sea, their distinctively different odors are due to extremely similar molecules. The main component of oil of cloves is eugenol; the fragrant compound in oil of nutmeg is isoeugenol. These two aromatic molecules—aromatic in both smell and chemical structure—differ only in the position of a double bond:

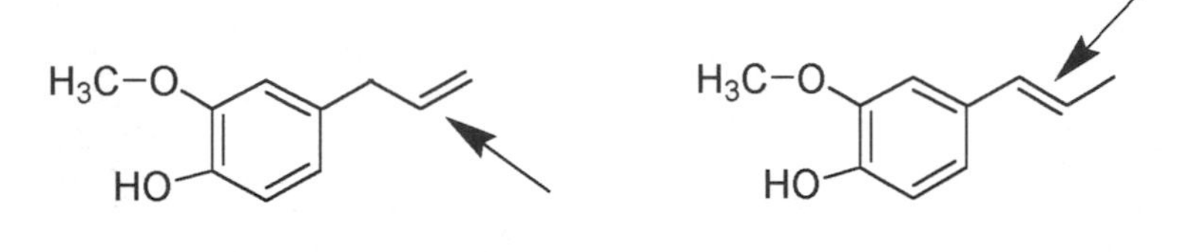

Eugenol (from cloves) *Isoeugenol (from nutmeg)*

The sole difference in these two compounds—the double bond position—is arrowed.

The similarities between the structures of these two compounds and of zingerone (from ginger) are also obvious. Again the smell of ginger is quite distinctive from that of either cloves or nutmeg.

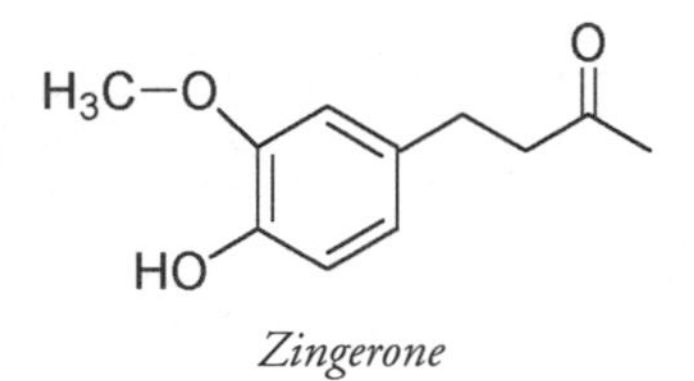

Zingerone

Plants do not produce these highly scented molecules for our benefit. As they cannot retreat from grazing animals, from sap-sucking and leaf-eating insects, or from fungal infestations, plants protect themselves with chemical warfare involving molecules such as eugenol and isoeugenol, as well as piperine, capsaicin, and zingerone. These are natural pesticides—very potent molecules. Humans can consume such compounds in small amounts since the detoxification process that occurs in our livers is very efficient. While a massive dose of a particular compound could theoretically overpower one of the liver's many metabolic pathways, it's reassuring to know that ingesting enough pepper or cloves to do this would be quite difficult.

Even at a distance from a clove tree, the wonderful smell of eugenol is apparent. The compound is found in many parts of the plant, in addition to the dried flower buds that we're familiar with. As long ago as 200 B.C., in the time of the Han dynasty, cloves were used as breath

Drying cloves on the street in Northern Sulawesi, Indonesia.
(Photo by Penny Le Couteur)

sweeteners for courtiers in the Chinese imperial court. Oil of clove was valued as a powerful antiseptic and a remedy for toothache. It is still sometimes used as a topical anesthetic in dentistry.

The nutmeg spice is one of two produced by the nutmeg tree, the other being mace. Nutmeg is ground from the shiny brown seed, or nut, of the apricotlike fruit, whereas mace comes from the red-colored covering layer, or aril, surrounding the nut. Nutmeg has long been used medicinally, in China to treat rheumatism and stomach pains, and in Southeast Asia for dysentery or colic. In Europe, as well as being considered an aphrodisiac and a soporific, nutmeg was worn in a small bag around the neck to protect against the Black Death, which swept Europe with regularity after its first recorded occurrence in 1347. While epidemics of other diseases (typhus, smallpox) periodically visited parts of Europe, the plague was the most feared. It occurred in three forms. The *bubonic* form manifested in painful buboes or swellings in the groin and armpits; internal hemorrhaging and neurological decay was fatal in 50 to 60 percent of cases. Less frequent but more virulent was the *pneumonic* form. *Septicemic* plague, where overwhelming amounts of bacilli invade the blood, was always fatal, often in less than a day.

It's entirely possible that the molecules of isoeugenol in fresh nutmeg did act as a deterrent to the fleas that carry bubonic plague bacteria. And other molecules in nutmeg may also have insecticidal properties. Quantities of two other fragrant molecules, myristicin and elemicin, occur in both nutmeg and mace. The structures of these two compounds are very similar to each other and to those molecules we have already seen in nutmeg, cloves, and peppers.

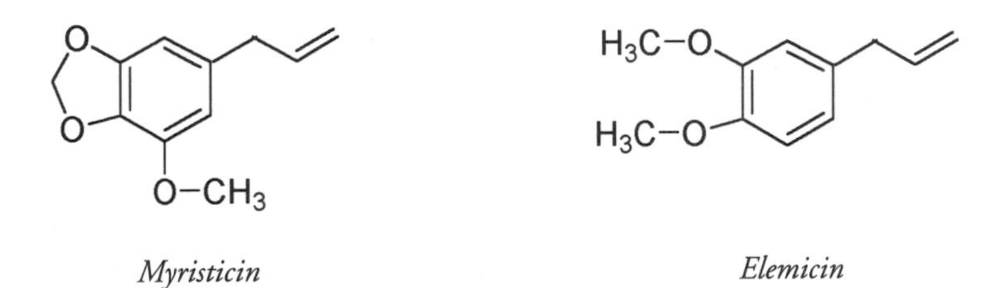

Myristicin *Elemicin*

Besides being a talisman against the plague, nutmeg was considered to be the "spice of madness." Its hallucinogenic properties—likely from the molecules myristicin and elemicin—were known for centuries. A 1576 report told of "a pregnant English lady, having eaten ten or twelve nutmegs, became deliriously inebriated." The accuracy of this tale is doubtful, especially the number of nutmegs consumed, as present-day accounts of ingestion of only one nutmeg describe nausea, profuse sweating, heart palpitations, and vastly elevated blood pressure, along with *days* of hallucinations. This is somewhat more than a delirious inebriation; death has been attributed to consumption of far fewer than twelve nutmegs. Myristicin in large quantities can also cause liver damage.

In addition to nutmeg and mace, carrots, celery, dill, parsley, and black pepper all contain trace amounts of myristicin and elemicin. We don't generally consume the huge quantities of these substances necessary for their psychedelic effects to be felt. And there is no evidence that myristicin and elemicin are psychoactive in themselves. It is possible that they are converted, by some as-yet-unknown metabolic pathway in our body, to traces of compounds that would be analogs of amphetamines.

The chemical rationale for this scenario depends on the fact that another molecule, safrole, with a structure that differs from myristicin only by a missing OCH_3, is the starting material for the illicit manufacture of the compound with the full chemical name of 3,4-methylenedioxy-N-methylamphetamine, abbreviated to MDMA and also called Ecstasy.

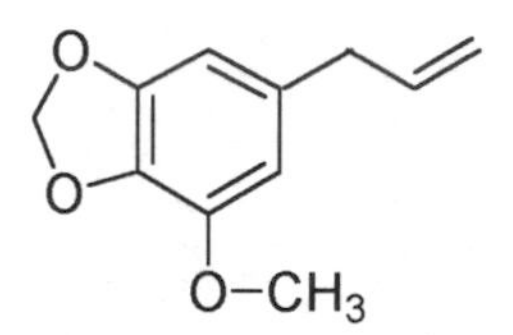

Myristicin

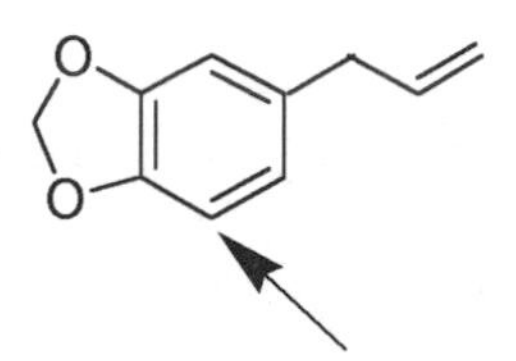

Safrole. The position of the missing OCH_3 is arrowed.

The transformation of safrole to Ecstasy can be shown as:

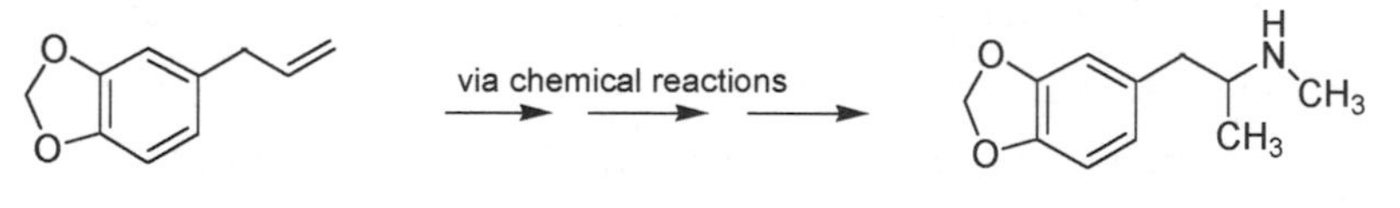

Safrole

3,4-methylenedioxy-N-methylamphetamine or MDMA (Ecstasy)

Safrole comes from the sassafras tree. Traces can also be found in cocoa, black pepper, mace, nutmeg, and wild ginger. Oil of sassafras, extracted from the root of the tree, is about 85 percent safrole and was once used as the main flavoring agent in root beer. Safrole is now deemed a carcinogen and, along with oil of sassafras, has been banned as a food additive.

NUTMEG AND NEW YORK

The clove trade was dominated by the Portuguese during most of the sixteenth century, but they never achieved a complete monopoly. They reached agreements on trading and building forts with the sultans of the islands of Ternate and Tidore, but these alliances proved temporary. The Moluccans continued to sell cloves to their traditional Javanese and Malayan trading partners.

In the next century the Dutch, who had more ships, more men, better guns, and a much harsher colonization policy, became masters of the spice trade, mainly through the auspices of the all-powerful Dutch East India Company—the Vereenigde Oostindische Compagnie, or VOC—established in 1602. The monopoly was not easily accomplished or sustained. It took until 1667 for the VOC to obtain complete control over the Moluccas, evicting the Spanish and Portuguese from their few remaining outposts and ruthlessly crushing opposition from the local people.

To fully consolidate their position, the Dutch needed to dominate the nutmeg trade in the Banda Islands. A 1602 treaty was supposed to have given the VOC sole rights to purchase all the nutmeg produced on the islands, but though the treaty was signed by the village chiefs, the concept of exclusivity was either not accepted or (maybe) not understood by the Bandanese, who continued selling their nutmeg to other traders at the highest offered price—a concept they did understand.

The response by the Dutch was ruthless. A fleet of ships, hundreds of men, and the first of a number of large forts appeared in the Bandas, all designed to control the trade in nutmeg. After a series of attacks, counterattacks, massacres, renewed contracts, and further broken treaties, the Dutch acted even more decisively. Groves of nutmeg trees were destroyed except around where the Dutch forts had been built. Bandanese villages were burned to the ground, the headmen were executed, and the remaining population was enslaved under Dutch settlers brought in to oversee nutmeg production.

The lone remaining threat to the VOC's complete monopoly was the continued presence of the English on Run, the most remote of the Banda Islands, where years before the headmen had signed a trade treaty with the English. This small atoll, where nutmeg trees were so numerous that they clung to the cliffs, became the scene of much bloody fighting. After a brutal siege, a Dutch invasion, and more destruction of nutmeg groves, with the 1667 Treaty of Breda the English surrendered all claims to the island of Run in exchange for a formal declaration renouncing Dutch rights to the island of Manhattan. New Amsterdam became New York, and the Dutch got nutmeg.

In spite of all their efforts, the Dutch monopoly in the trade in nutmeg and cloves did not last. In 1770 a French diplomat smuggled clove seedlings from the Moluccas to the French colony of Mauritius. From Mauritius they spread all along the East African coast and especially to Zanzibar, where cloves quickly became the major export.

Nutmeg, on the other hand, proved notoriously difficult to cultivate outside its original home on the Banda Islands. The tree requires rich, moist, well-drained soil and hot, humid conditions away from the sun and strong winds. Despite the difficulty competitors had in establishing nutmeg growth elsewhere, the Dutch took the precaution of dipping whole nutmegs in lime (calcium hydroxide or slaked lime) before export, to prevent any possibility of sprouting. Eventually the British managed to introduce nutmeg trees into Singapore and the West Indies. The Caribbean island of Grenada became known as "the Nutmeg Isle" and is now the major producer of the spice.

The great worldwide trade in spices would doubtless have continued were it not for the advent of refrigeration. With pepper, cloves, and nutmeg no longer needed as preservatives, the huge demand for piperine, eugenol, isoeugenol, and the other fragrant molecules of these once exotic spices has gone. Today pepper and other spices still grow in India, but they are not major exports. The islands of Ternate and Tidore and the Banda group, now part of Indonesia, are more remote than ever. No longer frequented by great sailing ships seeking to load their hulls with cloves and nutmeg, these small islands slumber in the hot sun, visited only by occasional tourists who explore the crumbling old Dutch forts or dive the pristine coral reefs.

The lure of spices is in the past. We still enjoy them for the rich, warm flavor their molecules impart to our food, but we rarely think of the fortunes they built, the conflicts they provoked, and the amazing feats of exploration they inspired.

2. ASCORBIC ACID

THE AGE OF DISCOVERY was fueled by molecules of the spice trade, but it was the lack of another, quite different molecule that almost ended it. Over 90 percent of his crew didn't survive Magellan's 1519–1522 circumnavigation of the world—in large part due to scurvy, a devastating disease caused by a deficiency of the ascorbic acid molecule, dietary vitamin C.

Exhaustion and weakness, swelling of the arms and legs, softening of the gums, excessive bruising, hemorrhaging from the nose and mouth, foul breath, diarrhea, muscle pain, loss of teeth, lung and kidney problems—the list of symptoms of scurvy is long and horrible. Death generally results from an acute infection such as pneumonia or some other respiratory ailment or, even in young people, from heart failure. One symptom, depression, occurs at an early stage, but whether it is an effect of the actual disease or a response to the other symptoms is not clear. After all, if you were constantly exhausted and had sores that did not heal, painful and bleeding gums, stinking breath, and diarrhea, and you knew that there was worse to come, would you not be depressed, too?

Scurvy is an ancient disease. Changes in bone structure in Neolithic remains are thought to be compatible with scurvy, and hieroglyphs from ancient Egypt have been interpreted as referring to it. The word *scurvy* is said to be derived from Norse, the language of the seafaring Viking warriors who, starting in the ninth century, raided the Atlantic coast of Europe from their northern homelands in Scandinavia. A lack of vitamin-rich fresh fruit and vegetables would have been common on board ships and in northern communities during winter. The Vikings supposedly made use of scurvy grass, a form of Arctic cress, on their way to America via Greenland. The first real descriptions of what was probably scurvy date from the Crusades in the thirteenth century.

Scurvy at Sea

In the fourteenth and fifteenth centuries, as longer voyages were made possible by the development of more efficient sets of sails and fully rigged ships, scurvy became commonplace at sea. Oar-propelled galleys, such as those used by the Greeks and Romans, and the small sailing boats of Arab traders had stayed fairly close to the coast. These vessels were not seaworthy enough to withstand the rough waters and huge swells of the open ocean. Consequently, they would seldom venture far from the coast, and supplies could be replenished every few days or weeks. Access to fresh food on a regular basis meant that scurvy was seldom a major problem. But in the fifteenth century, long ocean voyages in large sailing ships heralded not only the Age of Discovery but also reliance on preserved food.

Bigger ships had to carry cargo and arms, a larger crew to handle the more complicated rigging and sails, and food and water for months at sea. An increase in the number of decks and men and the amount of supplies inevitably translated into cramped sleeping and living conditions for the crew, poor ventilation, and a subsequent increase in infectious diseases and respiratory conditions. Consumption (tuberculosis) and the "bloody flux" (a pernicious form of diarrhea) were common as, no doubt, were body and head lice, scabies, and other contagious skin conditions.

The standard sailor's food did nothing to improve his health. Two major factors dictated the seafaring diet. Firstly, aboard wooden ships it was extremely difficult to keep anything, including food, dry and mold free. Water was absorbed through wooden hulls, as the only waterproofing material available was pitch, a dark-colored, sticky resin obtained as a by-product of charcoal manufacture, applied to the outside of the hull. The inside of the hull, particularly where ventilation was poor, would have been extremely humid. Many accounts of sailing journeys describe incessant dampness, as mold and mildew grew on clothing, on leather boots and belts, on bedding, and on books. The standard sailor's fare was salted beef or pork and ship's biscuits known as hardtack, a mixture of flour and water without salt that was baked rock hard and used as a substitute for bread. Hardtack had the desirable characteristic of being relatively immune to mildew. It was baked to such a degree of hardness that it remained edible for decades, but it was extremely difficult to bite into, especially for those whose gums were inflamed by the onset of scurvy. Typically, ship's biscuits were weevil infested, a circumstance that was actually welcomed by sailors as the weevil holes increased porosity and made the biscuits easier to break and chew.

The second factor governing diet on wooden ships was the fear of fire. Wooden construction and liberal use of highly combustible pitch meant that constant diligence was necessary to prevent fire at sea. For this reason the only fire permitted on board was in the galley and then only in relatively calm weather. At the first sign of foul weather, galley fires would be extinguished until the storm was over. Cooking was often not possible for days at a time. Salted meat could not be simmered in water for the hours necessary to reduce its saltiness; nor could ship's biscuits be made at least somewhat palatable by dunking them in hot stew or broth.

At the outset of a voyage provisions would be taken on board: butter, cheese, vinegar, bread, dried peas, beer, and rum. The butter was soon rancid, the bread moldy, the dried peas weevil infested, the cheeses hard, and the beer sour. None of these items provided vitamin C, so signs of scurvy were often evident after as little as six weeks out of port. Was it any wonder that the navies of European countries had to resort to the press-gang as a means of manning their ships?

Scurvy's toll on the lives and health of sailors is recorded in the logs of early voyages. By the time the Portuguese explorer Vasco da Gama sailed around the southern tip of Africa in 1497, one hundred of his 160-member crew had died from scurvy. Reports exist of the discovery of ships adrift at sea with entire crews dead from the disease. It is estimated that for centuries scurvy was responsible for more death at sea than all other causes; more than the combined total of naval battles, piracy, shipwrecks, and other illnesses.

Astonishingly, preventives and remedies for scurvy during these years were known—but largely ignored. As early as the fifth century, the Chinese were growing fresh ginger in pots on board their ships. The idea that fresh fruit and vegetables could alleviate symptoms of scurvy was, no doubt, available to other countries in Southeast Asia in contact with Chinese trading vessels. It would have been passed on to the Dutch and been reported by them to other Europeans as, by 1601, the first fleet of the English East India Company is known to have collected oranges and lemons at Madagascar on their way to the East. This small squadron of four ships was under the command of Captain James Lancaster, who carried bottled lemon juice with him on his flagship, the *Dragon*. Anyone who showed signs of scurvy was dosed with three teaspoons of lemon juice every morning. On arrival at the Cape of Good Hope, none of the men on board the *Dragon* was suffering from scurvy, but the toll on the other three ships was significant. Despite Lancaster's instructions and example, nearly a quarter of the total crew of this expedition died from scurvy—and not one of these deaths was on his flagship.

Some sixty-five years earlier the crew members on French explorer Jacques Cartier's second expedition to Newfoundland and Quebec were badly affected by a severe outbreak of scurvy, resulting in many deaths. An infusion of needles of the spruce tree, a remedy suggested by the local Indians, was tried with seemingly miraculous results. Almost overnight the symptoms were said to lessen and the disease rapidly disappeared. In 1593 Sir Richard Hawkins, an admiral of the British navy, claimed that within his own experience at least ten thousand men had died at sea from scurvy, but that lemon juice would have been an immediately effective cure.

There were even published accounts of successful treatments of scurvy. In 1617, John Woodall's *The Surgeon's Mate* described lemon juice as being prescribed for both cure and prevention. Eighty years later Dr. William Cockburn's *Sea Diseases, or the Treatise of their Nature, Cause and Cure* recommended fresh fruits and vegetables. Other suggestions such as vinegar, salt water, cinnamon, and whey were quite useless and may have obscured the correct action.

It was not until the middle of the following century that the effectiveness of citrus juice was proven in the first controlled clinical studies of scurvy. Although the numbers involved were very small, the conclusion was obvious. In 1747, James Lind, a Scottish naval surgeon at sea in the *Salisbury,* chose twelve of the crew suffering from scurvy for his experiment. He selected men whose symptoms seemed as similar as possible. He had them all eat the same diet: not the standard salted meat and hardtack, which these patients would have found very difficult to chew, but sweetened gruel, mutton broth, boiled biscuits, barley, sago, rice, raisins, currants, and wine. Lind added various supplements to this carbohydrate-based regime. Two of the sailors each received a quart of cider daily. Two others were dosed with vinegar, and another unfortunate pair received diluted elixir of vitriol (or sulfuric acid). Two more were required to drink half a pint of seawater daily, and another two were fed a concoction of nutmeg, garlic, mustard seed, gum myrrh, cream of tartar, and barley water. The lucky remaining pair was issued daily two oranges and one lemon each.

The results were sudden and visible and what we would expect with today's knowledge. Within six days the men who received the citrus fruit were fit for duty. Hopefully, the other ten sailors were then taken off their seawater, nutmeg, or sulfuric acid regimes and also supplied with lemons and oranges. Lind's results were published in *A Treatise of Scurvy,* but it was another forty years before the British navy began the compulsory issue of lemon juice.

If an effective treatment for scurvy was known, why wasn't it acted upon and used routinely? Sadly, the remedy for scurvy, though proven, seems to have not been recognized or believed. A widely held theory blamed scurvy on a diet of either too much salted meat or not enough fresh meat rather than a lack of fresh fruit and vegetables. Also, there was a logistical problem: it was difficult to keep fresh citrus fruit or juice for weeks at a time. Attempts were made to concentrate and preserve lemon juice, but such procedures were time consuming, costly, and perhaps not very effective, as we now know that vitamin C is easily destroyed by heat and light and that long-term storage reduces the amount in fruits and vegetables

Because of expense and inconvenience, naval officers, physicians, the British admiralty, and shipowners could see no way of growing sufficient greens or citrus fruit on heavily manned vessels. Precious cargo space would have to be used for this purpose. Fresh or preserved citrus fruit was expensive, especially if it was to be allocated daily as a preventive measure. Economy and the profit margin ruled—although, in hindsight, it does seem that this was a false economy. Ships had to be manned above capacity to allow for a 30, 40 or even 50 percent death rate from scurvy. Even without a high death rate, the effectiveness of a crew suffering from scurvy would have been remarkably low. And then there was the humane factor—rarely considered during these centuries.

Another element was the intransigence of the average crew. They were used to eating the standard ship's fare, and although they complained about the monotonous diet of salt meat and ship's biscuit when they were at sea, what they wanted in port was lots of fresh meat, fresh bread, cheese, butter, and good beer. Even if fresh fruit and vegetables were available, the majority of the crew would not have been interested in a quick stir-fry of tender crunchy greens. They wanted meat and more meat—boiled, stewed, or roasted. The officers, who generally came from a higher social class, where a wider and more varied diet was common, would have found eating fruit and vegetables in port to be normal and probably highly acceptable. It would not have been unusual for them to be interested in trying exotic new foodstuffs to be found in the locales where they made landfall. Tamarinds, limes, and other fruits high in vitamin C would have been used in the local cuisine that they, unlike the crew, might try. Scurvy was thus usually less of a problem among a ship's officers.

Cook: Hundreds–Scurvy: Nil

James Cook of the British Royal Navy was the first ship's captain to ensure that his crews remained scurvy free. Cook is sometimes associated with the discovery of antiscorbutics, as scurvy-curing foods are called, but his true achievement lay in the fact that he insisted on maintaining high levels of diet and hygiene aboard all his vessels. The result of his meticulous standards was an extraordinarily good level of health and a low mortality rate among his crew. Cook entered the navy at the relatively late age of twenty-seven, but his previous nine years of experience sailing as a merchant seaman mate in the North Sea and the Baltic, his intelligence, and his innate seamanship combined to ensure his rapid promotion within the naval ranks. His first experience with scurvy came aboard the *Pembroke,* in 1758, on his initial voyage across the Atlantic Ocean to Canada to challenge the French hold on the St. Lawrence River. Cook was alarmed by the devastation this common affliction caused and appalled that the deaths of so many crew, the dangerous reduction of working efficiency, and even actual loss of ships were generally accepted as inevitable.

His experience exploring and mapping around Nova Scotia, the Gulf of St. Lawrence, and Newfoundland and his accurate observations of the eclipse of the sun greatly impressed the Royal Society, a body founded in 1645 with the aim of "improving natural knowledge." He was granted command of the ship *Endeavour* and instructed to explore and chart the southern oceans, to investigate new plants and animals, and to make astronomical observations of the transit of planets across the sun.

Less known but nonetheless compelling reasons for this voyage and for Cook's subsequent later voyages were political. Taking possession in the name of Britain of already discovered lands; claiming of new lands still to be discovered, including Terra Australis Incognita, the great southern continent; and the hopes of finding a Northwest Passage were all on the minds of the admiralty. That Cook was able to complete so many of these objectives depended to a large degree on ascorbic acid.

Consider the scenario on June 10, 1770, when the *Endeavour* ran aground on coral of the Great Barrier Reef just south of present-day Cooktown, in northern Queensland, Australia. It was a near catastrophe. The ship had struck at high water; a resulting hole in the hull necessitated drastic measures. In order to lighten the ship, the entire crew heaved overboard everything that could be spared. For twenty-three hours straight they manned the pumps as seawater leaked inexorably into the hold, hauling desperately on cables and anchor in an attempt to plug the hole by fothering, a temporary method of mending a hole by drawing a heavy sail under the hull. Incredible effort, superb seamanship, and good fortune prevailed. The ship eventually slid off the reef and was beached for repairs. It had been a very close call—one that an exhausted, scurvy-inflicted crew could not have summoned the energy to answer.

A healthy, well-functioning crew was essential for Cook to accomplish what he did on his voyages. This fact was recognized by the Royal Society when it awarded him its highest honor, the Copley gold medal, not for his navigational feats but for his demonstration that scurvy was not an inevitable companion on long ocean voyages. Cook's methods were simple. He insisted on maintaining cleanliness throughout the ship, especially in the tight confines of the seamen's quarters. All hands were required to wash their clothes regularly, to air and dry their bedding when the weather permitted, to fumigate between decks, and in general to live up to the meaning of the term *shipshape.* When it was not possible to obtain the fresh fruit and vegetables he thought necessary for a balanced diet, he required that his men eat the sauerkraut he had included in the ship's provisions. Cook touched land at every possible opportunity to replenish stores and gather local grasses (celery grass, scurvy grass) or plants from which he brewed teas.

This diet was not at all popular with the crew, accustomed as they were to the standard seamen's fare and reluctant to try anything new. But Cook was adamant. He and his officers also adhered to this diet, and it was by his example, authority, and determination that his regimen was followed. There is no record that Cook had anyone flogged for refusing to eat sauerkraut or celery grass, but the crew knew the captain would not hesitate to prescribe the lash for opposing his rules. Cook also made use of a more subtle approach. He records that a "Sour Kroutt" prepared from local plants was initially made available only to the officers; within a week the lower ranks were clamoring for their share.

Success no doubt helped convince Cook's crew that their captain's strange obsession with what they ate was worthwhile. Cook never lost a single man to scurvy. On his first voyage of almost three years, one-third of his crew died after contracting malaria or dysentery in Batavia (now Jakarta) in the Dutch East Indies (now Indonesia). On his second voyage from 1772 to 1775, he lost one member of his crew to illness—but not to scurvy. Yet on that trip the crew of his companion vessel was badly affected by the problem. The commander, Tobias Furneaux, was severely reprimanded and instructed yet again by Cook on the need for preparation and administration of antiscorbutics. Thanks to vitamin C, the ascorbic acid molecule, Cook was able to compile an impressive list of accomplishments: the discovery of the Hawaiian Islands and the Great Barrier Reef, the first circumnavigation of New Zealand, the first charting of the coast of the Pacific Northwest, and first crossing of the Antarctic Circle.

A SMALL MOLECULE IN A BIG ROLE

What is this small compound that had such a big effect on the map of the world? The word *vitamin* comes from a contraction of two words, *vital* (necessary) and *amine* (a nitrogen-containing organic compound—it was originally thought that all vitamins contained at least one nitrogen atom). The C in vitamin C indicates that it was the third vitamin ever identified.

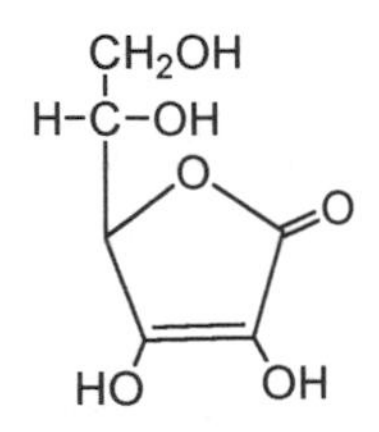

Structure of ascorbic acid (or vitamin C)

This system of naming has numerous flaws. The B vitamins and vitamin H are the only ones that actually do contain nitrogen. The original B vitamin was later discovered to consist of more than one compound, hence vitamin B_1, vitamin B_2, etc. Also, several supposedly different vitamins were found to be the same compound, and thus there is no vitamin F or vitamin G.

Among mammals, only primates, guinea pigs, and the Indian fruit bat require vitamin C in their diet. In all other vertebrates—the family dog or cat, for example—ascorbic acid is made in the liver from the simple sugar glucose by a series of four reactions, each catalyzed by an enzyme. Thus for these animals ascorbic acid is not a dietary necessity. Presumably, somewhere along the evolutionary path humans lost the ability to synthesize ascorbic acid from glucose, apparently by losing the genetic material that enabled us to make *gulonolactone oxidase,* the enzyme necessary for the final step in this sequence.

A similar set of reactions, in a somewhat different order, is the basis for the modern synthetic method (also from glucose) for the industrial preparation of ascorbic acid. The first step is an oxidation reaction, meaning that oxygen is added to a molecule, or hydrogen is removed, or possibly both. In the reverse process, known as reduction, either oxygen is removed from a molecule, or hydrogen is added, or again possibly both.

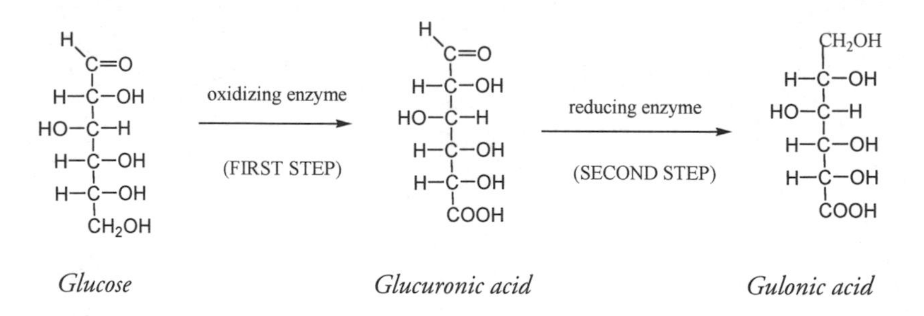

Glucose *Glucuronic acid* *Gulonic acid*

The second step involves reduction at the opposite end of the glucose molecule from that of the first reaction, forming a compound known as *gulonic acid.* The next part of the sequence, the third step, involves gulonic acid forming a cyclic or ring molecule in the form of a *lactone.* A final oxidation step then produces the double bond of the ascorbic acid molecule. It is the enzyme for this fourth and last step that humans are missing.

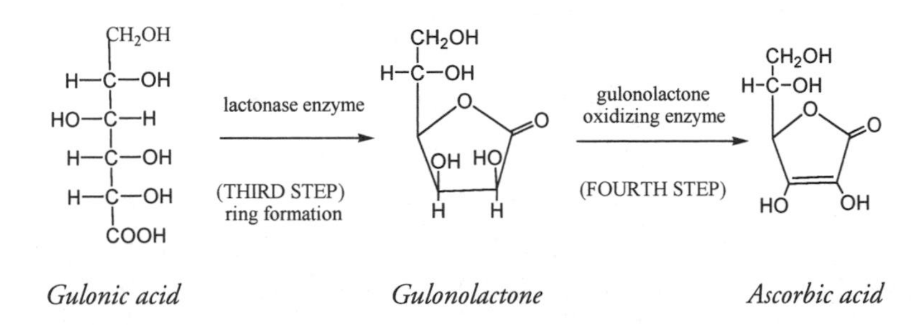

Gulonic acid *Gulonolactone* *Ascorbic acid*

The initial attempts to isolate and identify the chemical structure of vitamin C were unsuccessful. One of the major problems is that although ascorbic acid is present in reasonable amounts in citrus juices, separating it from the many other sugars and sugarlike substances that are also present in these juices is very difficult. It's not surprising, therefore, that the isolation of the first pure sample of ascorbic acid was not from plants but from an animal source.

In 1928, Albert Szent-Györgyi, a Hungarian doctor and biochemist working at Cambridge University in England, extracted less than a gram of crystalline material from bovine adrenal cortex, the inner fatty part of a pair of endocrine glands situated near a cow's kidneys. Present at only about 0.03 percent by weight in his source, the compound was not initially recognized as vitamin C. Szent-Györgyi thought he had isolated a new sugarlike hormone and suggested the name *ignose,* the *ose* part being the ending used for names of sugars (like *glucose* and *fructose*) and the *ig* part signifying that he was ignorant of the substance's structure. When Szent-Györgyi's second suggestion for a name, *Godnose,* was also rejected by the editor of the *Biochemical Journal* (who obviously did not share his sense of humor), he settled for the more sedate name *hexuronic acid.* Szent-Györgyi's sample had been pure enough for accurate chemical analysis to show six carbon atoms in the formula,

$C_6H_8O_6$, hence the *hex* of hexuronic acid. Four years later it was shown that hexuronic acid and vitamin C were, as Szent-Györgyi had come to suspect, one and the same.

The next step in understanding ascorbic acid was to determine its structure, a task that today's technology could accomplish relatively easily using very small amounts but was nearly impossible in the absence of large quantities in the 1930s. Once again Szent-Györgyi was in luck. He discovered that Hungarian paprika was particularly rich in vitamin C and, more important, was particularly lacking in other sugars that had made the compound's isolation in fruit juice such a problem. After only one week's work he had separated over a kilogram of pure vitamin C crystals, more than sufficient for his collaborator, Norman Haworth, professor of chemistry at the University of Birmingham, to begin the successful determination of the structure of what Szent-Györgyi and Haworth had now termed *ascorbic acid.* In 1937 the importance of this molecule was recognized by the scientific community. Szent-Györgyi was awarded the Nobel Prize for medicine for his work on vitamin C, and Haworth the Nobel Prize for chemistry.

Despite more than sixty years of further work, we are still not completely sure of all the roles that ascorbic acid plays in the body. It is vital for the production of collagen, the most abundant protein in the animal kingdom, found in connective tissues that bind and support other tissues. Lack of collagen, of course, explains some of the early symptoms of scurvy: the swelling of limbs, softening of gums, and loosening of teeth. As little as ten milligrams a day of ascorbic acid is said to be sufficient to keep the symptoms of scurvy at bay, although at that level subclinical scurvy (vitamin C deficiency at the cellular level but no gross symptoms) probably exists. Research in areas as varied as immunology, oncology, neurology, endocrinology, and nutrition is still discovering the involvement of ascorbic acid in many biochemical pathways.

Controversy as well as mystery has long surrounded this small molecule. The British navy delayed implementing James Lind's recommendations by a scandalous forty-two years. The East India Company purportedly withheld antiscorbutic foods on purpose in order to keep its sailors weak and controllable. At present there are debates on whether megadoses of vitamin C play a role in treatment of a variety of conditions. Linus Pauling was recognized in 1954 with the Nobel Prize in chemistry for his work on the chemical bond and again in 1962 with the Nobel Peace Prize for his activities opposing the testing of nuclear weapons. In 1970 this double Nobel laureate released the first of a number of publications on the role of vitamin C in medicine, recommending high doses of ascorbic acid for the prevention and cure of colds, flu, and cancer. Despite his eminence as a scientist, the medical establishment has not generally accepted Pauling's views.

The RDA (Recommended Daily Allowance) of vitamin C for an adult is generally given as sixty milligrams per day, about that found in a small orange. The RDA has varied over time and in different countries, perhaps indicating our lack of understanding of the complete physiological role of this not-so-simple molecule. It is agreed that a higher RDA is necessary during pregnancy and breast-feeding. The highest RDA is recommended for older people, a time when vitamin C intake is often reduced through poor diet or lack of interest in cooking and eating. Scurvy today is not unknown among the elderly.

A daily dose of 150 milligrams of ascorbic acid generally corresponds to a saturation level, and further intake does little to increase the ascorbic acid content of blood plasma. As excess vitamin C is eliminated through the kidneys, it has been claimed that the only good done by megadoses is to provide profits for pharmaceutical companies. It does seem, however, that higher doses may be necessary under circumstances such as infection, fever, wound recovery, diarrhea, and a long list of chronic conditions.

Research continues into the role of vitamin C in more than forty disease states; bursitis, gout, Crohn's disease, multiple sclerosis, gastric ulcers, obesity, osteoarthritis, Herpes simplex infections, Parkinson's, anemia, coronary heart disease, autoimmune diseases, miscarriages, rheumatic fever, cataracts, diabetes, alcoholism, schizophrenia, depression, Alzheimer's, infertility, cold, flu, and cancer, to name just some of them. When you look at this list, you may see why this molecule has sometimes been described as "youth in a bottle," although research results do not as yet support all the miracles that have been claimed.

Over fifty thousand tons of ascorbic acid are manufactured annually. Produced industrially from glucose, synthetic vitamin C is absolutely identical in every way to its natural counterpart. There is no physical or chemical difference between natural and synthetic ascorbic acid, so there is no reason to buy an expensive version marketed as "natural vitamin C, gently extracted from the pure rose hips of the rare *Rosa macrophylla,* grown on the pristine slopes of the lower Himalayas." Even if the product did originate at this source, if it is vitamin C, it is exactly the same as vitamin C that has been manufactured by the ton from glucose.

This is not to say that manufactured vitamin pills can replace the natural vitamins in foods. Swallowing a seventy-milligram ascorbic acid pill may not produce quite the same benefits as the seventy milligrams of vitamic C obtained from eating an average-sized orange. Other substances found in fruits and vegetables, such as those responsible for their bright colors, may help the absorption of vitamin C or in some way, as yet unknown, enhance its effect.

The main commercial use of vitamin C today is as a food preservative, where it acts as an antioxidant and an antimicrobial agent. In recent years food preservatives have come to be seen as bad. "Preservative free" is shouted from many a food package. Yet without preservatives much of our food supply would taste bad, smell bad, be inedible, or even kill us. The loss of chemical preservatives would be as great a disaster to our food supply as would the cessation of refrigeration and freezing.

It is possible to safely preserve fruit in the canning process at the temperature of boiling water, as fruit is usually acidic enough to prevent the growth of the deadly microbe *Clostridium botulinum.* Lower-acid-content vegetables and meats must be processed at higher temperatures to kill this common microorganism. Ascorbic acid is often used in home canning of fruit as an antioxidant to prevent browning. It also increases acidity and protects against botulism, the name given to the food poisoning resulting from the toxin produced by the microbe. *Clostridium botulinum* does not survive inside the human body. It is the toxin it produces in improperly canned food that is dangerous, although only if eaten. Tiny amounts of the purified toxin injected under the skin interrupt nerve pulses and induce muscle paralysis. The result is a temporary erasing of wrinkles—the increasingly popular Botox treatment.

Although chemists have synthesized many toxic chemicals, nature has created the most deadly. Botulinum toxin A, produced by *Clostridium botulinum,* is the most lethal poison known, one million times more deadly than dioxin, the most lethal man-made poison. For botulinum toxin A, the lethal dose that will kill 50 percent of a test population (the LD_{50}) is 3×10^{-8} mg per kg. A mere 0.00000003 milligrams of botulinum toxin A per kilogram of body weight of the subject is lethal. For dioxin, the LD_{50} is 3×10^{-2} mg per kg, or 0.03 milligrams per kilogram of body weight. It has been estimated that one ounce of botulinum toxin A could kill 100 million people. These numbers should surely make us rethink our attitudes toward the perceived evils of preservatives.

SCURVY ON ICE

Even in the early twentieth century a few Antarctic explorers still supported theories that putrefaction of preserved food, acid intoxication of the blood, and bacterial infections were the cause of scurvy. Despite the fact that compulsory lemon juice had virtually eliminated scurvy from the British navy in the early 1800s, despite observations that Eskimos in the polar regions who ate the vitamin C–rich fresh meat, brain, heart, and kidneys of seals never suffered from scurvy, and despite the experience of numerous explorers whose antiscorbutic precautions included as much fresh food as possible in the diet, the British naval commander Robert Falcon Scott persisted in his belief that scurvy was caused by tainted meat. The Norwegian explorer Roald Amundsen, on the other hand, took the threat of scurvy seriously and based the diet for his successful South Pole expedition on fresh seal and dog meat. His 1911 return journey to the pole, some fourteen hundred miles, was accomplished without sickness or accident. Scott's men were not so fortunate. Their return journey, after reaching the South Pole in January 1912, was slowed by what is now thought to be the Antarctic's worst weather in years. Symptoms of scurvy, brought on by several months on a diet devoid of fresh food and vitamin C, may have greatly hampered their efforts. Only eleven miles from a food and fuel depot they found themselves too exhausted to continue. For Commander Scott and his companions, just a few milligrams of ascorbic acid might have changed their world.

Had the value of ascorbic acid been recognized earlier, the world today might be a very different place. With a healthy crew Magellan might not have bothered to stop in the Philippines. He could have gone on to corner the Spice Islands clove market for Spain, sail triumphantly upriver to Seville, and enjoy the honors due to the first circumnavigator of the globe. A Spanish monopoly of the clove and nutmeg markets might have thwarted the establishment of the Dutch East India Company—and changed modern-day Indonesia. If the Portuguese, the first European explorers to venture these long distances, had understood the secret of ascorbic acid, they might have explored the Pacific Ocean centuries before James Cook. Portuguese might now be the language spoken in Fiji and Hawaii, which might have joined Brazil as colonies in a far-flung Portuguese Empire. Maybe the great Dutch navigator Abel

Janszoon Tasman, with the knowledge of how to prevent scurvy on his voyages of 1642 and 1644, would have landed on and formally laid claim to the lands known as New Holland (Australia) and Staten Land (New Zealand). The British, coming later to the South Pacific, would have been left with a much smaller empire and far less influence in the world, even to this day. Such speculation leads us to conclude that ascorbic acid deserves a prominent place in the history—and geography—of the world.

3. GLUCOSE

THE NURSERY RHYME phrase "Sugar and spice and everything nice" pairs sugar with spices—a classic culinary matching that we appreciate in such treats as apple pie and gingerbread cookies. Like spice, sugar was once a luxury affordable only by the rich, used as a flavoring in sauces for meat and fish dishes that today we would consider savory rather than sweet. And like spice molecules, the sugar molecule affected the destiny of countries and continents as it ushered in the Industrial Revolution, changing commerce and cultures around the world.

Glucose is a major component of sucrose, the substance we mean when we refer to sugar. Sugar has names specific to its source, such as cane sugar, beet sugar, and corn sugar. It also comes in a number of variations: brown sugar, white sugar, berry sugar, castor sugar, raw sugar, demerara sugar. The glucose molecule, present in all these kinds of sugar, is fairly small. It has just six carbon, six oxygen, and twelve hydrogen atoms, altogether the same number of atoms as in the molecules responsible for the tastes of nutmeg and cloves. But just as in those spice molecules, it is the spatial arrangements of the atoms of the glucose molecule (and other sugars) that result in a taste—a sweet taste.

Sugar can be extracted from many plants; in tropical regions it is usually obtained from sugarcane and in temperate regions from sugar beets. Sugarcane (*Saccharum officinarum*) is variously described as originating in the South Pacific or southern India. Sugarcane cultivation spread through Asia and to the Middle East, eventually reaching northern Africa and Spain. Crystalline sugar extracted from cane reached Europe with the first of the returning Crusaders during the thirteenth century. For the next three centuries it remained an exotic commodity, treated in much the same way as spices: the center for the sugar trade developed initially in Venice along with the burgeoning spice trade. Sugar was used in medicine to disguise the often-nauseating taste of other ingredients, to act as a binding agent for drugs, and as a medicine in itself.

By the fifteenth century sugar was more readily available in Europe, but it was still expensive. An increase in the demand for sugar and lower prices coincided with a decrease in the supply of honey, which had previously been the sweetening agent in Europe and much of the rest of the world. By the sixteenth century sugar was rapidly becoming the sweetener of choice for the masses. It became even more popular with the seventeenth- and eighteenth-century discoveries of the preservation of fruit by sugar and the making of jams, jellies, and marmalades. In England in 1700 the estimated yearly per capita consumption of sugar was about four pounds. By 1780 this had risen to twelve pounds and, in the 1790s, to sixteen pounds, much of it probably consumed in the newly popular drinks of tea, coffee, and chocolate. Sugar was also being used in sweet treats: syrup-covered nuts and seeds, marzipan, cakes, and candies. It had become a staple food, a necessity rather than a luxury, and consumption continued to rise through the twentieth century.

Between 1900 and 1964 world sugar production increased by 700 percent, and many developed countries reached a per capita annual consumption of one hundred pounds. This figure has dropped somewhat in recent years with the increasing use of artificial sweeteners and concerns over high-calorie diets.

SLAVERY AND SUGAR CULTIVATION

Without the demand for sugar, our world today would probably be a lot different. For it was sugar that fueled the slave trade, bringing millions of black Africans to the New World, and it was profit from the sugar trade that by the beginning of the eighteenth century helped spur economic growth in Europe. Early explorers of the New World brought back reports of tropical lands that were ideal for the cultivation of sugar. It took little time for Europeans, eager to overcome the sugar monopoly of the Middle East, to start growing sugar in Brazil and then in the West Indies. Sugarcane cultivation is labor intensive, and two possible sources of workers—native populations of the New World (already decimated by newly introduced diseases such as smallpox, measles, and malaria) and indentured servants from Europe—could not supply even a fraction of the needed workforce. The colonists of the New World looked toward Africa.

Until this time the slave trade from western Africa was mainly limited to the domestic markets of Portugal and Spain, an outgrowth of the trans-Saharan trade of the Moorish people around the Mediterranean. But the need for workers in the New World drastically increased what had been to that point a minor practice. The prospect of deriving great wealth from sugar cultivation was enough for England, France, Holland, Prussia, Denmark, and Sweden (and eventually Brazil and the United States) to become part of a massive system of transporting millions of Africans from their homes. Sugar was not the only commodity that relied on slave labor, but it was probably the major one. According to some estimates, around two-thirds of African slaves in the New World labored on sugar plantations.

The first slave-grown sugar from the West Indies was shipped to Europe in 1515, just twenty-two years after Christopher Columbus had, on his second voyage, introduced sugarcane to the island of Hispaniola. By the middle of the sixteenth century Spanish and Portuguese settlements in Brazil, Mexico, and many Caribbean islands were producing sugar. The annual slave shipment from Africa to these plantations numbered around ten thousand. Then in the seventeenth century the English, French, and Dutch colonies in the West Indies began growing sugarcane. The rapidly expanding demand for sugar, the growing technology of sugar processing, and the development of a new alcoholic drink, rum, from the by-products of sugar refining contributed to an explosive rise in the number of people dispatched from Africa to work the sugarcane fields.

It is impossible to establish the exact numbers of slaves that were loaded onto sailing vessels off the west coast of Africa and later sold in the New World. Records are incomplete and possibly fraudulent, reflecting attempts to get around the laws that belatedly tried to improve conditions aboard these transport ships by regulating the number of slaves that could be carried. As late as the 1820s more than five hundred human beings were being packed, on Brazilian slave ships, into an area less than nine hundred feet square and three feet high. Some historians calculate that upward of fifty million Africans were shipped to the Americas over the three and a half centuries of the slave trade. This figure does not include those who would have been killed in slaving raids, or those who died on the trip from the interior of the continent to the African coasts, or those who did not survive the horrors of the sea voyage that came to be known as the middle passage.

The middle passage refers to the second side of the trade triangle known as the Great Circuit. The first leg of this triangle was the trip from Europe to the coast of Africa, predominantly the west coast of Guinea, bringing manufactured goods to exchange for slaves. The third leg was the passage from the New World back to Europe. The slave ships by that point would have exchanged their human cargo for ore from mines and produce from the plantations, generally rum, cotton, and tobacco. Each leg of the triangle was hugely profitable, especially for Britain: by the end of the eighteenth century the value of British income derived from the West Indies was much greater than the value of income from trade with the rest of the world. Sugar and sugar products, in fact, were the source of the enormous increase in capital and the rapid economic expansion necessary to fuel the British and later the French Industrial Revolution of the late eighteenth and early nineteenth centuries.

SWEET CHEMISTRY

Glucose is the most common of the simple sugars, which are sometimes called *monosaccharides* from the Latin word *saccharum* for sugar. The *mono* prefix refers to one unit, as opposed to the two-unit *disaccharides* or the many-unit *polysaccharides*. The structure of glucose can be drawn as a straight chain

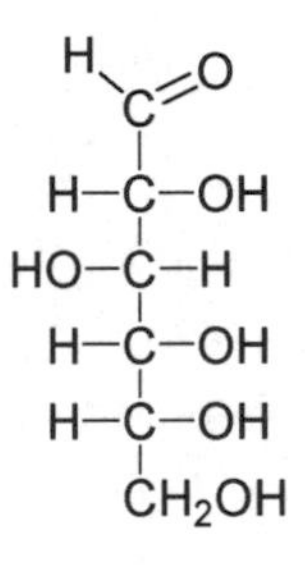

Glucose

or as a slight adaptation of this chain, where each intersection of vertical and horizontal lines represents a carbon atom. A set of conventions that need not concern us give numbers to the carbon atoms, with carbon number 1 always drawn at the top. This is known as a Fischer projection formula, after Emil Fischer, a German chemist who in 1891 determined the actual structure of glucose and a number of other related sugars. Though the scientific tools and techniques available to Fischer at the time were very rudimentary, his results still stand today as one of the most elegant examples of chemical logic. He was awarded the 1902 Nobel Prize in chemistry for his work on sugars.

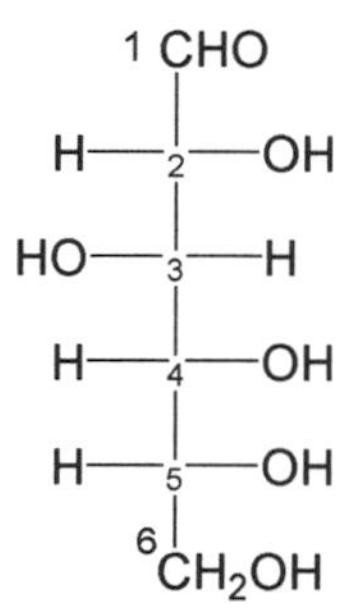

Fischer projection formula for glucose, showing the numbering of the carbon chain

Although we can still draw sugars such as glucose in this straight chain form, we now know that they normally exist in a different form—cyclic (ring) structures. The drawings of these cyclic structures are known as Haworth formulas, after Norman Haworth, the British chemist whose 1937 Nobel Prize recognized his work on vitamin C and on structures of carbohydrates (see Chapter 2). The six-membered ring of glucose consists of five carbon atoms and one oxygen atom. Its Haworth formula, shown below, indicates by number how each carbon atom corresponds to the carbon atom shown in the previous Fischer projection formula.

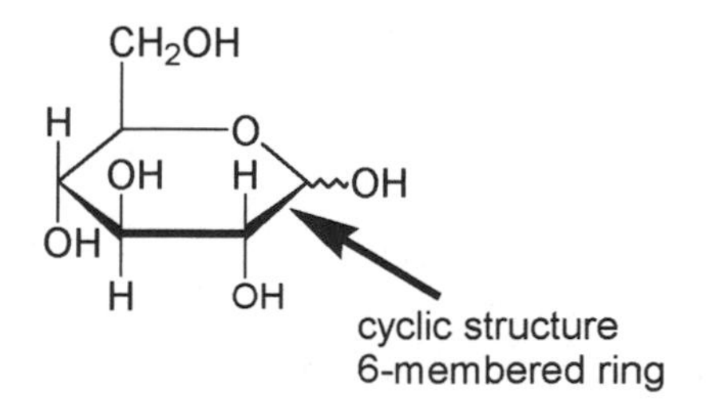

Haworth formula of glucose, showing all the hydrogen atoms

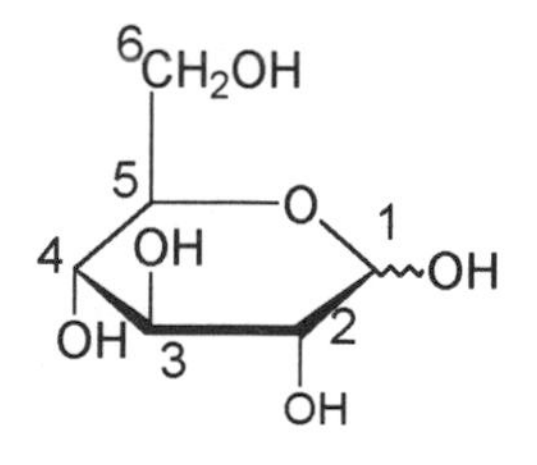

Haworth formula of glucose, without all the H atoms but showing carbon atoms numbered

There are actually two versions of glucose in the cyclic form, depending on whether the OH at carbon number 1 is above or below the ring. This might seem a very minor distinction, but it is worth noting as it has very important consequences for the structures of more complicated molecules that contain glucose units, such as complex carbohydrates. If the OH at carbon number 1 is below the ring it is known as alpha (α)-glucose. If it is above the ring, it is beta (β)-glucose.

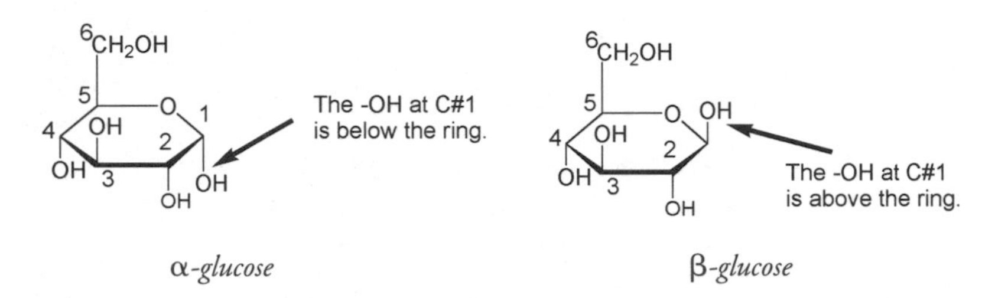

When we use the word *sugar,* we are referring to sucrose. Sucrose is a disaccharide, so called because it is composed of two simple monosaccharide units: one a glucose unit and the other a fructose unit. Fructose, or fruit sugar, has the same formula as glucose; $C_6H_{12}O_6$, and also the same number and type of atoms (six carbon, twelve hydrogen, and six oxygen) found in glucose. But fructose has a different structure. Its atoms are arranged in a different order. The chemical definition of this is that fructose and glucose are *isomers.* Isomers are compounds that have the same chemical formula (same number of each atom) but different arrangements of these atoms.

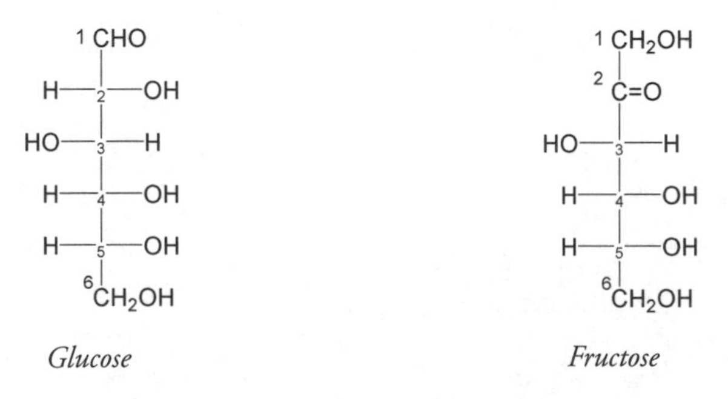

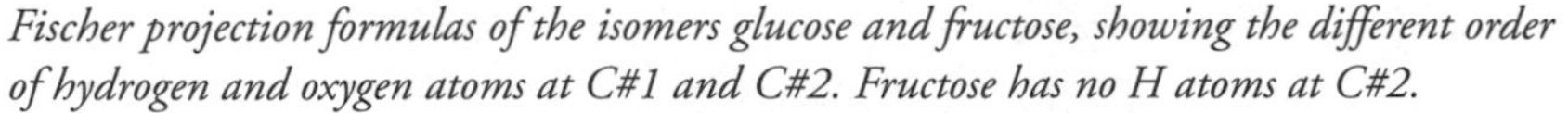

Fischer projection formulas of the isomers glucose and fructose, showing the different order of hydrogen and oxygen atoms at C#1 and C#2. Fructose has no H atoms at C#2.

Fructose exists mainly in the cyclic form, but it looks a bit different from glucose since fructose forms a five-membered ring, shown below as a Haworth formula, rather than the six-membered ring of glucose. As with glucose, there are α and β forms of fructose, but as it is carbon number 2 that joins the ring oxygen in fructose, it is around this carbon atom that we designate OH below the ring as α and OH above the ring as β.

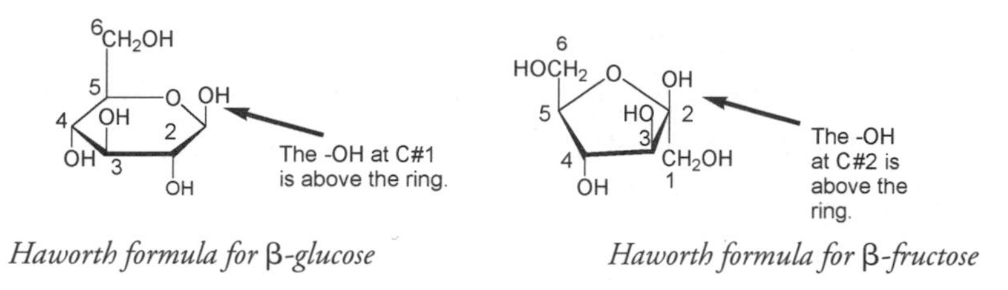

Haworth formula for β-glucose *Haworth formula for β-fructose*

Sucrose contains equal amounts of glucose and fructose but not as a mixture of two different molecules. In the sucrose molecule one glucose and one fructose are joined together through the removal of a molecule of water (H_2O) between the OH at carbon number 1 of α-glucose and the OH on carbon number 2 of β-fructose.

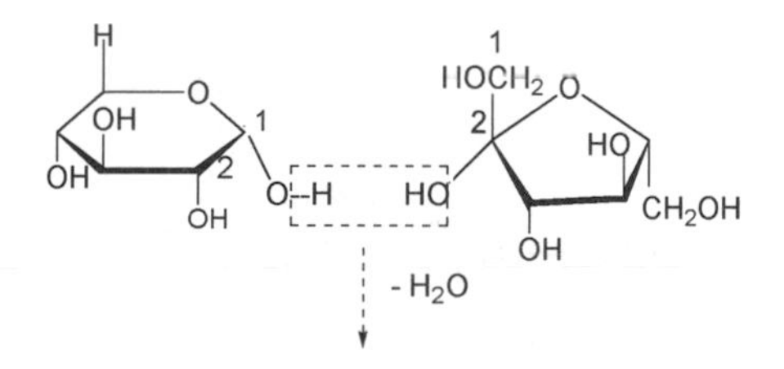

Removal of a molecule of H_2O between glucose and fructose forms sucrose. The fructose molecule has been turned 180° and inverted in these diagrams.

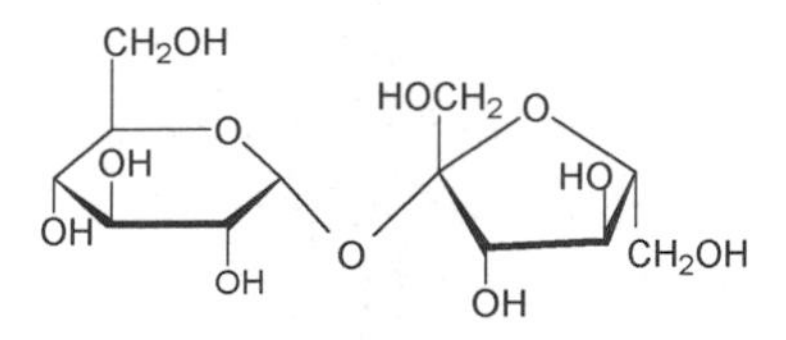

Structure of the sucrose molecule

Fructose is largely found in fruit but also in honey, which is about 38 percent fructose and 31 percent glucose, with another 10 percent of other sugars including sucrose. The remainder is mainly water. Fructose is sweeter than sucrose or glucose, so because of its fructose component, honey is sweeter than sugar. Maple syrup is approximately 62 percent sucrose with only 1 percent of each of fructose and glucose.

Lactose, also called milk sugar, is a disaccharide formed from one unit of glucose and one unit of another monosaccharide, galactose. Galactose is an isomer of glucose; the only difference is that in galactose the OH group at carbon number 4 is above the ring and not below the ring as it is in glucose.

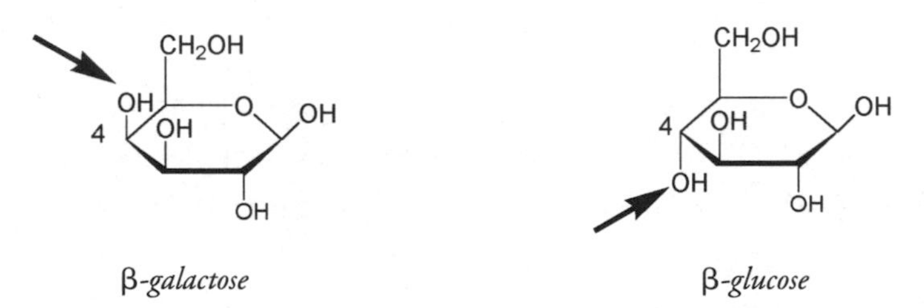

β-galactose with arrow showing C#4 OH above the ring compared to β-glucose where the C#4 OH is below the ring. These two molecules combine to form lactose.

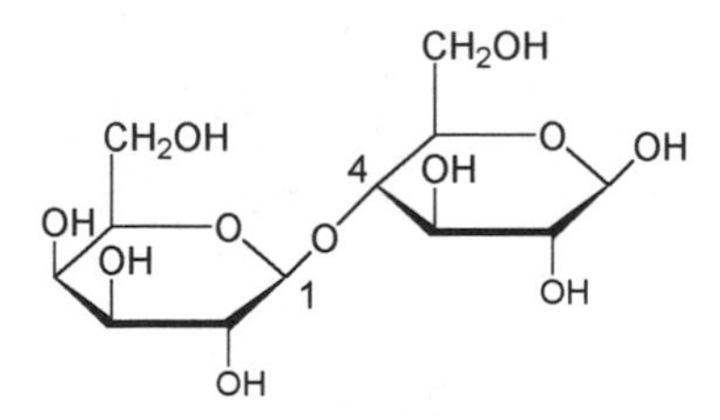

Structure of the lactose molecule

Galactose on the left is joined through C#1 to C#4 of glucose on the right.

Again, having an OH above or below the ring may seem like a very minor difference, but for those people who suffer from lactose intolerance, it is not. To digest lactose and other disaccharides or larger sugars, we need specific enzymes that initially break down these complex molecules into simpler monosaccharides. In the case of lactose, the enzyme is called *lactase* and is present in only small amounts in some adults. (Children generally produce greater amounts of lactase than adults.) Insufficient lactase makes the digestion of milk and milk products difficult and causes the symptoms associated with lactose intolerance: abdominal bloating, cramps, and diarrhea. Lactose intolerance is an inherited trait, easily treated with over-the-counter preparations of the lactase enzyme. Adults and children (but not babies) from certain ethnic groups, such as some African tribes, are missing the lactase enzyme completely. For these people, powdered milk and other milk products, often found in food aid programs, are indigestible and even harmful.

The brain of a normal healthy mammal uses only glucose for fuel. Brain cells are dependent on a minute-to-minute supply from the bloodstream, as there are essentially no fuel reserves or storage in the brain. If blood glucose level falls to 50 percent of the normal level, some symptoms of brain dysfunction appear. At 25 percent of the normal level, possibly from an overdose of insulin—the hormone that maintains the level of glucose in the blood—a coma may result.

SWEET TASTE

What makes all these sugars so appealing is that they taste sweet, and humans like sweetness. Sweetness is one of the four principal tastes; the other three are sourness, bitterness, and saltiness. Achieving the ability to distinguish among these tastes was an important evolutionary step. Sweetness generally implies "good to eat." A sweet taste indicates that fruit is ripe, whereas sour tells us there are still lots of acids present, and the unripe fruit may cause a stomachache. A bitter taste in plants often indicates the presence of a type of compound known as an alkaloid. Alkaloids are often poisonous, sometimes in only very small amounts, so the ability to detect traces of an alkaloid is a distinct advantage. It has even been suggested that the extinction of the dinosaurs might have been due to their inability to detect the poisonous alkaloids found in some of the flowering plants that evolved toward the end of the Cretaceous period, about the time the dinosaurs disappeared, although this is not the generally accepted theory of dinosaur extinction.

Humans do not seem to have an inborn liking for bitterness. In fact, their preference is probably just the opposite. Bitterness invokes a response involving secretion of extra saliva. This is a useful reaction to something poisonous in the mouth, allowing one to spit it out as completely as possible. Many people do, however, learn to appreciate, if not like, the bitter taste. Caffeine in tea and coffee and quinine in tonic water are examples of this phenomenon, although many of us still rely on having sugar in these drinks. The term *bittersweet,* connoting pleasure mixed with sadness, conveys our ambivalence about bitter tastes.

Our sense of taste is located in the taste buds, specialized groups of cells found mainly on the tongue. Not all parts of the tongue detect taste the same way or to the same degree. The front tip of the tongue is the most sensitive to sweetness, while sourness is detected most strongly on the sides of the tongue toward the back. You can test this easily for yourself by touching a sugar solution to the side of the tongue and then to the tip of the tongue. The tip of the tongue will definitely detect the sweet sensation more strongly. If you try the same thing with lemon juice, the result will be even more obvious. Lemon juice on the very tip of the tongue does not seem very sour, but put a freshly cut slice of lemon on the side of the tongue, and you will discover where the sourness reception area is the strongest. You can continue this experiment: bitterness is detected most strongly on the middle of the tongue but back from the tip, and the salty sensation is greatest just to each side of the tip.

Sweetness has been investigated far more than any of the other tastes, no doubt because, as in the days of the slave trade, it is still big business. The relationship between chemical structure and sweetness is complicated. One simple model, known as the A-H,B Model, suggests that a sweet taste depends on an arrangement of a group of atoms within a molecule. These atoms (A and B in the diagram) have a particular geometry, allowing atom B to be attracted to the hydrogen atom attached to atom A. This results in the short-term binding of the sweet molecule to a protein molecule of a taste receptor, causing a generation of a signal (transmitted through nerves) informing the brain, "This is sweet." A and B are usually oxygen or nitrogen atoms, although one of them may also be a sulfur atom.

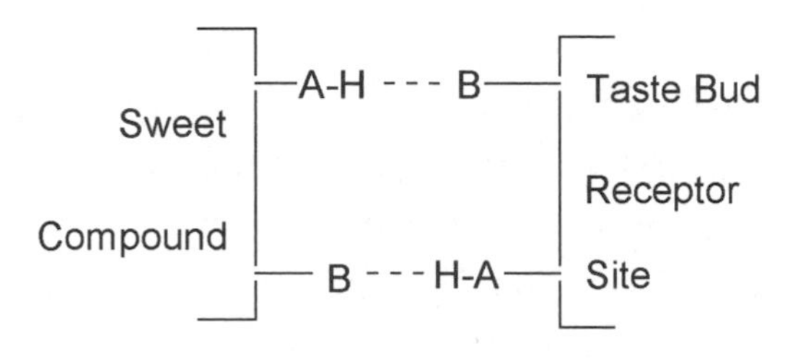

The A-H,B Model of Sweetness

There are many sweet compounds other than sugar, and not all of them are good to eat. Ethylene glycol, for example, is the major component of antifreeze used in car radiators. The solubility and flexibility of the ethylene glycol molecule, as well as the distance between its oxygen atoms (similar to the distance between oxygen atoms in sugars), account for its sweet taste. But it is very poisonous. A dose of as little as one tablespoon can be lethal for humans or family pets.

Interestingly, it is not ethylene glycol but what the body turns it into that is the toxic agent. Oxidation of ethylene glycol by enzymes in the body produces oxalic acid.

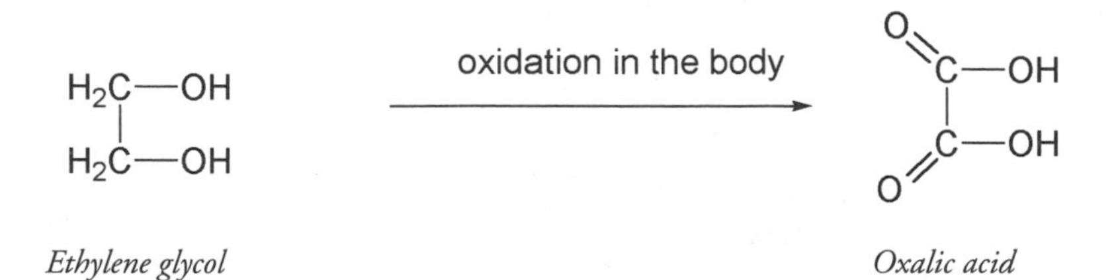

Ethylene glycol *Oxalic acid*

Oxalic acid occurs naturally in a number of plants, including some that we eat, such as rhubarb and spinach. We usually consume these foods in moderate amounts, and our kidneys can cope with the traces of oxalic acid from such sources. But if ethylene glycol is swallowed, the sudden appearance of a large amount of oxalic acid can cause kidney damage and death. Eating spinach salad and rhubarb pie at the same meal will not hurt you. It would probably be difficult to consume enough spinach and rhubarb to do any harm, except perhaps if you are prone to kidney stones, which build up over some years. Kidney stones consist mainly of calcium oxalate, the insoluble calcium salt of oxalic acid; those prone to kidney stones are often advised to avoid foods high in oxalates. For the rest of us, moderation is the best advice.

A compound that has a very similar structure to ethylene glycol and also tastes sweet is glycerol, but glycerol in moderate amounts is safe to consume. It is used as an additive in many prepared foods because of its viscosity and high water solubility. The term *food additive* has had a bad press in recent years, implying that food additives are essentially nonorganic, unhealthy, and unnatural. Glycerol is definitely organic, is nontoxic, and occurs naturally in products such as wine.

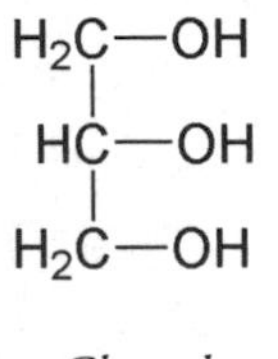

Glycerol

When you swirl a glass of wine, the "legs" that form on the glass are due to the presence of glycerol increasing the viscosity and smoothness characteristic of good vintages.

SWEET NOTHING

There are numerous other nonsugars that taste sweet, and some of these compounds are the basis for the billion-dollar artificial sweetener industry. As well as having a chemical structure that in some way mimics the geometry of sugars, allowing it to fit and bind to the sweetness receptor, an artificial sweetener needs to be water soluble and nontoxic and, often, not metabolized in the human body. These substances are usually hundreds of times sweeter than sugar.

The first of the modern artificial sweeteners to be developed was saccharin. Saccharin is a fine powder. Those who work with it sometimes detect a sweet taste if they accidentally touch their fingers to their mouth. It is so sweet that only a very small amount triggers the sweetness response. This is evidently what happened in 1879, when a chemistry student at Johns Hopkins University in Baltimore noticed an unusual sweetness in the bread he was eating. He returned to his laboratory bench to systematically taste the compounds that he had been using in that day's experiments—a risky but common practice with new molecules in those days—and discovered that saccharin was intensely sweet.

Saccharin has no calorific value, and it did not take long (1885) for this combination of sweetness and no calories to be commercially exploited. Originally intended as a replacement for sugar in the diet of diabetic patients, it quickly became an accepted sugar substitute for the general population. Concern about possible toxicity and the problem of a metallic aftertaste led to the development of other artificial sweeteners, such as cyclamate and aspartame. As you can see, the structures of these are all quite different and are very different from sugars, yet they all have the appropriate atoms, along with the specific atomic position, geometry, and flexibility that is necessary for sweetness.

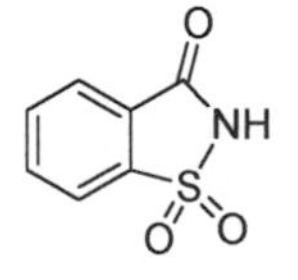

Saccharin

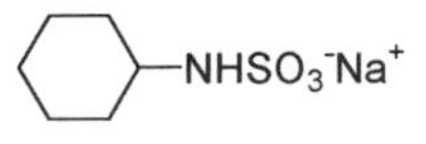

Sodium cyclamate

Aspartame

No artificial sweetener is completely free of problems. Some decompose on heating and so can be used only in soft drinks or cold foods; some are not particularly soluble; and others have a detectable side taste along with their sweetness. Aspartame, although synthetic, is composed of two naturally occurring amino acids. It is metabolized by the body, but as it is over two hundred times sweeter than glucose, a lot less is needed to produce a satisfactory level of sweetness. Those with the inherited condition known as PKU (phenylketonuria), an inability to metabolize the amino acid phenylalanine, one of the breakdown products of aspartame, are told to avoid this particular artificial sweetener.

A new sweetener that was approved by the U.S. Food and Drug Administration in 1998 approaches the problem of creating artificial sweetness in a different way. Sucralose has a very similar structure to that of sucrose except for two factors. The glucose unit, on the left-hand side in the diagram, is replaced by galactose, the same unit as in lactose. Three chlorine atoms (Cl) replace three of the OH groups: one on the galactose unit and the other two on the right-hand fructose unit, as indicated. The three chlorine atoms do not affect the sweetness of this sugar, but they do stop the body from metabolizing it. Hence sucralose is a noncalorific sugar.

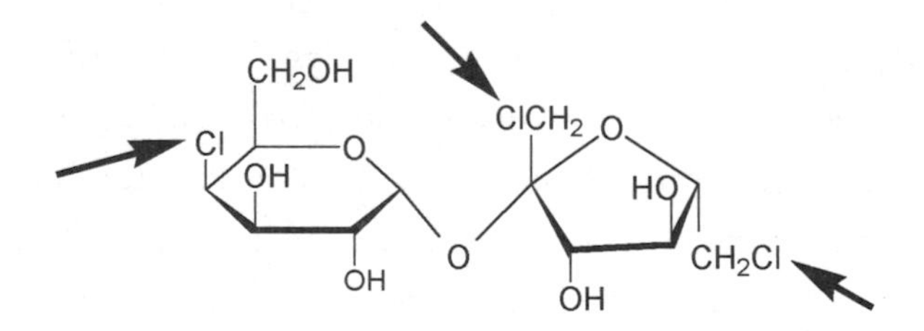

Sucralose structure, showing the three Cl atoms (arrows) replacing three OHs

Natural nonsugar sweeteners are now being sought from plant sources containing "high-potency sweeteners"—compounds that can be as much as a thousand times sweeter than sucrose. For centuries indigenous people have known about plants that have a sweet taste; the South American herb *Stevia rebaudiana;* roots of the licorice plant *Gly-*

cyrrhiza glabra; Lippia dulcis, a Mexican member of the verbena family; and rhizomes from *Selliguea feei,* a fern from West Java, are examples. Sweet compounds from natural sources have shown potential for commercial application, but problems with small concentrations, toxicity, low water solubility, unacceptable aftertaste, stability, and variable quality still need to be overcome.

While saccharin has been used for more than a hundred years, it was not the first substance to be used as an artificial sweetener. That distinction probably belongs to lead acetate, $Pb(C_2H_3O_2)_2$, which was used to sweeten wine in the days of the Roman Empire. Lead acetate, known as sugar of lead, would sweeten a vintage without causing further fermentation, which would have occurred with the addition of sweeteners like honey. Lead salts are known to be sweet, and many are insoluble, but all are poisonous. Lead acetate is very soluble, and its toxicity was obviously not known to the Romans. This should give us pause to think, if we long for the good old days when food and drink were uncontaminated with additives.

The Romans also stored wine and other beverages in lead containers and supplied their houses with water through lead pipes. Lead poisoning is cumulative. It affects the nervous system and the reproductive system as well as other organs. Initial symptoms of lead poisoning are vague but include restless sleep, loss of appetite, irritation, headaches, stomachaches, and anemia. Brain damage, leading to gross mental instability and paralysis, develops. Some historians have attributed the fall of the Roman Empire to lead poisoning, as many Roman leaders, including the Emperor Nero, are reported to have exhibited these symptoms. Only the wealthy, aristocratic, ruling Roman class had water piped to their houses and used lead vessels for storing wine. Ordinary people would have fetched their water and stored their wine in other containers. If lead poisoning did indeed contribute to the fall of the Roman Empire, it would be yet another example of a chemical that changed the course of history.

Sugar—the desire for its sweetness—shaped human history. It was profit from the huge sugar market developing in Europe that motivated the shipping of African slaves to the New World. Without sugar there would have been a much-reduced slave trade; without slaves there would have been a much-reduced sugar trade. Sugar started the massive buildup of slavery, and sugar revenues sustained it. The wealth of West African states—their people—was transferred to the New World to build wealth for others.

Even after slavery was abolished, the desire for sugar still affected human movement around the globe. At the end of the nineteenth century large numbers of indentured laborers from India went to the Fijian Islands to work in the sugarcane fields. As a result, the racial composition of this Pacific island group changed so completely that the native Melanesians were no longer a majority. After three coups in recent years Fiji is still a country of political and ethnic unrest. The racial makeup of the population of other tropical lands also owes much to sugar. Many of the ancestors of present-day Hawaii's largest ethnic group emigrated from Japan to work in the Hawaiian sugarcane fields.

Sugar continues to shape human society. It is an important trade commodity; vagaries in weather and pest infestations affect the economies of sugar-growing countries and stock markets around the world. An increase in sugar price has a ripple effect throughout the food industry. Sugar has been used as a political tool; for decades purchase of Cuban sugar by the USSR supported the economy of Fidel Castro's Cuba.

We have sugar in much of what we drink and much of what we eat. Our children prefer sugary treats. We tend to offer sweet foods when we entertain; offering hospitality to guests no longer means breaking a simple loaf of bread. Sugar-laden treats and candies are associated with major holidays and celebrations in cultures around the world. Levels of consumption of the glucose molecule and its isomers, many times higher than in previous generations, are reflected in health problems such as obesity, diabetes, and dental caries. In our everyday lives we continue to be shaped by sugar.

4. CELLULOSE

SUGAR PRODUCTION promoted the buildup of the slave trade to the Americas, but sugar was not alone in sustaining it for over three centuries. The cultivation of other crops for the European market also depended on slavery. One of these crops was cotton. Raw cotton shipped to England could be made into the cheap manufactured goods that were sent to Africa in exchange for the slaves shipped to plantations in the New World, especially to the southern United States. Profit from sugar was the first fuel for this trade triangle, and it supplied the initial capital for the growing British industrialization. But it was cotton and the cotton trade that launched rapid economic expansion in Britain in the late eighteenth century and early nineteenth century.

Cotton and the Industrial Revolution

The fruit of the cotton plant develops as a globular pod known as a boll, which contains oily seeds within a mass of cotton fibers. There is evidence that cotton plants, members of the *Gossypium* genus, were cultivated in India and Pakistan and also in Mexico and Peru some five thousand years ago, but the plant was unknown in Europe until around 300 B.C. when soldiers from the army of Alexander the Great returned from India with cotton robes. Arab traders brought cotton plants to Spain during the Middle Ages. The cotton plant is frost tender and needs moist but well-drained soil and long hot summers, not the conditions found in the temperate regions of Europe. Cotton had to be imported to Britain and other northern countries.

Lancashire, in England, became the center of the great industrial complex that grew up around cotton manufacture. The damp climate of the region helped cotton fibers stick together, which was perfect for the manufacture of cotton, as it meant less likelihood of threads breaking during the spinning and weaving processes. Cotton mills in drier climates suffered higher production costs due to this factor. As well, Lancashire had land available for building mills, land for housing the thousands who were needed to work in the cotton industry, abundant soft water for the bleaching, dyeing, and printing of cotton, and a plentiful supply of coal, a factor that became very important as steam power arrived.

In 1760, England imported 2.5 million pounds of raw cotton. Less than eighty years later the country's cotton mills were processing more than 140 times this amount. This increase had an enormous effect on industrialization. Demand for cheap cotton yarns led to mechanical innovation, and eventually all stages of cotton processing became mechanized. The eighteenth century saw the development of the mechanical cotton gin for separating the cotton fiber from the seeds, carding machines to prepare the raw fiber, the spinning jenny and spinning throstles for drawing out the fiber and twisting it into a thread, and various versions of mechanical shuttles for weaving. Soon these machines, initially powered by humans, were run by animals or by water wheels. The invention of the steam engine by James Watt led to the gradual introduction of steam as the main power source.

The social consequences of the cotton trade were enormous. Large areas of the English Midlands were transformed from a farming district with numerous small trading centers into a region of almost three hundred factory towns and villages. Working and living conditions were terrible. Very long hours were required of workers, under a factory system of strict rules and harsh discipline. While not quite the slavery that existed on the cotton plantations on the other side of the Atlantic, the cotton trade brought servitude, squalor, and misery to the many thousands who worked in the dusty, noisy, and dangerous cotton factories. Wages were often paid in overpriced goods—workers had no say in this practice. Housing conditions were deplorable; in areas around the factories buildings were crowded together along narrow, dark, and poorly drained lanes. Factory workers and their families were crammed into these cold, damp, and dirty accommodations, often two or three families to a house with another family in the cellar. Less than half of children born under these conditions survived to their fifth birthday. Some authorities were concerned, not because of the appallingly high infant mortality rate but because these children died "before they can be engaged in factory labor, or in any other labor whatsoever." When children did reach an age to work in the cotton mills, where their small size allowed them to crawl underneath machines and their nimble fingers to repair breaks in the threads, they were often beaten to keep them awake for the twelve to fourteen hours of the working day.

Indignation over ill treatment of children and other abuses generated a widespread humanitarian movement pushing for laws governing work hours, child labor, and factory safety and health, from which much of our present-day industrial legislation evolved. The conditions encouraged many factory workers to take an active role in the trade union movement and numerous other movements for social, political, and educational reforms. Change did not come easily, however. Factory owners and their shareholders wielded enormous political power and were reluctant to accept any decrease in the huge profits from the cotton trade that might result from the cost of improving working conditions.

A pall of dark smoke from the hundreds of cotton mills was a permanent fixture over the city of Manchester, which grew and flourished along with the cotton trade. Cotton profits were used to further industrialize the region. Canals and railways were built to transport raw materials and coal to the factories and finished products to the nearby port of Liverpool. Demand grew for engineers, mechanics, builders, chemists, and artisans—those with the technical skills needed by a vast manufacturing enterprise with products and services as diverse as dyestuffs, bleaches, iron foundries, metalworks, glassmaking, shipbuilding, and railway manufacturing.

Despite legislation enacted in England in 1807 abolishing the slave trade, industrialists did not hesitate to import slave-grown cotton from the American South. Raw cotton, from other cotton-producing countries such as Egypt and India as well as the United States, was Britain's largest import during the years between 1825 and 1873, but the processing of cotton declined when raw cotton supplies were cut off during World War I. The industry in Britain never recovered to its former levels because cotton-growing countries, installing more modern machinery and able to use less expensive local labor, became important producers—and significant consumers—of cotton fabric.

The sugar trade had provided the original capital for the Industrial Revolution, but much of the prosperity of nineteenth-century Britain was based on the demand for cotton. Cotton fabric was cheap and attractive, ideal for clothing and household furnishings. It could mix with other fibers without problems and was easy to wash and sew. Cotton quickly replaced the more expensive linen as the plant fiber of choice for a vast number of ordinary people. The huge increase in demand for raw cotton in Europe, especially in England, led to a great expansion of slavery in America. Cotton cultivation was very labor intensive. Agricultural mechanization, pesticides, and herbicides were much later inventions, so cotton plantations relied on the human labor supplied by slaves. In 1840 the slave population of the United States was estimated at 1.5 million. Just twenty years later, when raw cotton exports accounted for two-thirds of the total value of U.S. exports, there were four million slaves.

CELLULOSE, A STRUCTURAL POLYSACCHARIDE

Like other plant fibers, cotton consists of over 90 percent cellulose, which is a polymer of glucose and a major component of plant cell walls. The term *polymer* is often associated with synthetic fibers and plastics, but there are also many naturally occurring polymers. The word comes from two Greek words, *poly* meaning "many" and *meros* meaning "parts"—or units—so a polymer consists of many units. Polymers of glucose, also known as polysaccharides, can be classified on the basis of their function in a cell. Structural polysaccharides, like cellulose, provide a means of support for the organism; storage polysaccharides supply a way of storing glucose until it is needed. The units of structural polysaccharides are β-glucose units; those of storage polysaccharides are α-glucose. As discussed in Chapter 3, β refers to the OH group on carbon number 1 above the glucose ring. The structure of α-glucose has the OH at carbon number 1 below the ring.

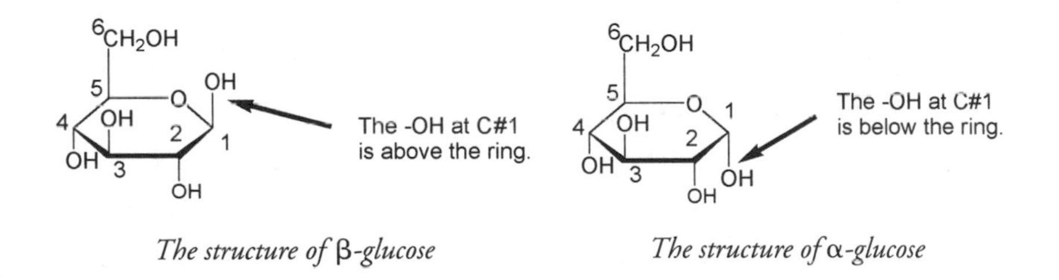

The structure of β-glucose *The structure of α-glucose*

The difference between α- and β-glucose may seem small, but it is responsible for enormous differences in function and role among the various polysaccharides derived from each version of glucose: above the ring, structural; and below the ring, storage. That a very small change in the structure of a molecule can have profound consequences for properties of the compound is something that occurs again and again in chemistry. The α and β polymers of glucose demonstrate this observation extremely well.

In both structural and storage polysaccharides, the glucose units are joined to each other through carbon number 1 on a glucose molecule and carbon number 4 on the adjacent glucose molecule. This joining occurs by the removal of a molecule of water formed from an H of one of the glucose molecules and an OH from the other glucose molecule. The process is known as condensation—thus these polymers are known as condensation polymers.

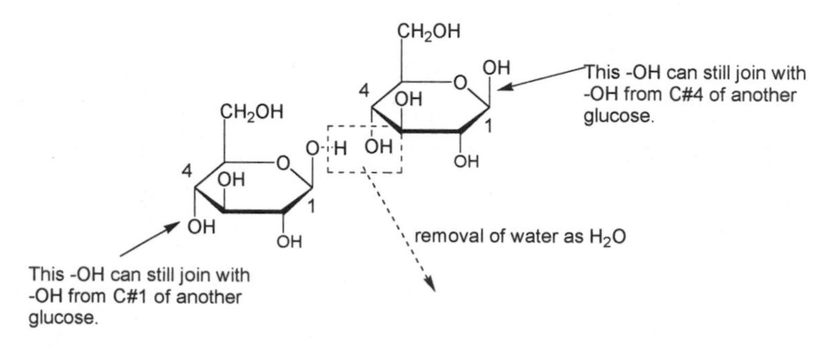

Condensation (loss of a molecule of water) between two β-glucose molecules. Each molecule can repeat this process again at its opposite end.

Each end of the molecule is able to join to another by condensation, forming long continuous chains of glucose units with the remaining OH groups distributed around the outside of the chain.

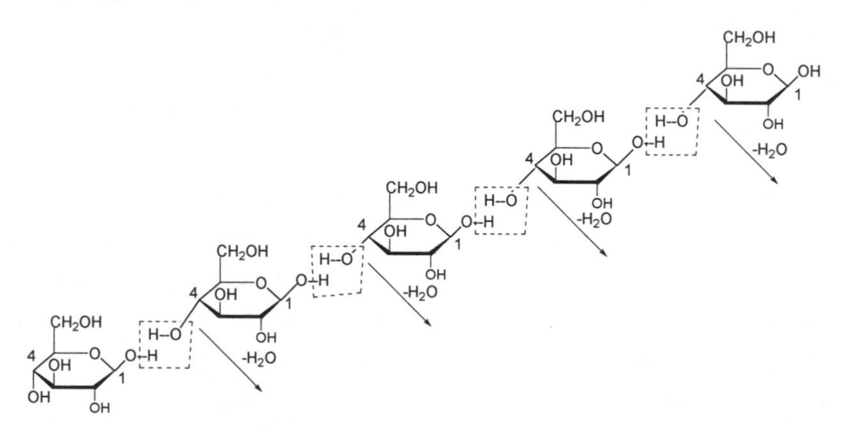

Elimination of a molecule of H_2O between C#1 of one β-glucose and C#4 of the next forms a long polymer chain of cellulose. The diagram shows five β-glucose units.

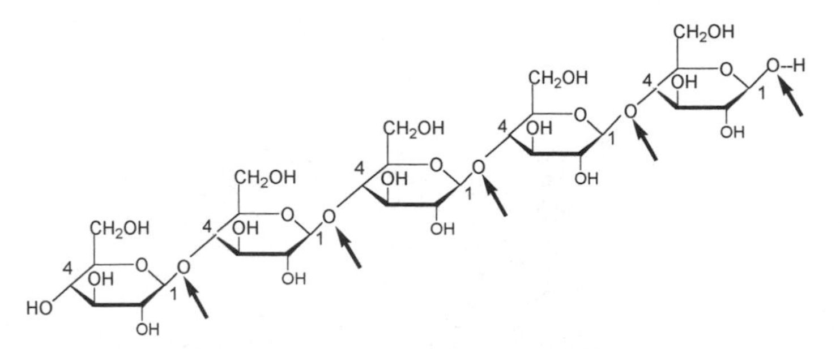

Structure of part of a long cellulose chain. The O attached at each C#1, as indicated with an arrow, is β, i.e., is above the ring to its left.

Many of the traits that make cotton such a desirable fabric arise from the unique structure of cellulose. Long cellulose chains pack tightly together, forming the rigid, insoluble fiber of which plant cell walls are constructed. X-ray analysis and electron microscopy, techniques used to determine physical structures of substances, show that the cellulose chains lie side by side in bundles. The shape a β linkage confers on the structure allows the cellulose chains to pack closely enough to form these bundles, which then twist together to form fibers visible to the naked eye. On the outside of the bundles are the OH groups that have not taken part in the formation of the long cellulose chain, and these OH groups can attract water molecules. Thus cellulose can take up water, accounting for the high absorbency of cotton and other cellulose-based products. The statement that "cotton breathes" has nothing to do with the passage of air but everything to do with the absorbency of water by cotton. In hot weather perspiration from the body is absorbed by cotton garments as it evaporates, cooling us down. Clothes made from nylon or polyester don't absorb moisture, so perspiration is not "wicked" away from the body, leading to an uncomfortable humid state.

Another structural polysaccharide is *chitin,* a variation of cellulose found in the shells of crustaceans such as crabs, shrimps, and lobsters. Chitin, like cellulose, is a β-polysaccharide. It differs from cellulose

Fields of cellulose—a cotton field. (Photo by Peter Le Couteur)

only at the carbon number 2 position on each β-glucose unit, where the OH is replaced by an amide ($NHCOCH_3$) group. So each unit of this structural polymer is a glucose molecule where $NHCOCH_3$ replaces

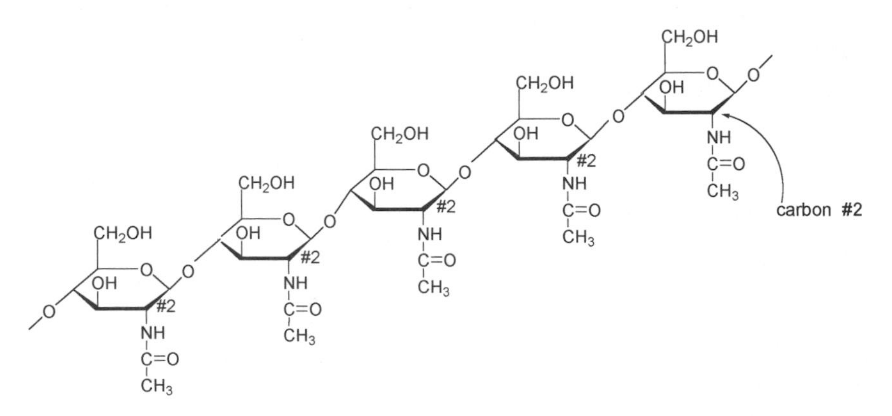

Part of the structure of the polymer chitin found in crustacean shells. At C#2 the OH of cellulose has been substituted by $NHCOCH_3$.

OH at carbon number 2. The name of this molecule is N-acetyl glucosamine. This may not seem terribly interesting, but if you suffer from arthritis or other joint ailments, you may already know the name. N-acetyl glucosamine and its closely related derivative glucosamine, both manufactured from crustacean shell, have provided relief for many arthritis victims. They are thought to stimulate the replacement of or to supplement cartilaginous material in the joints.

Humans and all other mammals lack the digestive enzymes needed to break down β linkages in these structural polysaccharides, and so we cannot use them as a food source, despite the billions and billions of glucose units available as cellulose in the plant kingdom. But there are bacteria and protozoa that do produce the enzymes necessary to split the β linkage and are thus able to break down cellulose into its component glucose molecules. The digestive systems of some animals include temporary storage areas where these microorganisms live, enabling their hosts to obtain nourishment. For example, horses have a cecum—a large pouch where the small and large intestines connect—for this purpose. Ruminants, a group that includes cattle and sheep, have a four-chambered stomach, one part of which contains the symbiotic bacteria. Such animals also periodically regurgitate and rechew their cud, another digestive system adaptation designed to improve access to the β linkage enzyme.

With rabbits and some other rodents, the necessary bacteria live in the large intestine. As the small intestine is where most nutrients are absorbed and the large intestine comes after the small intestine, these animals obtain the products from the cleaving of the β link by eating their feces. As the nutrients pass through the alimentary canal a second time, the small intestine is now able to absorb the glucose units released from cellulose during the first passage. This may seem, to us, a thoroughly distasteful method of coping with the problem of the orientation of an OH group, but it obviously works well for these rodents. Some insects, including termites, carpenter ants, and other wood-eating pests, harbor microorganisms that allow them to access cellulose for food, with sometimes disastrous results for human homes and buildings. Even though we cannot metabolize cellulose, it is still very important in our diet. Plant fiber, consisting of cellulose and other indigestible material, helps move waste products along the digestive tract.

STORAGE POLYSACCHARIDES

Though we lack the enzyme that breaks down a β linkage, we do have a digestive enzyme that splits an α linkage. The α configuration is found in the storage polysaccharides, starch and glycogen. One of our major dietary sources of glucose, starch is found in roots, tubers, and seeds of many plants. It consists of two slightly different polysaccharide molecules, both polymers of α-glucose units. Twenty to 30 percent of starch is made up of amylose, an unbranched chain of several thousand glucose units joined between carbon number 4 on one glucose and carbon number 1 on the next glucose. The only difference between amylose and cellulose is that in amylose the linkages are α and in cellulose they are β. But the roles played by cellulose and amylose polysaccharides are vastly different.

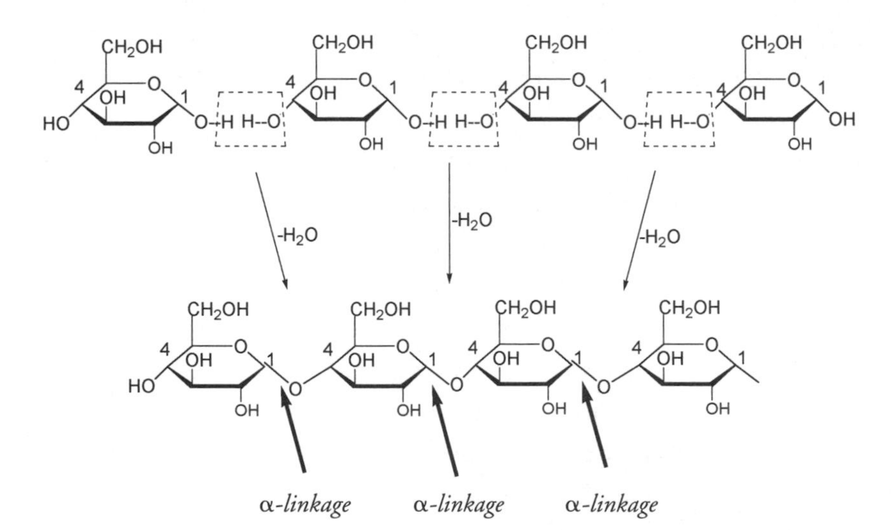

Part of the amylose chain formed from loss of H_2O molecules between α-glucose units. The linkages are α as the –O is below the ring for C#1.

Amylopectin forms the remaining 70 to 80 percent of starch. It also consists of long chains of α-glucose units joined between carbons number 1 and number 4, but amylopectin is a branched molecule with cross-linkages, between the carbon number 1 of one glucose unit and carbon number 6 of another glucose unit, occurring every twenty to twenty-five glucose units. The presence of up to a million glucose units in interconnecting chains makes amylopectin one of the largest molecules found in nature.

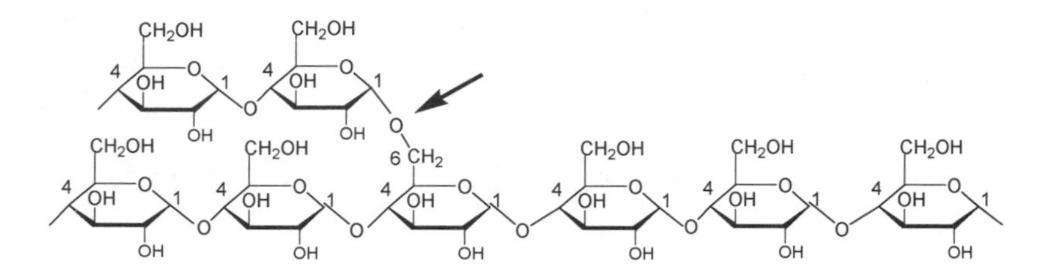

Part of the structure of amylopectin. The arrow shows the C#1 to C#6 α-cross-linkage responsible for the branching of amylopectin.

In starches, the α-linkage is responsible for other important properties besides our ability to digest them. Chains of amylose and amylopectin form into the shape of a helix rather than the tightly packed linear structure of cellulose. When water molecules have enough energy, they are able to penetrate into the more open helical coils; thus starch is water soluble whereas cellulose is not. As all cooks know, the water solubility of starch is strongly temperature dependent. If a suspension of starch and water is heated, granules of starch absorb more and more water until, at a certain temperature, the starch molecules are forced apart, resulting in a mesh of long molecules interspersed in the liquid. This is known as a gel. The cloudy suspension then becomes clear, and the mixture starts to thicken. Thus cooks use starch sources such as flour, tapioca, and cornstarch to thicken sauces.

The storage polysaccharide in animals is glycogen, formed mainly in the cells of the liver and skeletal muscle. Glycogen is a very similar molecule to amylopectin, but where amylopectin has carbon number 1 to carbon number 6 α-cross-linkages only every twenty or twenty-five glucose units, glycogen has these α-cross-links every ten glucose units. The resulting molecule is highly branched. This has a very important consequence for animals. An unbranched chain has only two ends, but a highly branched chain, with the same overall number of glucose units, has a large number of ends. When energy is needed quickly, many glucose units can be removed simultaneously from these many ends. Plants, unlike animals, do not need sudden bursts of energy to escape from predators or chase a prey, so fuel storage as the lesser branched amylopectin and unbranched amylose is sufficient for a plant's lower metabolic rate. This small chemical difference, relating only to the number and not to the type of cross-link, is the basis for one of the fundamental differences between plants and animals.

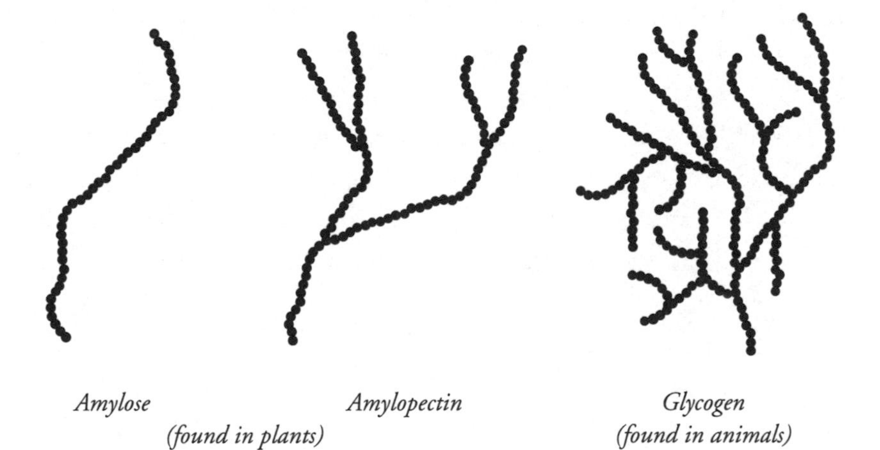

Amylose *Amylopectin*
(found in plants)

Glycogen
(found in animals)

The different branching in starch (amylose and amylopectin) compared with glycogen. The greater the branching, the greater the number of chain ends for enzymes to break down the linkages and the faster glucose can be metabolized.

CELLULOSE MAKES A BIG BANG

Although there is a very large amount of storage polysaccharide in the world, there is a lot more of the structural polysaccharide, cellulose. By some accounts half of all organic carbon is tied up in cellulose. An estimated 10^{14} kilograms (about a 100 billion tons) of cellulose is biosynthesized and degraded annually. As it is not only an abundant but also a replenishable resource, the possibility of using cellulose as a cheap and readily available starting material for new products long interested chemists and entrepreneurs.

By the 1830s it was found that cellulose would dissolve in concentrated nitric acid and that this solution, when poured into water, formed a highly flammable and explosive white powder. Commercialization of this compound had to wait until 1845 and a discovery by Friedrich Schönbein of Basel, Switzerland. Schönbein was experimenting with mixtures of nitric and sulfuric acids in the kitchen of his home, against the wishes of his wife, who perhaps understandably had strictly forbidden the use of her residence for such activities. On this particular day his wife was out, and Schönbein spilled some of the acid mixture. Anxious to clean up the mess quickly, he grabbed the first thing that came to hand—his wife's cotton apron. He mopped up the spill and then hung the apron over the stove to

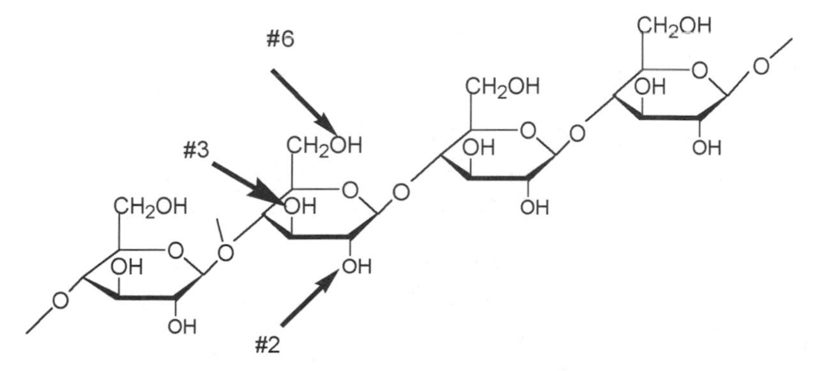

The structure of part of a cellulose molecule. The arrows show where nitration can take place at the OH on C#2, 3, and 6 of each of the glucose units

dry. Before long, with an extremely loud bang and a great flash, the apron exploded. How Schönbein's wife reacted when she came home to find her husband continuing his kitchen experiments on cotton and the nitric acid mix is not known. What is recorded is what Schönbein called his material—*schiessbaumwolle,* or guncotton. Cotton is 90 percent cellulose, and we now know that Schönbein's guncotton was nitrocellulose, the compound formed when the nitro group (NO_2) replaces the H of OH at a number of positions on the cellulose molecule. Not all these positions are necessarily nitrated, but the more nitration on cellulose, the more explosive is the guncotton produced.

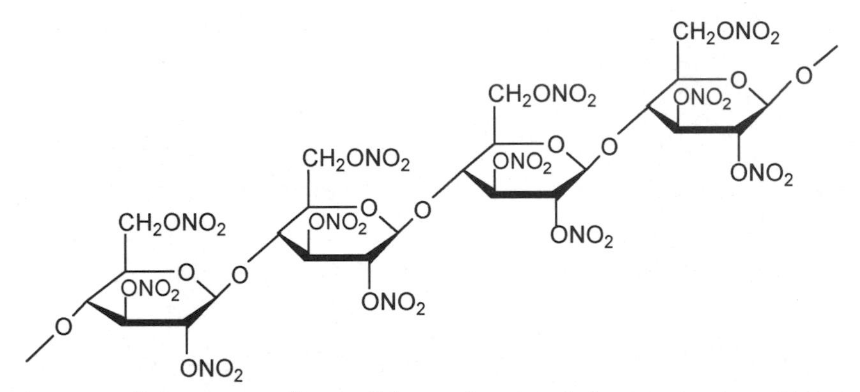

A portion of the structure of nitrocellulose or "guncotton" showing nitration;—NO_2 is substituted for -H at every possible OH position on each glucose unit of the cellulose.

Schönbein, recognizing the potential profit from his discovery, established factories to manufacture nitrocellulose, hoping it would become an alternative to gunpowder. But nitrocellulose can be an extremely dangerous compound unless it is kept dry and handled with proper care. At the time the destabilizing effect of residual nitric acid on the material was not understood, and thus a number of factories were accidentally destroyed by violent explosions, putting Schönbein out of business. It was not until the late 1860s, when proper methods were found to clean guncotton of excess nitric acid, that it could be made stable enough for use in commercial explosives.

Later, control of this nitration process led to different nitrocelluloses, including a higher-nitrate-content guncotton and the lower-nitrate-content materials collodion and celluloid. Collodion is a nitrocellulose mixed with alcohol and water and was used extensively in early photography. Celluloid, a nitrocellulose mixed with camphor, was one of the first successful plastics and was originally used as film for moving pictures. Another cellulose derivative, cellulose acetate, was found to be less flammable than nitrocellulose and quickly replaced it for many uses. The photography business and the movie industry, today enormous commercial enterprises, owe their beginnings to the chemical structure of the versatile cellulose molecule.

Cellulose is insoluble in almost all solvents but does dissolve in an alkaline solution of one organic solvent, carbon disulfide, forming a derivative of cellulose called cellulose xanthate. Cellulose xanthate is in the form of a viscous colloidal dispersion and was given the trade name of viscose. When viscose is forced through tiny holes and the resulting filament is treated with acid, the cellulose is regenerated in the form of fine threads that can be woven into a fabric known commercially as rayon. A similar process, where the viscose is extruded through a narrow slit, produces sheets of cellophane. Rayon and cellophane are usually considered to be synthetic textiles, but they are not totally man-made in the sense that they are just somewhat different forms derived from naturally occurring cellulose.

Both the α polymer of glucose (starch) and the β polymer (cellulose) are essential components of our diet and as such have had, and always will have, an indispensable function in human society. But it is the non-dietary roles of cellulose and its various derivatives that have created milestones in history. Cellulose, in the form of cotton, was responsible for two of the most influential events of the nineteenth century: the Industrial Revolution and the American Civil War. Cotton was the star of the Industrial Revolution, transforming the face of England through rural depopulation, urbanization, rapid industrialization, innovation and invention, social change, and prosperity. Cotton evoked one of the greatest crises in the history of the United States; slavery was the most important issue in the Civil War between abolitionist North and the southern states, whose economic system was based on slave-grown cotton.

Nitrocellulose (guncotton) was one of the very first explosive organic molecules made by man, and its discovery marked the start of a number of modern industries originally based on nitrated forms of cellulose: explosives, photography, and the movie business. The synthetic textile industry, with its beginnings from rayon—a different form of cellulose—has played a significant role in shaping the economy over the last century. Without these applications of the cellulose molecule, our world would be a very different place.

5. NITRO COMPOUNDS

SCHÖNBEIN'S WIFE'S exploding apron was not the first example of a man-made explosive molecule, nor would it be the last. When chemical reactions are very rapid, they can have an awesome power. Cellulose is only one of the many molecules we have altered to take advantage of the capacity for explosive reaction. Some of these compounds have been of enormous benefit; others have caused widespread destruction. Through their very explosive properties, these molecules have had a marked effect on the world.

Although the structures of explosive molecules vary widely, most often they contain a nitro group. This small combination of atoms, one nitrogen and two oxygens, NO_2, attached at the right position, has vastly increased our ability to wage war, changed the fate of nations, and literally allowed us to move mountains.

Gunpowder—The First Explosive

Gunpowder (or black powder), the first explosive mixture ever invented, was used in ancient times in China, Arabia, and India. Early Chinese texts refer to "fire-chemical" or "fire-drug." Its ingredients were not recorded until early in A.D. 1000, and even then the actual proportions required of the component nitrate salt, sulfur, and carbon were not given. Nitrate salt (called saltpeter or "Chinese snow") is potassium nitrate, chemical formula KNO_3. The carbon in gunpowder was in the form of wood charcoal and gives the powder its black color.

Gunpowder was initially used for firecrackers and fireworks, but by the middle of the eleventh century flaming objects—used as weapons and known as fire arrows—were launched by gunpowder. In 1067 the Chinese placed the production of sulfur and saltpeter under government control.

We have no certainty as to when gunpowder arrived in Europe. The Franciscan monk Roger Bacon, born in England and educated at Oxford University and the University of Paris, wrote of gunpowder around 1260, a number of years before Marco Polo's return to Venice with stories of gunpowder in China. Bacon was also a physician and an experimentalist, knowledgeable in the sciences that we would now call astronomy, chemistry, and physics. He was also fluent in Arabic, and it is likely that he learned about gunpowder from a nomadic tribe, the Saracens, who acted as middlemen between the Orient and the West. Bacon must have been aware of the destructive potential of gunpowder, as his description of its composition was in the form of an anagram that had to be deciphered to reveal the ratio: seven parts saltpeter, five parts charcoal, and five parts sulfur. His puzzle remained unsolved for 650 years before finally being decoded by a British army colonel. By then gunpowder had, of course, been in use for centuries.

Present-day gunpowder varies somewhat in composition but contains a larger proportion of saltpeter than Bacon's formulation. The chemical reaction for the explosion of gunpowder can be written as

$$4KNO_{3(s)} + 7C_{(s)} + S_{(s)} \rightarrow 3CO_{2(g)} + 3CO_{(g)} + 2N_{2(g)} + K_2CO_{3(s)} + K_2S_{(s)}$$

potassium nitrate, carbon, sulfur, carbon dioxide, carbon monoxide, nitrogen, potassium carbonate, potassium sulfide

This chemical equation tells us the ratios of substances reacting and the ratios of the products obtained. The subscript (s) means the substance is a solid, and (g) means it is a gas. You can see from the equation that all the reactants are solids, but eight molecules of gases are formed: three carbon dioxide, three carbon monoxide, and two nitrogens. It is the hot, expanding gases produced from the rapid burning of gunpowder that propel a cannonball or bullet. The solid potassium carbonate and sulfide formed are dispersed as tiny particles, the characteristic dense smoke of exploding gunpowder.

Thought to have been produced somewhere around 1300 to 1325, the first firearm, the firelock, was a tube of iron loaded with gunpowder, which was ignited by the insertion of a heated wire. As more sophisticated firearms developed (the musket, the flintlock, the wheellock), the need for different rates of burning of gunpowder became apparent. Sidearms needed faster-burning powder, rifles a slower-burning powder, and cannons and rockets an even slower burn. A mixture of alcohol and water was used to produce a powder that caked and could be crushed and screened to give fine, medium, and coarse fractions. The finer the powder, the faster the burn, so it was possible to manufacture gunpowder that was appropriate for the various applications. The water used for manufacture was frequently supplied as urine from workers in the gunpowder mill; the urine of a heavy wine drinker was believed to create particularly potent gunpowder. Urine from a clergyman, or better yet a bishop, was also considered to give a superior product.

EXPLOSIVE CHEMISTRY

The production of gases and their consequent fast expansion from the heat of the reaction is the driving force behind explosives. Gases have a much greater volume than do similar amounts of solids or liquids. The destructive power of an explosion is due to the shock wave caused by the very rapid increase in volume as gases form. The shock wave for gunpowder travels around a hundred meters per second, but for "high" explosives (TNT or nitroglycerin, for example) it can be up to six thousand meters per second.

All explosive reactions give off large amounts of heat. Such reactions are said to be highly exothermic. The large amounts of heat act dramatically to increase the volume of the gases—the higher the temperature the larger the volume of gas. Heat comes from the energy difference between the molecules on each side of the explosive reaction equation. The molecules produced (on the right of the equation) have less energy tied up in their chemical bonds than the starting molecules (on the left). The compounds that form are more stable. In explosive reactions of nitro compounds, the extremely stable nitrogen molecule, N_2, is formed. The stability of the N_2 molecule is due to the strength of the triple bond that holds the two nitrogen atoms together.

$$N \equiv N$$

Structure of the N_2 molecule

That this triple bond is very strong means that a lot of energy is needed to break it. Conversely, when the N_2 triple bond is made, a lot of energy is released, which is exactly what is wanted in an explosive reaction.

Besides production of heat and of gases, a third important property of explosive reactions is that they must be extremely rapid. If the explosive reaction were to occur slowly, the resulting heat would dissipate and the gases would diffuse into the surroundings without the violent pressure surge, damaging shock wave, and high temperatures characteristic of an explosion. The oxygen required for such a reaction has to come from the molecule that is exploding. It cannot come from the air, because oxygen from the atmosphere is not available quickly enough. Thus nitro compounds, in which nitrogen and oxygen are bonded together, are often explosive, while other compounds containing both nitrogen and oxygen, but not bonded together, are not.

This can be seen using isomers as an example, isomers being compounds that have the same chemical formula but different structures. *Para*-nitrotoluene and *para*-aminobenzoic acid both have seven carbon atoms, seven hydrogen atoms, one nitrogen atom, and two oxygen atoms for identical chemical formulae of $C_7H_7NO_2$, but these atoms are arranged differently in each molecule.

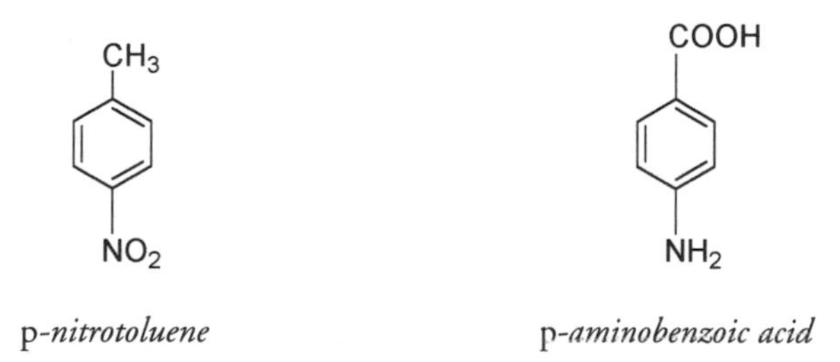

p-*nitrotoluene* p-*aminobenzoic acid*

Para- or *p*-nitrotoluene (the *para* just tells you that the CH_3 and NO_3 groups are at opposite ends of the molecule) can be explosive, whereas *p*-aminobenzoic acid is not at all explosive. In fact you have probably rubbed it over your skin in the summer; it is PABA, the active ingredient in many sunscreen products. Compounds such as PABA absorb ultraviolet light at the very wavelengths that have been found to be most damaging to skin cells. Absorption of ultraviolet light at particular wavelengths depends on the presence in the compound of alternating double and single bonds, possibly also with oxygen and nitrogen atoms attached. Variation in the number of bonds or atoms of this alternating pattern changes the wavelength of absorption. Other compounds that absorb at the required wavelengths can be used as sunscreens provided they also do not wash off easily in water, have no toxic or allergic effects, no unpleasant smell or taste, and do not decompose in the sun.

The explosiveness of a nitrated molecule depends on the number of nitro groups attached. Nitrotoluene has only one nitro group. Further nitration can add two or three more nitro groups, resulting in di- or trinitrotoluenes respectively. While nitrotoluene and dinitrotoluene can explode, they do not pack the same power as the high-explosive trinitrotoluene (TNT) molecule.

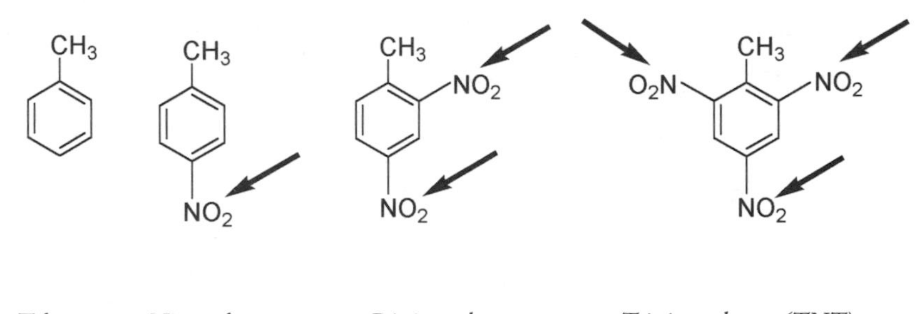

Toluene *Nitrotoluene* *Dinitrotoluene* *Trinitrotoluene (TNT)*

The nitro groups are indicated by arrows.

Advances in explosives came about in the nineteenth century when chemists began studying the effects of nitric acid on organic compounds. Only a few years after Friedrich Schönbein destroyed his wife's apron with his experiments, an Italian chemist, Ascanio Sobrero, of Turin, prepared another highly explosive nitro molecule. Sobrero had been studying the effects of nitric acid on other organic compounds. He dripped glycerol, also known as glycerin and readily obtained from animal fat, into a cooled mixture of sulfuric and nitric acids and poured the resulting mixture into water. An oily layer of what is now known as nitroglycerin separated out. Using a procedure that was normal in Sobrero's time but unthinkable today, he tasted the new compound and recorded his comments: "a trace placed on the tongue but not swallowed gives rise to a most pulsating, violent headache, accompanied by great weakness of the limbs."

Later investigations into the severe headaches suffered by workers in the explosives industry showed that these headaches were due to the dilation of blood vessels caused by handling nitroglycerin. This discovery resulted in the prescription of nitroglyerin for treatment of the heart disease angina pectoris.

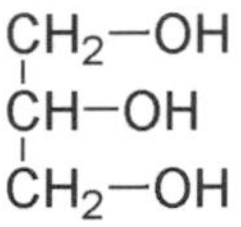

$$\begin{array}{l} CH_2-OH \\ | \\ CH-OH \\ | \\ CH_2-OH \end{array} \qquad\qquad \begin{array}{l} CH_2-O-NO_2 \\ | \\ CH-O-NO_2 \\ | \\ CH_2-O-NO_2 \end{array}$$

Glycerol (glycerin) 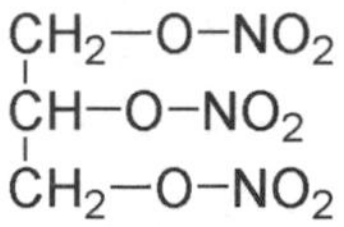 *Nitroglycerin*

For angina sufferers, dilation of previously constricted blood vessels supplying the heart muscle allows an adequate flow of blood and relieves the pain of angina. We now know that in the body nitroglycerin releases the simple molecule nitric oxide (NO), which is responsible for the dilation effect. Research on this aspect of nitric oxide led to the development of the anti-impotence drug Viagra, which also depends on the blood-vessel-dilating effect of nitric oxide.

Other physiological roles of nitric oxide include maintaining blood pressure, acting as a messenger molecule carrying signals between cells, establishing long-term memory, and aiding digestion. Drugs for treating high blood pressure in newborns and for treating shock victims have been developed from these investigations. The 1998 Nobel Prize for medicine was awarded to Robert Furchgott, Louis Ignarro, and Ferid Murad for the discovery of the role played by nitric oxide in the body. Yet in one of chemistry's many ironic twists, Alfred Nobel, whose nitroglycerin-derived fortune would be used to establish the Nobel prizes, personally refused nitroglycerin treatment for the chest pains from his heart disease. He did not believe it would work—only that it would cause headaches.

Nitroglycerin is a highly unstable molecule, exploding when heated or struck with a hammer. The explosive reaction

$$\underset{\text{nitroglycerin}}{4C_3H_5N_3O_{9(l)}} \longrightarrow \underset{\text{nitrogen}}{6N_{2(g)}} + \underset{\text{carbon dioxide}}{12CO_{2(g)}} + \underset{\text{water}}{10H_2O_{(g)}} + \underset{\text{oxygen}}{O_{2(g)}}$$

produces clouds of rapidly expanding gases and vast amounts of heat. In contrast to gunpowder, which produces six thousand atmospheres of pressure in thousandths of a second, an equal amount of nitroglycerin produces 270,000 atmospheres of pressure in millionths of a second. Gunpowder is relatively safe to handle, but nitroglycerin is very unpredictable and can spontaneously explode due to shock or heating. A safe and reliable way to handle and set off or "detonate" this explosive was needed.

NOBEL'S DYNAMITE IDEA

Alfred Bernard Nobel, born in 1833 in Stockholm, had the idea of employing—instead of a fuse, which just caused nitroglycerin to burn slowly—an explosion of a very small amount of gunpowder to detonate a larger explosion of nitroglycerin. It was a great idea; it worked, and the concept is still used today in the many controlled explosions that are routine in the mining and construction industries. Having solved the problem of producing a desired explosion, however, Nobel still faced the problem of preventing an undesired explosion.

Nobel's family had a factory that manufactured and sold explosives, which by 1864 had begun to manufacture nitroglycerin for commercial applications such as blasting tunnels and mines. In September of that year one of their laboratories in Stockholm blew up, killing five people, including Alfred Nobel's younger brother, Emil. Though the cause of the accident was never precisely determined, Stockholm officials banned the production of nitroglycerin. Not one to be deterred, Nobel built a new laboratory on pontoons and anchored it in Lake Mälaren, just beyond the Stockholm city limits. The demand for nitroglycerin increased rapidly as its advantages over the much less powerful gunpowder became known. By 1868, Nobel had opened manufacturing plants in eleven countries in Europe and had even expanded to the United States with a company in San Francisco.

Nitroglycerin was often contaminated by the acid used in the manufacturing process and tended to slowly decompose. The gases produced by this decomposition would pop the corks of the zinc cans in which the explosive was packed for shipping. As well, acid in the impure nitroglycerin would corrode the zinc, causing the cans to leak. Packing materials such as sawdust were used to insulate the cans and to absorb any leakages or spills, but such precautions were inadequate and did little to improve safety. Ignorance and misinformation frequently led to terrible accidents. Mishandling was common. In one case, nitroglycerin oil had even been used as a lubricant on the wheels of a cart transporting the explosive, obviously with disastrous results. In 1866 a shipment of nitroglycerin detonated in a Wells Fargo warehouse in San Francisco, killing fourteen people. In the same year a seventeen-thousand-ton steamship, the S.S. *European,* blew up while unloading nitroglycerin on the Atlantic coast of Panama, killing forty-seven people and causing more than a million dollars in damages. Also in 1866 explosions leveled nitroglycerin plants in Germany and Norway. Authorities around the world became concerned. France and Belgium banned nitroglycerin, and similar action was proposed in other countries, despite an increased worldwide demand for the use of the incredibly powerful explosive.

Nobel began to look for ways to stabilize nitroglycerin without losing its power. Solidification seemed an obvious method, so he experimented by mixing the oily liquid nitroglycerin with such neutral solids as sawdust, cement, and powdered charcoal. There has always been speculation as to whether the product we now know as "dynamite" was the result of a systematic investigation, as claimed by Nobel, or was more a fortuitous discovery. Even if the discovery was serendipitous, Nobel was astute enough to recognize that kieselguhr, a natural, fine, siliceous material that was occasionally substituted for sawdust packing material, could soak up spilled liquid nitroglycerin but remain porous. Kieselguhr, also known as diatomaceous earth, is the remains of tiny marine animals and has a number of other uses: as a filter in sugar refineries, as insulation, and as a metal polish. Further testing showed that mixing liquid nitroglycerin with about one-third of its weight of kieselguhr formed a plastic mass with the consistency of putty. The kieselguhr diluted the nitroglycerin; separation of the nitroglycerin particles slowed down the rate of their decomposition. The explosive effect could now be controlled.

Nobel named the nitroglycerin/kieselguhr mixture dynamite, from the Greek *dynamis* or power. Dynamite could be molded into any desired shape or size, was not readily subject to decomposition, and did not explode accidentally. By 1867, Nobel and Company, as the family firm was now called, began shipping dynamite, newly patented as Nobel's Safety Powder. Soon there were Nobel dynamite factories in countries around the world, and the Nobel family fortune was assured.

That Alfred Nobel, a munitions manufacturer, was also a pacifist may seem a contradiction, but then Nobel's whole life was full of contradictions. As a child he was sickly and was not expected to live to adulthood, but he outlasted his parents and brothers. He has been described in somewhat paradoxical terms as shy, extremely considerate, obsessed by his work, highly suspicious, lonely, and very charitable. Nobel firmly believed that the invention of a truly terrible weapon might act as a deterrent that could bring lasting peace to the world, a hope that over a century later and with a number of truly terrible weapons now available has still not been realized. He died in 1896, working alone at his desk in his home in San Remo, Italy. His enormously wealthy estate was left to provide yearly prizes for research in chemistry, physics, medicine, literature, and peace. In 1968 the Bank of Sweden, in memory of Alfred Nobel, established a prize in the field of economics. Although now called a Nobel Prize, it was not part of the original endowment.

WAR AND EXPLOSIVES

Nobel's invention could not be used as a propellant for projectiles, as guns cannot withstand the tremendous explosive force of dynamite. Military leaders still wanted a more powerful explosive than gunpowder, one that did not produce clouds of black smoke, was safe to handle, and allowed for quick loading. From the early 1880s various formulations of nitrocellulose (guncotton), or nitrocellulose mixed with nitroglycerin had been used as "smokeless powder" and are still today the basis of firearm explosives. Cannons and other heavy artillery are not as restricted in the choice of propellant. By World War I, munitions contained mainly picric acid and trinitrotoluene. Picric acid, a bright yellow solid first synthesized in 1771, was used originally as an artificial dye for silk and wool. It is a triple-nitrated phenol molecule and relatively easy to make.

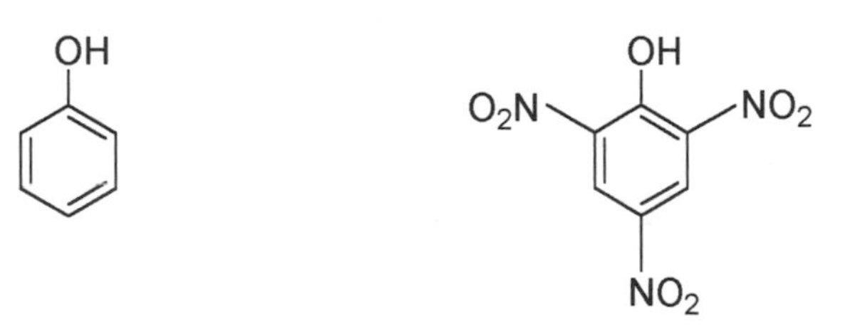

Phenol *Trinitrophenol or picric acid*

In 1871 it was found that picric acid could be made to explode if a sufficiently powerful detonator was used. It was first employed in shells by the French in 1885, then by the British during the Boer War of 1899–1902. Wet picric acid was difficult to detonate, however, leading to misfiring under rainy or humid conditions. It was also acidic and would react with metals to form shock-sensitive "picrates." This shock sensitivity caused shells to explode on contact, preventing them from penetrating thick armor plate.

Chemically similar to picric acid, trinitrotoluene, known as TNT from the initials of *tri, nitro,* and *toluene,* was better suited for munitions.

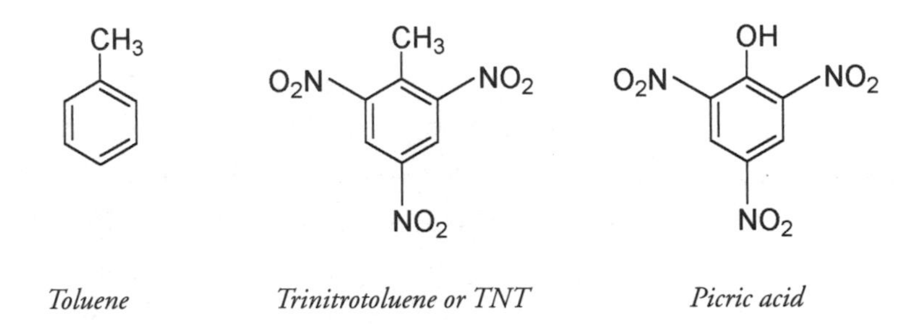

Toluene *Trinitrotoluene or TNT* *Picric acid*

It was not acidic, was not affected by the damp, and had a relatively low melting point so it could be readily melted and poured into bombs and shells. Being harder to detonate than picric acid, it could take a greater impact and thus had better armor-penetrating ability. TNT has a lower ratio of oxygen to carbon than nitroglycerin, so its carbon is not converted completely to carbon dioxide nor its hydrogen to water. The reaction can be represented as

$$\underset{\text{TNT}}{2C_7H_5N_3O_{6(s)}} \longrightarrow \underset{\text{carbon dioxide}}{6CO_{2(g)}} + \underset{\text{hydrogen}}{5H_{2(g)}} + \underset{\text{nitrogen}}{3N_{2(g)}} + \underset{\text{carbon}}{8C_{(s)}}$$

Carbon produced in this reaction causes the large amount of smoke that is associated with the explosions of TNT compared to those of nitroglycerin and guncotton.

At the beginning of World War I, Germany, using TNT-based munitions, had a definite advantage over the French and British, who were still using picric acid. A crash program to start producing TNT, aided by large quantities shipped from manufacturing plants in the United

States, allowed Britain to rapidly develop similar quality shells and bombs containing this pivotal molecule.

Another molecule, ammonia (NH_3), became even more crucial during World War I. While not a nitro compound, ammonia is the starting material for making the nitric acid, HNO_3, which is needed to make explosives. Nitric acid has probably been known for a long time. Jabir ibn Hayyan, the great Islamic alchemist who lived around A.D. 800, would have known about nitric acid and probably made it by heating saltpeter (potassium nitrate) with ferrous sulfate (then called green vitriol because of its green crystals). The gas produced by this reaction, nitrogen dioxide (NO_2), was bubbled into water to form a dilute solution of nitric acid.

Nitrates are not commonly found in nature, as they are very soluble in water and tend to be dissolved away, but in the extremely arid deserts of northern Chile huge deposits of sodium nitrate (so-called Chile saltpeter) have been mined for the past two centuries as a source of nitrate for direct preparation of nitric acid. Sodium nitrate is heated with sulfuric acid. The nitric acid that is produced is driven off because it has a lower boiling point than sulfuric acid. It is then condensed and collected in cooling vessels.

During World War I supplies of Chile saltpeter to Germany were cut off by a British naval blockade. Nitrates were strategic chemicals, necessary for manufacture of explosives, so Germany needed to find another source.

While nitrates may not be plentiful, the two elements, nitrogen and oxygen, that make up nitrates exist in the world in a generous supply. Our atmosphere is composed of approximately 20 percent oxygen gas and 80 percent nitrogen gas. Oxygen (O_2) is chemically reactive, combining readily with many other elements, but the nitrogen molecule

(N_2) is relatively inert. At the beginning of the twentieth century, methods of "fixing" nitrogen—that is, removing it from the atmosphere by chemical combination with other elements—were known but not very advanced.

For some time Fritz Haber, a German chemist, had been working on a process to combine nitrogen from the air with hydrogen gas to form ammonia.

$$N_{2(g)} + 3H_{2(g)} \longrightarrow 2NH_{3(g)}$$

nitrogen hydrogen ammonia

Haber was able to solve the problem of using inert atmospheric nitrogen by working with reaction conditions that produced the highest yield of ammonia for the lowest possible cost: high pressure, temperatures of around 400 to 500°C, and removal of the ammonia as soon as it formed. Much of Haber's work involved finding a catalyst to increase the rate of this particularly slow reaction. His experiments were aimed at producing ammonia for the fertilizer industry. Two-thirds of the world's fertilizer needs were at that time being supplied from the saltpeter deposits in Chile; as these deposits became depleted, a synthetic route to ammonia was needed. By 1913 the world's first synthetic ammonia plant had been established in Germany, and when the British blockade later cut nitrate supply from Chile, the Haber process, as it is still known, was quickly expanded to other plants to supply ammonia not only for fertilizers but also for ammunition and explosives. The ammonia thus produced is reacted with oxygen to form nitrogen dioxide, the precursor of nitric acid. For Germany, with ammonia for fertilizers and nitric acid to make explosive nitro compounds, the British blockade was irrelevant. Nitrogen fixation had become a vital factor in waging war.

The 1918 Nobel Prize for chemistry was awarded to Fritz Haber for his role in the synthesis of ammonia, which ultimately led to increased fertilizer production and the consequent greater ability of agriculture to feed the world's population. The announcement of this award aroused a storm of protest because of the role Fritz Haber had played in Germany's gas warfare program in World War I. In April 1915 cylinders of chlorine gas had been released over a three-mile front near Ypres, Belgium. Five thousand men had been killed and another ten thousand suffered devastating effects on their lungs from chlorine exposure. Under Haber's leadership of the gas warfare program, a number of new substances, including mustard gas and phosgene, were also tested and used. Ultimately gas warfare was not the deciding factor in the outcome of the war, but in the eyes of many of his peers Haber's earlier great innovation—so crucial to world agriculture—did not compensate for the appalling result of the exposure of thousands to poisonous gases. Many scientists considered awarding the Nobel Prize to Haber under these circumstances to be a travesty.

Haber saw little difference between conventional and gas warfare and was greatly upset by the controversy. In 1933, as director of the prestigious Kaiser Wilhelm Institute for Physical Chemistry and Electrochemistry, he was ordered by the Nazi government of Germany to dismiss all Jewish workers on his staff. In an unusual act of courage for those times, Haber refused, citing in his letter of resignation that "for more than forty years I have selected my collaborators on the basis of their intelligence and their character and not on the basis of their grandmothers, and I am not willing for the rest of my life to change this method that I have found so good."

Today, worldwide annual production of ammonia, still made by Haber's process, is about 140 million tons, much of it used for ammonium nitrate (NH_4NO_3), probably the world's most important fertilizer. Ammonium nitrate is also used for blasting in mines, as a mixture of 95 percent ammonium nitrate and 5 percent fuel oil. The explosive reaction

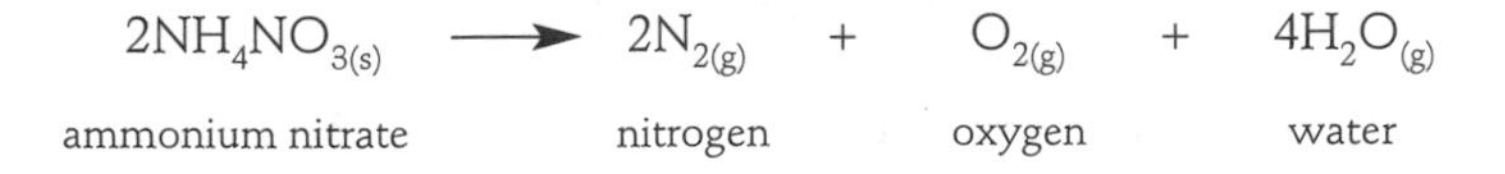

$$2NH_4NO_{3(s)} \longrightarrow 2N_{2(g)} + O_{2(g)} + 4H_2O_{(g)}$$

ammonium nitrate — nitrogen — oxygen — water

produces oxygen gas as well as nitrogen and steam. The oxygen gas oxidizes the fuel oil in the mixture, increasing the energy released by the blast.

Ammonium nitrate is considered a very safe explosive when properly handled, but it has been responsible for a number of disasters as a result of improper safety procedures or deliberate bombings by terrorist organizations. In 1947, in the port of Texas City, Texas, a fire broke out in the hold of a ship as it was being loaded with paper bags of ammonium nitrate fertilizer. In an attempt to stop the fire, the ship's crew closed the hatches, which had the unfortunate effect of creating the conditions of heat and compression needed to detonate ammonium nitrate. More than five hundred people were killed in the ensuing explosion. More recent disasters involving ammonium nitrate bombs planted by terrorists include the incidents at the World Trade Center in New York City in 1993 and at the Alfred P. Murrah Federal Building in Oklahoma City in 1995.

One of the more recently developed explosives, pentaerythritoltetranitrate (abbreviated to PETN), is regrettably also favored by terrorists because of the very same properties that have made it so useful for legitimate purposes. PETN can be mixed with rubber to make what is called a plastic explosive, which can be pressed into any shape. PETN may have a complicated chemical name, but its structure is not that complicated. It is chemically similar to nitroglycerin but has five carbons instead of three and one more nitro group.

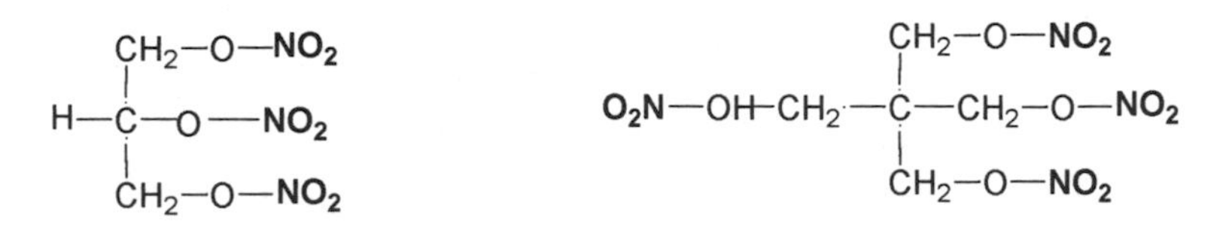

Nitroglycerin (left) and pentaerythritoltetranitrate (PETN) (right). The nitro groups are bolded.

Easily detonated, shock sensitive, very powerful, and with little odor so that even trained dogs find it difficult to detect, PETN may have become the explosive of choice for airplane bombings. It gained fame as a component of the bomb that brought down Pan Am flight 103 over Lockerbie, Scotland, in 1988. Further notoriety has resulted from the 2001 "Shoebomber" incident, in which a passenger on an American Airlines flight from Paris attempted to set off PETN hidden in the soles of his sneakers. Disaster was averted only due to quick action by crew and passengers.

The role of explosive nitro molecules has not been confined to wars and terrorism. There is evidence that the power of the saltpeter, sulfur, and charcoal mixture was used in mining in northern Europe by the early 1600s. The Malpas Tunnel (1679) of the Canal du Midi in France, the original canal linking the Atlantic Ocean to the Mediterranean Sea, was just the first of many major canal tunnels built with the help of gunpowder. The 1857–1871 building of the Mont Cenis or Fréjus railway tunnel, through the French Alps, was the largest use of explosive molecules of the time, changing the face of travel in Europe by allowing easy passage from France to Italy. The new explosive nitroglycerin was first used in construction in the Hoosac railway tunnel (1855–1866) at North Adams in Massachusetts. Major engineering feats have been accomplished with the aid of dynamite: the 1885 completion of the Canadian Pacific Railway, allowing passage through the Canadian Rockies; the eighty-kilometer-long Panama Canal, which opened in 1914; and the 1958 removal of the navigational hazard Ripple Rock off the west coast of North America—still the largest-ever nonnuclear explosion.

In 218 B.C., the Carthaginian general Hannibal made his way through the Alps with his vast army and his forty elephants for an assult on the heart of the Roman Empire. He used the standard but extremely slow road-building method of the day: rock obstacles were heated by bonfires, then doused with cold water to crack them apart. Had Hannibal possessed explosives, a rapid passage through the Alps might have allowed him an eventual victory at Rome, and the fate of the whole western Mediterranean would have been very different.

From Vasco da Gama's defeat of the rulers of Calicut, through the conquest of the Aztec empire by Hernán Cortés and a handful of Spanish conquistadors, to the British army's Light Cavalry Brigade charge of Russian field batteries in the 1854 Battle of Balaklava, explosive-propelled weapons have had the advantage over bows and arrows, spears, and swords. Imperialism and colonialism—systems that have molded our world—depended on the power of armaments. In war and in peace, from destroying to constructing, for worse or for better, explosive molecules have changed civilization.

6. SILK AND NYLON

EXPLOSIVE MOLECULES may seem very remote from the images of luxury, softness, suppleness, and sheen conjured up by the word *silk.* But explosives and silk have a chemical connection, one that led to the development of new materials, new textiles, and by the twentieth century, a whole new industry.

Silk has always been prized as a fabric by the wealthy. Even with the wide choice of both natural and man-made fibers available today, it is still considered irreplaceable. The properties of silk that have long made it so desirable—its caressing feel, its warmth in cool weather and coolness in hot weather, its wonderful luster, and the fact that it takes dyes so beautifully—are due to its chemical structure. Ultimately it was the chemical structure of this remarkable substance that opened trade routes between the East and the rest of the known world.

THE SPREAD OF SILK

The history of silk goes back more than four and a half millennia. Legend has it that around 2640 B.C., Princess Hsi-ling-shih, the chief concubine of the Chinese emperor Huang-ti, found that a delicate thread of silk could be unwound from an insect cocoon that had fallen into her tea. Whether this story is myth or not, it is true that the production of silk started in China with the cultivation of the silkworm, *Bombyx mori,* a small gray worm that feeds only on the leaves of the mulberry tree, *Morus alba.*

Common in China, the silkworm moth lays around five hundred eggs over a five-day period and then dies. One gram of these tiny eggs produces more than a thousand silkworms, all of which together devour about thirty-six kilograms of ripe mulberry leaves to produce around two hundred grams of raw silk. The eggs must initially be kept at 65°F, then gradually rise to the hatching temperature of 77°F. The worms are kept in clean, well-ventilated trays, where they eat voraciously and shed their skins several times. After a month they are moved to spinning trays or frames to start spinning their cocoons, a process that takes several days. A single continuous strand of silk thread is extruded from the worm's jaw, along with a sticky secretion that holds the threads together. The worm moves its head constantly in a figure-eight pattern, spinning a dense cocoon and gradually changing itself into a chrysalis.

To obtain the silk, the cocoons are heated to kill the chrysalis inside and then plunged into boiling water to dissolve the sticky secretion holding the threads together. Pure silk thread is then unwound from the cocoon and wound onto reels. The length of a silk thread from one cocoon can be anywhere from four hundred to more than three thousand yards.

The cultivation of silkworms and the use of the resulting silk fabric spread rapidly throughout China. Initially, silk was reserved for members of the imperial family and for nobility. Later, though its price remained high, even the common people were allowed to wear garments made of silk. Beautifully woven, lavishly embroidered, and wonderfully dyed, silk fabric was greatly prized. It was a highly valued commodity of trade and barter and even a form of currency—rewards and taxes were sometimes paid in silk.

For centuries, long after the opening of the trade routes through Central Asia collectively known as the Silk Road, the Chinese kept the details of silk production a secret. The path of the Silk Road varied over the centuries, depending mostly on the politics and safety of the regions along the way. At its longest it extended some six thousand miles from Peking (Beijing), in eastern China, to Byzantium (later Constantinople, now Istanbul) in present-day Turkey, and to Antioch and Tyre, on the Mediterranean, with major arteries diverting into northern India. Some parts of the Silk Road date back over four and a half millennia.

Trade in silk spread slowly, but by the first century B.C. regular shipments of silk were arriving in the West. In Japan sericulture began around A.D. 200 and developed independently from the rest of the world. The Persians quickly became middlemen in the silk trade. The Chinese, to maintain their monopoly on production, made attempts to smuggle silkworms, silkworm eggs, or white mulberry seeds from China punishable by death. But as the legend goes, in 552 two monks of the Nestorian church managed to return from China to Constantinople with hollowed-out canes that concealed silkworm eggs and mulberry seeds. This opened the door to silk production in the West. If the story is true, it is possibly the first recorded example of industrial espionage.

Sericulture spread throughout the Mediterranean and by the fourteenth century was a flourishing industry in Italy, especially in the north, where cities like Venice, Lucca, and Florence became renowned for beautiful heavy silk brocades and silk velvets. Silk exports from these areas to northern Europe are considered to be one of the financial bases of the Renaissance movement, which started in Italy about this time. Silk weavers fleeing political instability in Italy helped France become a force in the silk industry. In 1466, Louis XI granted tax exemptions for silk weavers in the city of Lyons, decreed the planting of mulberry trees, and ordered the manufacture of silk for the royal court. For the next five centuries European sericulture would be centered near Lyons and its surrounding area. Macclesfield and Spittalfield, in England, became major centers for finely woven silks when Flemish and French weavers, escaping religious persecution on the Continent, arrived in the late sixteenth century.

Various attempts to cultivate silk in North America were not commercially successful. But the spinning and weaving of silk, processes that could be readily mechanized, were developed. In the early part of the twentieth century the United States was one of the largest manufacturers of silk goods in the world.

THE CHEMISTRY OF SHEEN AND SPARKLE

Silk, like other animal fibers such as wool and hair, is a protein. Proteins are made from twenty-two different α-amino acids. The chemical structure for an α-amino acid has an amino group (NH_2) and an organic acid group (COOH) arranged as shown, with the NH_2 group on the carbon atom of the alpha carbon—that is, the carbon adjacent to the COOH group.

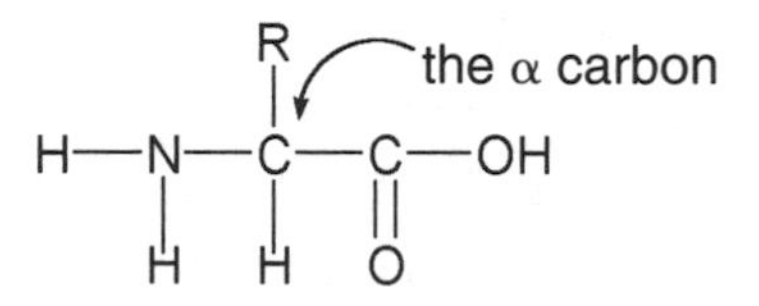

Generalized structure for an α-amino acid

This is often drawn more simply in its condensed version as

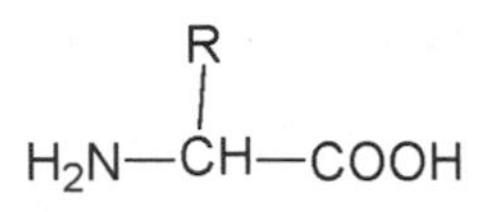

The condensed structure of the generalized amino acid structure

In these structures R represents a different group, or combination of atoms, for each amino acid. There are twenty-two different structures for R, and this is what makes the twenty-two amino acids. The R group is sometimes called the side group or the side chain. The structure of this side group is responsible for the special properties of silk—and indeed for the properties of any protein.

The smallest side group, and the only side group consisting of just one atom, is the hydrogen atom. Where this R group is H, the name of the amino acid is *glycine,* and the structure is as shown as follows.

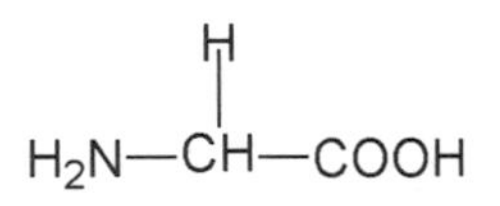

The amino acid glycine

Other simple side groups are CH_3 and CH_2OH, giving the amino acids *alanine* and *serine* respectively.

Alanine *Serine*

These three amino acids have the smallest side groups of all the amino acids, and they are also the most common amino acids in silk, constituting together about 85 percent of silk's overall structure. That the side groups in the silk amino acids are physically very small is an important factor in the smoothness of silk. In comparison, other amino acids have much larger, more complicated side groups.

Like cellulose, silk is a polymer—a macromolecule made up of repeating units. But unlike the cellulose polymer of cotton, in which the repeating units are exactly the same, the repeating units of protein polymers, amino acids, vary somewhat. The parts of the amino acid that form a polymer chain are all the same. It is the side group on each amino acid that differs.

Two amino acids combine by eliminating water between themselves, an H atom from the NH_2 or amino end and an OH from the COOH or acid end. The resulting link between the two amino acids is known as an *amide group.* The actual chemical bond between the carbon of one amino acid and the nitrogen of the other amino acid is known as a *peptide bond.*

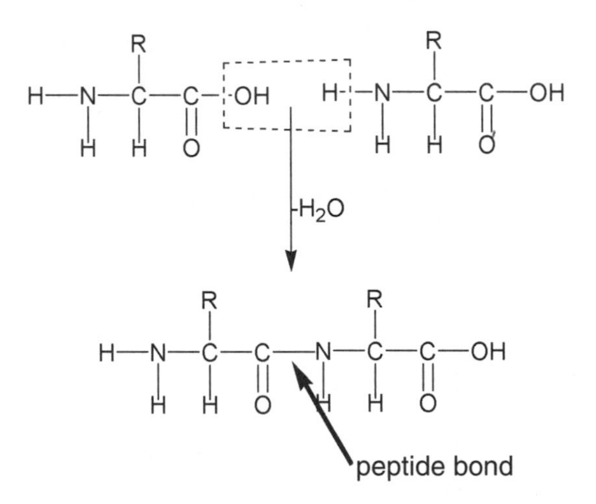

Of course, on one end of this new molecule there is still an OH that can be used to form another peptide bond with another amino acid, and on the other end there is an NH_2 (also written H_2N) that can form a peptide bond to yet another amino acid.

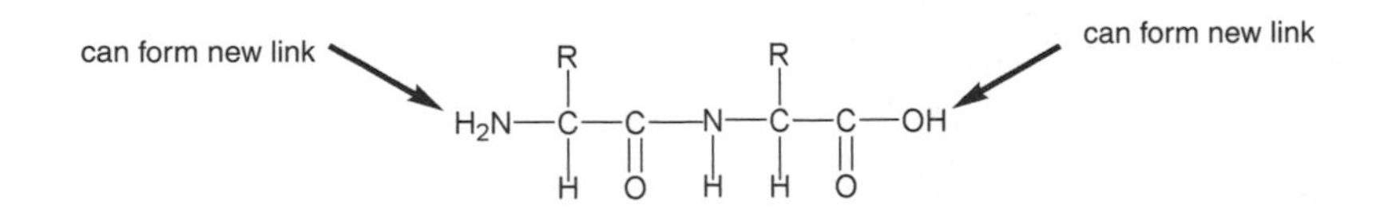

The amide group

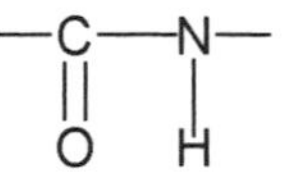

is usually shown in a more space-saving way as

—CO—NH—

If we add two more amino acids, there are now four amino acids joined through amide linkages.

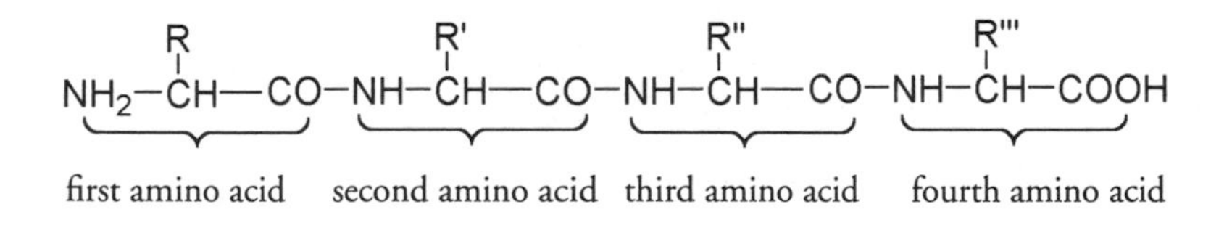

As there are four amino acids, there are four side groups, designated above as R, R', R'' and R'''. These side groups could all be the same, or some of them could be the same, or they could all be different. Despite having only four amino acids in the chain, a very large number of combinations is possible. R could be any of twenty-two amino acids, R' could also be any of the twenty-two, and so could R'' and R'''. This means there are 22^4 or 234,256 possibilities. Even a very small protein such as insulin, the hormone secreted by the pancreas that regulates glucose metabolism, contains 51 amino acids, so the number of combinations possible for insulin would be 22^{51} (2.9×10^{68}), or billions of billions.

It is estimated that 80 to 85 percent of the amino acids of silk is a repeating sequence of glycine-serine-glycine-alanine-glycine-alanine. A chain of the silk protein polymer has a zigzag arrangement with the side groups alternating on each side.

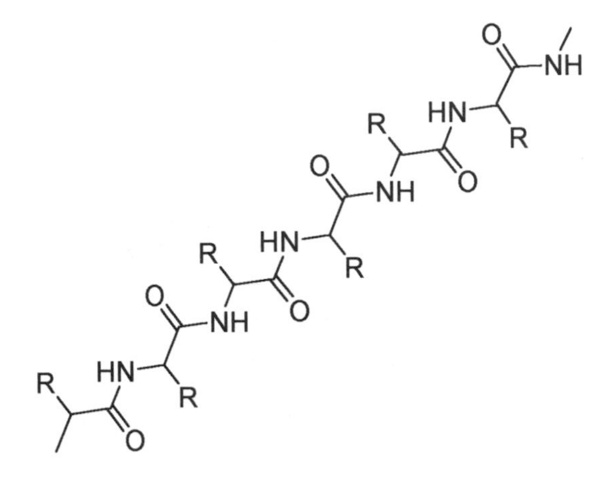

The silk protein chain is a zigzag; the R groups alternate on each side of the chain.

These chains of the protein molecule lie parallel with adjacent chains running in opposite directions. They are held to one another through cross-attractions between the molecular strands, as shown by dotted lines below.

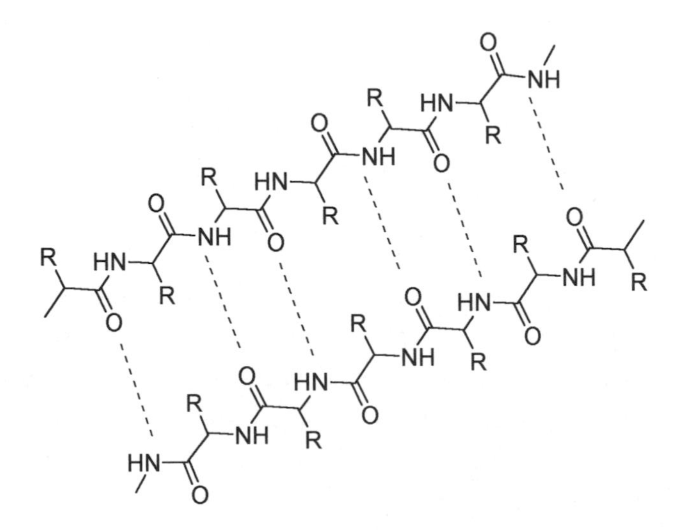

Attractions between side-by-side protein chains hold silk molecules together.

This produces a pleated sheet structure, where alternate R groups along the protein chain point either up or down. It can be shown as:

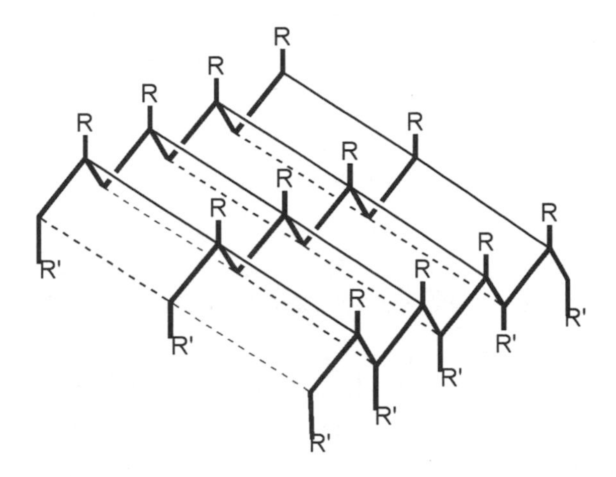

The pleated sheet structure. The bold lines represent the protein amino acid chains. Here R represents groups that are above the sheet, while R′ groups (where shown) are below. The narrow and dotted lines show the attractive forces holding the protein chains together.

The flexible structure resulting from the pleated sheet structure is resistant to stretching and accounts for many of the physical properties of silk. The protein chains fit together tightly; the small R groups on the surfaces are relatively similar in size, creating a uniform surface responsible for the smooth feel of silk. As well, this uniform surface acts as a good reflector of light, accounting for silk's characteristic luster. Thus many of the highly valued qualities of silk are due to the small side groups in its protein structure.

Connoisseurs of silk also appreciate the fabric's "sparkle," which is attributed to the fact that not all silk molecules are part of a regular pleated sheet structure. These irregularities break up reflected light, creating flashes of brightness. Often considered unsurpassed in its ability to absorb both natural and man-made dyes, silk is easy to color. This property is again due to the parts of the silk structure that are not included in the regular repeating sequence of pleated sheets. Among these remaining 15 to 20 percent or so of amino acids—those that are not glycine, alanine, or serine—are some whose side groups can easily chemically bond with dye molecules, producing the deep, rich, and colorfast hues for which silk is famous. It is this dual nature of silk—the repetitive small-side-group pleated sheet structure responsible for strength, sheen, and smoothness, combined with the more variable remaining amino acids, giving sparkle and ease of dyeing—that has for centuries made silk such a desirable fabric.

THE SEARCH FOR SYNTHETIC SILK

All of these properties make silk difficult to replicate. But because silk was so expensive and the demand for it so great, numerous attempts were made, beginning in the late nineteenth century, to produce a synthetic version. Silk is a very simple molecule—just a repetition of very similar units. But attaching these units together in the random and non-random combination found in natural silk is a very complicated chemical problem. Modern chemists are now able, on a very small scale, to replicate the set pattern of a particular protein strand, but the process is time consuming and exacting. A silk protein produced in the laboratory in this way would be many times more expensive than the natural article.

As the complexities of the chemical structure of silk were not understood until the twentieth century, early efforts to make a synthetic version were largely guided by fortunate accidents. Sometime in the late 1870s a French count, Hilaire de Chardonnet, while pursuing his favorite hobby of photography, discovered that a spilled solution of collodion—the nitrocellulose material used to coat photographic plates—had set to a sticky mass, from which he was able to pull out long threads resembling silk. This reminded Chardonnet of something he had seen a number of years previously: as a student, he had accompanied his professor, the great Louis Pasteur, to Lyons, in the south of France, to investigate a silkworm disease that was causing enormous problems for the French silk industry. Though unable to find the cause of the silkworm blight, Chardonnet had spent much time studying the silkworm and how it spun its silk fiber. With this in mind, he now tried forcing collodion solution through a set of tiny holes. He thus produced the first reasonable facsimile of silk fiber.

The words *synthetic* and *artificial* are often used interchangeably in everyday language and are given as synonyms in most dictionaries. But there is an important chemical distinction between them. For our purposes *synthetic* is taken to mean that a compound is man-made by chemical reactions. The product may be one that occurs in nature or it may not occur naturally. If it does occur in nature, the synthetic version will be chemically identical to that of the natural source. For example, ascorbic acid, vitamin C, can be synthesized in a laboratory or factory; synthetic vitamin C has exactly the same chemical structure as naturally occurring vitamin C.

The word *artificial* refers more to the properties of a compound. An artificial compound has a different chemical structure from that of another compound, but it has properties similar enough to mimic the other's role. For example, an artificial sweetener does not have the same structure as sugar, but it will have an important property—in this case, sweetness—in common. Artificial compounds are often man-made and are therefore synthetic, but they need not necessarily be synthetic. Some artificial sweeteners are naturally occurring.

What Chardonnet produced was artificial silk, not synthetic silk, although it was made synthetically. (By our definitions, synthetic silk would be man-made but chemically identical to real silk.) *Chardonnet silk,* as it came to be known, resembled silk in some of its properties but not in all of them. It was soft and lustrous, but unfortunately it was highly flammable—not a desirable property for a fabric. Chardonnet silk was spun from a solution of nitrocellulose, and as we have seen, nitrated versions of cellulose are flammable and even explosive, depending on the degree of nitration of the molecule.

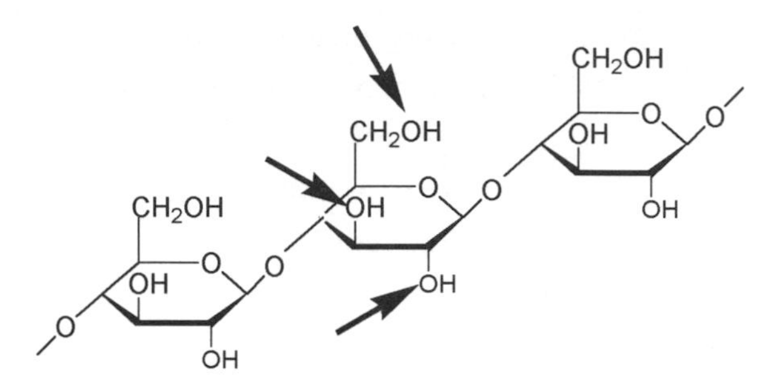

A portion of a cellulose molecule. The arrows on the middle glucose unit indicate the OH groups where nitration could take place on each glucose unit along the chain.

Chardonnet patented his process in 1885 and started manufacturing Chardonnet silk in 1891. But the flammability of the material proved to be its downfall. In one incident a cigar-smoking gentleman flicked ash on the Chardonnet silk dress of his dancing partner. The garment was reported to have disappeared in a flash of flame and a puff of smoke; there was no mention of the fate of the lady. Although this event and a number of other disasters at the Chardonnet factory led to its closure, Chardonnet did not give up on his artificial silk. By 1895 he was using a somewhat different process, involving a denitrating agent that produced a much safer cellulose-based artificial silk that was no more flammable than ordinary cotton.

Another method, developed in England, in 1901, by Charles Cross and Edward Bevan produced *viscose,* so named because of its high viscosity. When viscose liquid was forced through a spinnerette into an acid bath, cellulose was regenerated in the form of a fine filament called viscose silk. This process was used by both the American Viscose Company, formed in 1910, and by the Du Pont Fibersilk Company (later to become the Du Pont Corporation), formed in 1921. By 1938, 300 million pounds of viscose silk were being produced annually, supplying a growing demand for the new synthetic fabrics with the desired silky gloss so reminiscent of silk.

The viscose process is still in use today, as the principal means of making what are now called rayons—artificial silks, like viscose silk, in which the threads are composed of cellulose. Although it is still the same polymer of β-glucose units, the cellulose in rayon is regenerated under a slight tension, supplying a slight difference in twist to rayon threads that accounts for its high luster. Rayon, pure white in color and still having the same chemical structure, can be dyed to any number of tints and shades in the same manner as cotton. But it also has a number of drawbacks. While the pleated sheet structure of silk (flexible but resistant to stretching) makes it ideal for hosiery, the cellulose of rayon absorbs water, causing it to sag. This is not a desirable characteristic when used for stockings.

NYLON—A NEW ARTIFICIAL SILK

A different type of artificial silk was needed, one that had rayon's good characteristics without its weaknesses. Noncellulose-based nylon arrived on the scene in 1938, created by an organic chemist hired by the Du Pont Fibersilk Company. By the late 1920s Du Pont had become interested in the plastic materials coming into the market. Wallace Carothers, a thirty-one-year-old organic chemist at Harvard University, was offered the opportunity to perform independent research for Du Pont on a virtually unlimited budget. He began work in 1928 at the new Du Pont laboratory dedicated to basic research—itself a highly unusual concept, as within the chemical industry the practice of basic research was normally left to the universities.

Carothers decided that he wanted to work on polymers. At that time most chemists thought that polymers were actually groups of molecules clumped together and known as colloids; hence the name *collodion,* for the nitrocellulose derivative used in photography and in Chardonnet silk. Another opinion on the structure of polymers, championed by the German chemist Hermann Staudinger, was that these materials were extremely large molecules. The largest molecule synthesized up to that time—by Emil Fischer, the great sugar chemist—had a molecular weight of 4,200. In comparison, a simple water molecule has a molecular weight of 18, and a glucose molecule's molecular weight is 180. Within a year of starting work in the Du Pont laboratory, Carothers had made a polyester molecule with a molecular weight of over 5,000. He was then able to increase this value to 12,000, adding more evidence to the giant molecule theory of polymers, for which Staudinger was to receive the 1953 Nobel Prize in chemistry.

Carothers's first polymer initially looked as if it had some commercial potential, as its long threads glistened like silk and did not become stiff or brittle on drying. Unfortunately, it melted in hot water, dissolved in common cleaning solvents, and disintegrated after a few weeks. For four years Carothers and his coworkers prepared different types of polymers and studied their properties, before they finally produced nylon, the man-made fiber that comes the closest to having the properties of silk and that deserves to be described as "artificial silk."

Nylon is a polyamide, meaning that, as with silk, its polymer units are held together through amide linkages. But while silk has both an acid end and an amine end on each of its individual amino acid units, Carothers's nylon was made from two different monomer units—one with two acid groups and one with two amine groups—alternating in the chain. Adipic acid has acid groups COOH at both ends:

$$HOOC-CH_2-CH_2-CH_2-CH_2-COOH$$

Structure of adipic acid, showing the two acid groups at each end of the molecule. The acid group –COOH is written backward as HOOC– when it is shown on the left-hand side.

or written as a condensed structure (below):

$$HOOC-(CH_2)_4-COOH$$

The condensed structure of the adipic acid molecule

The other molecular unit, 1,6-diaminohexane, has a very similar structure to that of adipic acid except there are amino groups (NH_2) attached in place of the COOH acid groups. The structure and its condensed version are shown below:

$H_2N—CH_2—CH_2—CH_2—CH_2—CH_2—CH_2—NH_2$

Structure of 1,6-diaminohexane

$H_2N—(CH_2)_6—NH_2$

Condensed structure of 1,6-diaminohexane

The amide link in nylon, like the amide link in silk, is formed by eliminating a molecule of water between the ends of the two molecules, from the H atom from NH_2 and the OH from COOH. The resulting amide bond, shown as -CO-NH- (or in reverse order as -NH-CO-) joins the two different molecules. It is in this respect—having the same amide link—that nylon and silk are chemically similar. In the making of nylon both the amino ends of 1,6-diaminohexane react with the acid ends of different molecules. This continues with alternating molecules adding to each end of a growing nylon chain. Carothers's version of nylon became known as "nylon 66" because each monomer unit has six carbon atoms.

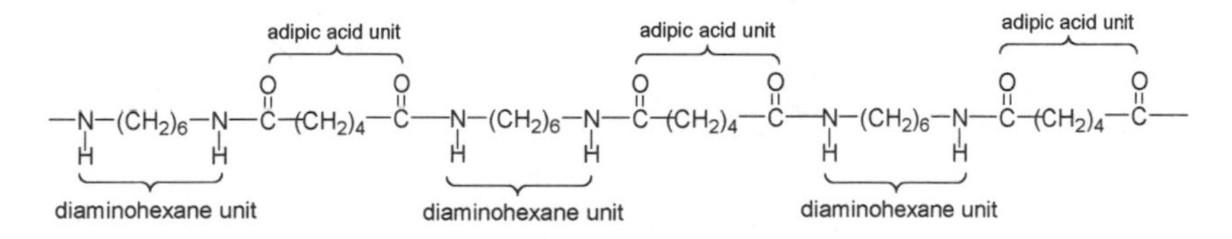

Structure of nylon, showing alternating molecules of adipic acid and 1,6-diaminohexane

The first commercial use of nylon, in 1938, was for toothbrush bristles. Then in 1939 nylon stockings were marketed for the first time. Nylon proved to be the ideal polymer for stockings. It had many of the desirable properties of silk; it did not sag and wrinkle like cotton or rayon; and most important, it was far less expensive than silk. Nylon hosiery was an enormous commercial success. In the year after they were introduced, some 64 million pairs of "nylons" were manufactured and sold. So overwhelming was the response to this product that the word *nylons* is now synonymous with women's hosiery. With its exceptional strength, durability, and lightness, nylon quickly found a use in many other products: fishing lines and nets, strings for tennis and badminton rackets, surgical sutures, and coatings for electrical wires.

During World War II Du Pont's main production of nylon shifted from the fine filaments used in hosiery to the coarser yarns needed for military products. Tire cords, mosquito netting, weather balloons, ropes, and other military items dominated the use of nylon. In aviation nylon proved to be an excellent substitute for silk parachute shrouds. After the end of the war production in nylon plants was quickly converted back to civilian products. By the 1950s nylon's versatility was apparent in its use in clothing, skiwear, carpets, furnishings, sails, and many other products. It was also found to be an excellent molding compound and became the first "engineering plastic," a plastic that is strong enough to be used as a replacement for metal. Ten million pounds of nylon were produced in 1953 for this use alone.

Unfortunately, Wallace Carothers did not live to see the success of his discovery. A victim of depression that became worse as he got older, he ended his life in 1937 by swallowing a vial of cyanide, unaware that the polymer molecule he had synthesized would play such a dominant role in the world of the future.

Silk and nylon share a similar legacy. It is more than just a comparable chemical structure and an eminent suitability for use in hosiery and parachutes. Both these polymers contributed—in their own way—to enormous changes in the economic prosperity of their times. Not only did the demand for silk open worldwide trade routes and new trade agreements; it also led to the growth of cities that depended on silk production or the silk trade and helped establish other industries, such as dyeing, spinning, and weaving, that developed alongside sericulture. Silk brought great wealth and great change to many parts of the globe.

Just as silk and silk production stimulated fashions—in clothing, fur-

Women rushed to buy—and wear—nylons after World War II when the polymer became available for hosiery again. (Photo courtesy of Du Pont)

nishings, and art—in Europe and Asia for centuries, the introduction of nylon and a wealth of other modern textiles and materials has had a vast influence on our world. Where once plants and animals furnished the starting materials for our clothing, the raw products for many fabrics now come from by-products of oil refining. As a commodity, oil has taken over a position that once belonged to silk. As was once the case with silk, the demand for oil has forged new trade agreements, opened new trade routes, encouraged the growth of some cities and the establishment of others, created new industries and new jobs, and brought great wealth and great change to many parts of the globe.

7. PHENOL

THE VERY FIRST totally man-made polymer was produced about twenty-five years before Du Pont's nylon. It was a somewhat random cross-linked material made from a compound whose chemical structure was similar to some of the spice molecules to which we attributed the Age of Discovery. This compound, phenol, started another age, the Age of Plastics. Linked to such diverse topics as surgical practices, endangered elephants, photography, and orchids, phenols have played a pivotal role in a number of advances that have changed the world.

STERILE SURGERY

In 1860 you would not have wanted to be a patient in a hospital—especially not to undergo an operation. Hospitals were dark, grimy, and airless. Patients were commonly given beds where the bedclothes were not changed after the previous occupant left—or more probably died. Surgical wards exuded an appalling stench from gangrene and sepsis. Equally appalling was the death rate from such bacterial infections; at least 40 percent of amputees died from so-called hospital disease. In army hospitals that number approached 70 percent.

Despite the fact that anesthetics had been introduced at the end of 1864, most patients agreed to surgery only as a last resort. Surgical wounds always became infected; accordingly, a surgeon would ensure that the stitches closing an operation site were left long, hanging down toward the ground, so that pus could drain away from the wound. When this happened, it was considered a positive sign, as chances were good that the infection would stay localized and not invade the rest of the body.

Of course, we now know why "hospital disease" was so prevalent and so lethal. It was actually a group of diseases caused by a variety of bacteria that easily passed from patient to patient or even from doctor to a series of patients under unsanitary conditions. When hospital disease became too rife, a physician usually closed down his surgical ward, sent the remaining patients elsewhere, and had the premises fumigated with sulfur candles, the walls whitewashed, and the floors scrubbed. For a while after these precautions infections would be under control—until another outbreak required further attention.

Some surgeons insisted on maintaining constant strict cleanliness, a regime involving lots of cooled boiled water. Others supported the miasma theory, the belief that a poisonous gas generated by drains and sewers was carried in the air and that once a patient was infected, this miasma was transferred through the air to other patients. This miasma theory probably seemed very reasonable at the time. The stench from drains and sewers would have been as bad as the smell of putrefying gangrenous flesh in surgery wards, which would further explain how patients treated at home rather than in a hospital often escaped infection altogether. Various treatments were prescribed to counteract miasma gases, including thymol, salicylic acid, carbon dioxide gas, bitters, raw carrot poultices, zinc sulfate, and boracic acid. The occasional success of any of these remedies was fortuitous and could not be replicated at will.

This was the world in which the physician Joseph Lister was practicing surgery. Born in 1827 to a Quaker family from Yorkshire, Lister completed his medical degree at University College in London, and by 1861 was a surgeon at the Royal Infirmary in Glasgow and a professor of surgery at the University of Glasgow. Though a new modern surgical block had been opened at the Royal Infirmary during Lister's tenure, hospital disease was just as much a problem there as it was elsewhere.

Lister believed its cause might not be a poisonous gas but something else in the air, something that could not be seen with the human eye, something microscopic. On reading a paper that described "The Germ Theory of Diseases," he immediately recognized its applicability to his own ideas. The paper had been written by Louis Pasteur, a professor of chemistry in Lille, northeastern France, and mentor to Chardonnet of Chardonnet silk fame. Pasteur's experiments on the souring of wine and milk had been presented to a gathering of scientists at the Sorbonne in Paris in 1864. Germs—microorganisms that could not be detected by the human eye—were considered by Pasteur to be everywhere. His experiments showed that such microorganisms could be eliminated by boiling, leading of course to our present-day pasteurization of milk and other foodstuffs.

As boiling patients and surgeons was not practical, Lister had to find some other way to safely eliminate germs on all surfaces. He settled on carbolic acid, a product made from coal tar that had been used successfully to treat stinking city drains and that had already been tried as a dressing on surgical wounds, without very positive results. Lister persevered and met with success in the case of an eleven-year-old boy who came to the Royal Infirmary with a compound fracture of the leg. At the time compound fractures were a dreaded injury. A simple fracture could be set without invasive surgery, but a compound fracture, where the sharp ends of broken bones had pierced the skin, was almost certain to become infected, despite a surgeon's skill at setting the bone. Amputation was a common outcome, and death from an uncontrollable persistent infection was likely.

Lister carefully cleaned the area in and around the boy's broken bone with lint soaked in carbolic acid. Then he prepared a surgical dressing consisting of layers of linen soaked in carbolic solution and covered with a thin metal sheet bent over the leg to reduce possible evaporation of the carbolic acid. This dressing was carefully taped in place. A scab soon formed, the wound healed rapidly, and infection never occurred.

Other patients had survived their infections from hospital disease, but this was a case in which infection had been prevented, not just outlasted. Lister treated subsequent compound fracture cases the same way, producing the same positive outcome, convincing him of the effectiveness of carbolic solutions. By August 1867 he was using carbolic acid as an antiseptic agent during all his surgical procedures, not just as a postoperative dressing. He improved his antiseptic techniques over the next decade, gradually convincing other surgeons, many of whom still refused to believe in the germ theory as "if you can't see them, they are not there."

Coal tar, from which Lister obtained his carbolic acid solution, was readily available as a waste product from the gaslight illumination of city streets and houses during the nineteenth century. The National Light and Heat Company had installed the first gas street lighting in Westminster, London, in 1814, and widespread use of gas lighting followed in other cities. Coal gas was produced by heating coal at high temperatures; it was a flammable mixture—about 50 percent hydrogen, 35 percent methane, and smaller amounts of carbon monoxide, ethylene, acetylene, and other organic compounds. It was piped to homes, factories, and streetlamps from local gasworks. As demand for coal gas grew, so did the problem of what to do with coal tar, the seemingly unimportant residue of the coal-gasification process.

Coal tar was a viscous, black, acrid-smelling liquid that would eventually prove an amazingly prolific source for a number of important aromatic molecules. Not until huge reservoirs of natural gas, consisting mainly of methane, were discovered in the early part of the twentieth century did the coal-gasification process and its accompanying production of coal tar decline. Crude carbolic acid, as first used by Lister, was a mixture distilled from coal tar at temperatures between 170°C and 230°C. It was a dark and very strong-smelling oily material that burned the skin. Lister was eventually able to obtain the main constitutent of carbolic acid, *phenol,* in its pure form as white crystals.

Phenol is a simple aromatic molecule consisting of a benzene ring, to which is attached an oxygen-hydrogen or OH group.

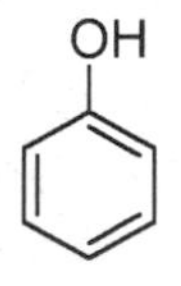

Phenol

It is somewhat water soluble and is very soluble in oil. Lister made use of these characteristics by developing what became known as the "carbolic putty poultice," a mixture of phenol with linseed oil and whitening (powdered chalk). The resultant paste (spread on a sheet of tinfoil) was placed poultice side down on the wound and acted like a scab, providing a barrier to bacteria. A less concentrated solution of phenol in water, usually about one part of phenol to between twenty and forty parts of water, was used to wash the skin around a wound, the surgical instruments, and the surgeon's hands, and it was also sprayed onto an incision during an operation.

Despite the effectiveness of his carbolic acid treatment, as shown by the recovery rates of his patients, Lister was not satisfied that he had achieved totally antiseptic conditions during surgical operations. He thought that every dust particle in the air bore germs, and in an effort to prevent these airborne germs from contaminating operations, he developed a machine that continually sprayed a fine mist of carbolic acid solution into the air, effectively drenching the whole area. Airborne germs are actually far less a problem than Lister had assumed. The real issue was microorganisms that came from the clothes, hair, skin, mouths, and noses of the surgeons, the other doctors, and the medical students who routinely assisted with or watched operations without taking any antiseptic precautions. Modern operating-room protocol of sterile masks, scrub suits, hair coverings, drapes, and latex gloves now solves this problem.

Lister's carbolic spray machine did help prevent contamination from microorganisms, but it had negative effects on the surgeons and others in the operating room. Phenol is toxic, and even in dilute solutions, it causes bleaching, cracking, and numbing of the skin. Inhaling phenolic spray can lead to illness; some surgeons refused to continue working when a phenol spray was in use. Despite these drawbacks Lister's techniques of antiseptic surgery were so effective and the positive results so obvious that by 1878 they were in use throughout the world. Phenol is rarely used today as an antiseptic; its harsh effect on the skin and its toxicity make it less useful than newer antiseptics that have been developed.

MANY-FACETED PHENOLS

The name *phenol* does not apply only to Lister's antiseptic molecule; it is applied to a very large group of related compounds that all have an OH group attached directly to a benzene ring. This may seem a bit confusing, as there are thousands or even hundreds of thousands of phenols, but only one "phenol." There are man-made phenols, like trichlorophenol and hexylresorcinols, with antibacterial properties that are used today as antiseptics.

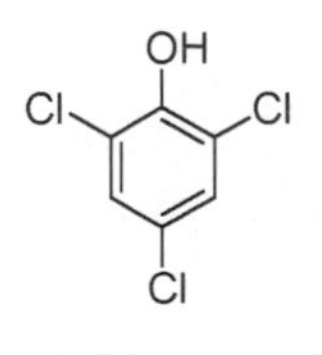

Trichlorophenol

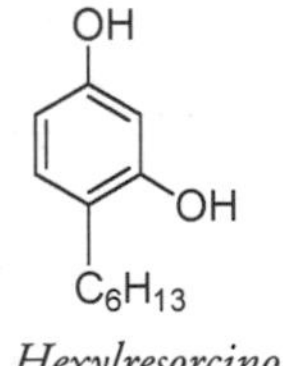

Hexylresorcinol

Picric acid, originally used as a dye—especially for silk—and later in armaments by the British in the Boer War and in the initial stages of World War I, is a triple-nitrated phenol and is highly explosive.

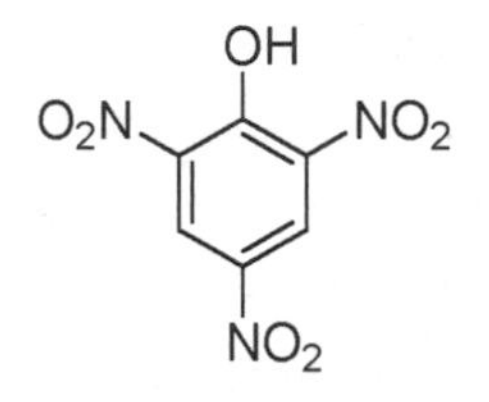

Trinitrophenol (picric acid)

Many different phenols occur in nature. The hot molecules—capsaicin from peppers and zingerone from ginger—can both be classified as phenols, and some of the highly fragrant molecules in spices—eugenol from cloves and isoeugenol from nutmeg—are members of the phenol family.

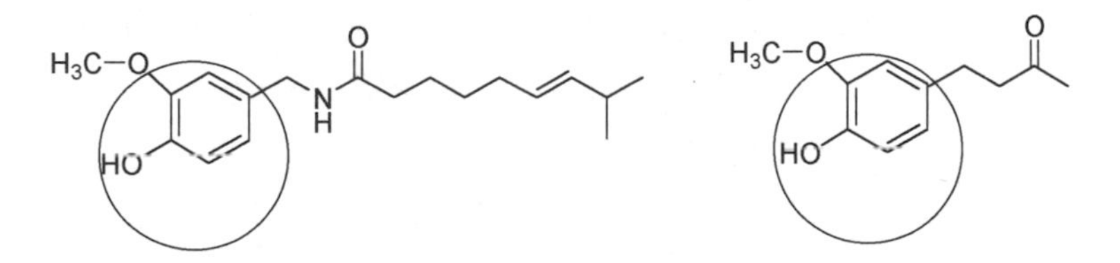

Capsaicin (left) and zingerone (right). The phenol part of each structure is circled.

Vanillin, the active ingredient in one of our most widely used flavoring compounds, vanilla, is also a phenol, with a very similar structure to that of eugenol and isoeugenol.

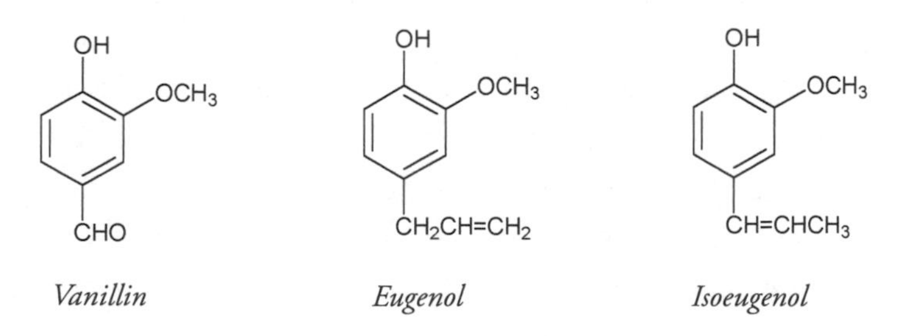

Vanillin *Eugenol* *Isoeugenol*

Vanillin is present in the dried fermented seedpods from the vanilla orchid (*Vanilla planifolia*), native to the West Indies and Central America but now grown around the world. These long, thin, fragrant seedpods are sold as vanilla beans, and up to 2 percent of their weight can be vanillin. When wine is stored in oak casks, vanillin molecules are leached from the wood, contributing to the changes that constitute the aging process. Chocolate is a mixture containing cacao and vanillin; custards, ice cream, sauces, syrups, cakes, and many other foods depend partly on vanilla for their flavor. Perfumes also incorporate vanillin for its heady and distinctive aroma.

We are only beginning to understand the unique properties of some naturally occurring members of the phenol family. Tetrahydrocannabinol (THC), the active ingredient in marijuana, is a phenol found in *Cannabis sativa,* the Indian hemp plant. Marijuana plants have been grown for centuries for the strong fibers found in the stem, which make excellent rope and a coarse cloth, and for the mildly intoxicating, sedative, and hallucinogenic properties of the THC molecule found—in some cannabis varieties—in all parts of the plant but most often concentrated in the female flower buds.

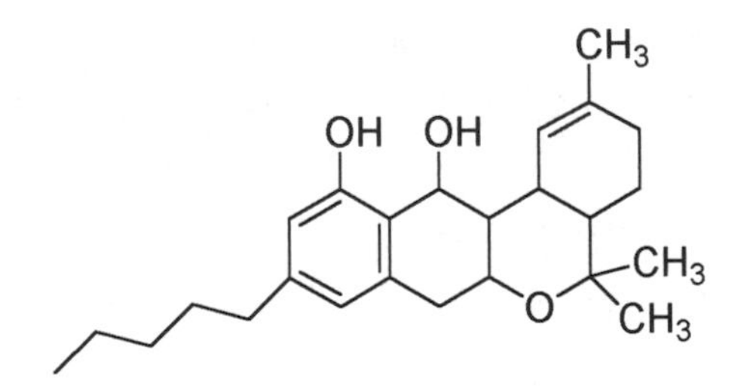

Tetrahydrocannabinol, the active ingredient of marijuana

Medicinal use of the tetrahydrocannabinol in marijuana to treat nausea, pain, and loss of appetite in patients suffering from cancer, AIDS, and other illnesses is now permitted in some states and countries.

Naturally occurring phenols often have two or more OH groups attached to the benzene ring. Gossypol is a toxic compound, classified as a polyphenol because it has six OH groups on four different benzene rings.

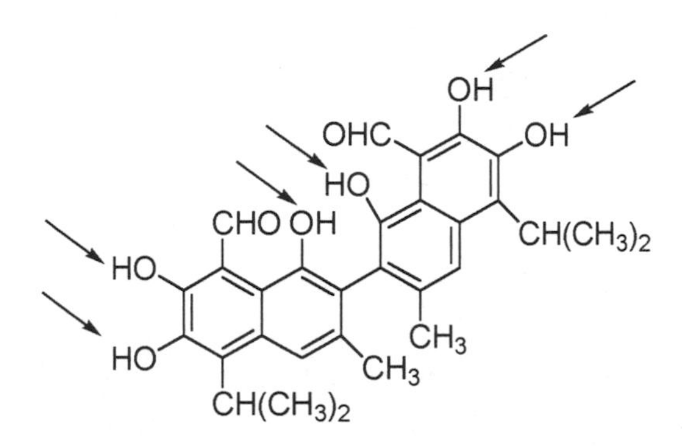

The gossypol molecule. The six phenol (OH) groups are indicated by arrows.

Extracted from seeds of the cotton plant, gossypol has been shown to be effective in suppressing sperm production in men, making it a possible candidate for a male chemical birth control method. The social implications of such a contraceptive could be significant.

The molecule with the complicated name of epigallocatechin-3-gallate, found in green tea, has even more phenolic OH groups.

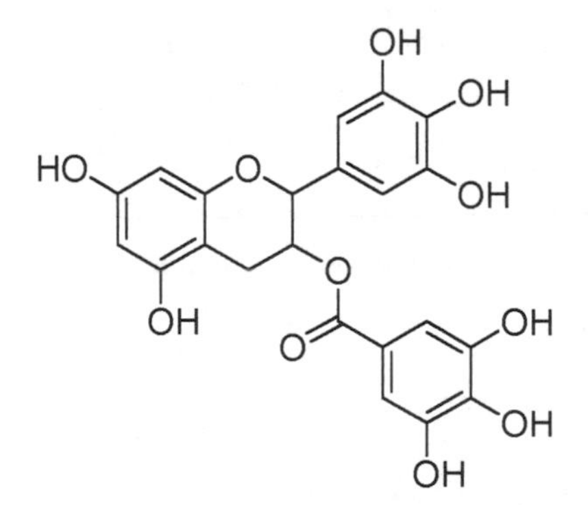

The epigallocatechin-3-gallate molecule in green tea has eight phenolic groups.

It has recently been credited with providing protection against various types of cancer. Other studies have shown that polyphenolic compounds in red wine inhibit the production of a substance that is a factor in hardening of the arteries, possibly explaining why in countries where a lot of red wine is consumed there is a lower incidence of heart disease, despite a diet rich in butter, cheese, and other foods high in animal fat.

PHENOL IN PLASTICS

As valuable as the many different derivatives of phenol are, however, it is the parent compound, phenol itself, that has brought about the greatest changes in our world. As useful and influential as the phenol molecule was in the development of antiseptic surgery, it had a vastly different and possibly even more important role in the growth of an entirely new industry. Around the same time Lister was experimenting with carbolic acid, the use of ivory from animals for items as diverse as combs and cutlery, buttons and boxes, chessmen and piano keys was rapidly increasing. As more and more elephants were killed for their tusks, ivory was becoming scarce and expensive. Concern for the reduction of the elephant population was most noticeable in the United States, not for the conservationist reasons that we espouse today but because of the exploding popularity of the game of billiards. Billiard balls require extremely high-quality ivory for the ball to roll true. They must be cut from the very center of a flaw-free animal tusk, and only one out of every fifty tusks provides the consistent density necessary.

In the last decades of the nineteenth century, as supplies of ivory dwindled, the idea of making an artificial material to replace it appeared sensible. The first artificial billiard balls were made from pressed mixtures of substances such as wood pulp, bone dust, and soluble cotton paste impregnated by or coated with a hard resin. The main component of these resins was cellulose, often a nitrated form of cellulose. A later and more sophisticated version used the cellulose-based polymer celluloid. The hardness and density of celluloid could be controlled during the manufacturing process. Celluloid was the first *thermoplastic* material—that is, one that could be melted and remolded many times in a process that was the forerunner of the modern injection-molding machine, a method of repeatedly reproducing objects inexpensively with unskilled labor.

A major problem with cellulose-based polymers is their flammability and, especially where nitrocellulose is involved, their tendency to explode. There is no record of exploding celluloid billiard balls, but celluloid was a potential safety hazard. In the movie business film was originally composed of a celluloid polymer made from nitrocellulose, using camphor as a plasticizer for improving flexibility. After a disastrous 1897 fire in a movie theater in Paris killed 120 people, projection booths were lined with tin to prevent the spread of fire if the film happened to ignite. This precaution, however, did nothing for the safety of the projectionist.

In the early 1900s, Leo Baekeland, a young Belgian immigrant to the

A growing scarcity of good quality ivory from elephant tusks was relieved by the development of phenolic resins like Bakelite. (Photo courtesy Michael Beugger)

United States, developed the first truly synthetic version of the material we now call *plastic.* This was revolutionary, since varieties of polymers that had been made up to this time had been composed, at least partly, of the naturally occurring material cellulose. With his invention Baekeland initiated the Age of Plastics. A bright and inventive chemist who had obtained his doctorate from the University of Ghent at the age of twenty-one, he could have settled for the security of an academic life. He chose instead to emigrate to the New World, where he believed the opportunities to develop and manufacture his own chemical inventions would be greater.

At first this choice seemed a mistake, as despite working diligently for a few years on a number of possible commercial products, he was on the verge of bankruptcy in 1893. Then, desperate for capital, Baekeland approached George Eastman, the founder of the photographic company Eastman Kodak, with an offer to sell him a new type of photographic paper he had devised. This paper was prepared with a silver chloride emulsion that eliminated the washing and heating steps of image development and boosted light sensitivity to such a level that it could be exposed by artificial light (gaslight in the 1890s). Amateur photographers could thus develop their photographs quickly and easily at home or send them to one of the new processing laboratories opening across the country.

While he was taking the train to meet with Eastman, Baekeland decided he would ask for $50,000 for his new photographic paper, since it was a great improvement over the celluloid product with associated fire hazards that was then being used by Eastman's company. If forced to compromise, Baekeland told himself he would accept no less than $25,000, still a reasonably large amount of money in those days. Eastman was so impressed with Baekeland's photographic paper, however, that he immediately offered him the then-enormous sum of $750,000. A stunned Baekeland accepted and used the money to establish a modern laboratory next to his home.

With his financial problem solved, Baekeland turned his attention to creating a synthetic version of shellac, a material that had been used for many years as a lacquer and wood preservative and is still used today. Shellac is obtained from an excretion of the female lac or *Laccifer lacca* beetle, a native of Southeast Asia. The beetles attach themselves to trees, suck the sap, and eventually become encased in a covering of their own secretion. After reproducing, the beetles die, and their cases or shells—hence the *shell* part of the word *shellac*—are gathered and melted. The liquid thus obtained is filtered to remove the bodies of the dead beetles. It takes fifteen thousand lac beetles six months to produce a single pound of shellac. As long as the shellac was used only as a thin coating, the price was affordable, but with increased use of shellac by the electrical industry, which was rapidly expanding at the beginning of the twentieth century, the demand for shellac soared. The cost of making an electrical insulator, even just using shellac-impregnated paper, was high, and Baekeland realized that artificial shellac would become a necessity for insulators in this increasing market.

Baekeland's first approach to the problem of making a shellac involved the reaction of phenol—the same molecule with which Lister had successfully transformed surgery—and formaldehyde, a compound derived from methanol (or wood alcohol) and in those days used extensively as an embalming agent by undertakers and for preserving animal specimens.

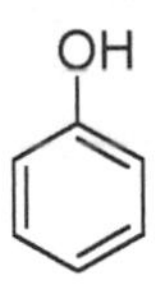

Phenol

Formaldehyde

Previous attempts to combine these compounds had produced discouraging results. Rapid, uncontrolled reactions led to insoluble and infusible materials that were too brittle and inflexible to be useful. But Baekeland recognized that such properties could be just what was needed in a synthetic shellac for electrical insulators, if only he could control the reaction so the material could be processed into a usable form.

By 1907, using a reaction in which he was able to control both heat and pressure, Baekeland had produced a liquid that rapidly hardened into a transparent, amber-colored solid conforming exactly to the shape of the mold or vessel into which it was poured. He named the material *Bakelite* and called the modified pressure cooker–like device used to produce it the *Bakelizer.* We can perhaps excuse the self-promotion inherent in these names when we consider that Baekeland had spent five years working with this one reaction in order to synthesize this substance.

While shellac would become distorted with heat, Bakelite retained its shape at high temperatures. Once set, it could not be melted and remolded. Bakelite was a *thermoset* material; that is, it was frozen in its shape forever, as opposed to a *thermoplastic* material like celluloid. This phenolic resin's unique thermoset property was due to its chemical structure: formaldehyde in Bakelite can react at three different places on the benzene ring of phenol, causing cross-links among the polymer chains. Rigidity in Bakelite is attributed to these very short cross-links attached to already rigid and planar benzene rings.

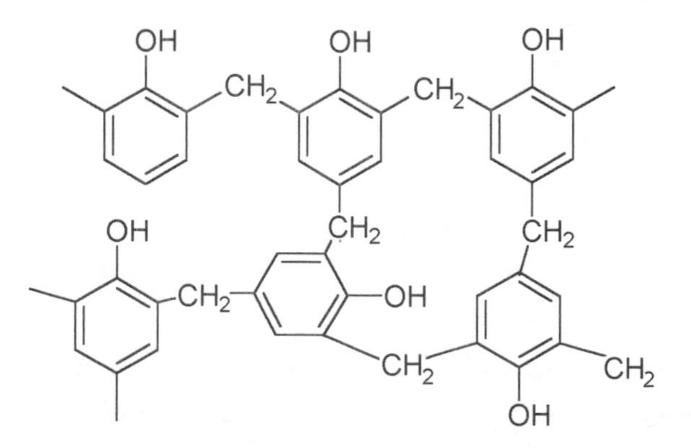

Schematic formula for Bakelite showing -CH2- cross-links between phenol molecules. These are only some possible ways of linking; in the actual material the linkages are random.

When used for electrical insulators, Bakelite's performance was superior to that of any other material. It was more heat resistant than shellac or any shellac-impregnated paper versions; it was less brittle than ceramic or glass insulators; and it had better electrical resistance than porcelain or mica. Bakelite did not react with sun, water, salt air, or ozone and was impervious to acid and solvents. It didn't easily crack, chip, discolor, fade, burn, or melt.

Subsequently, though not the original intent of its inventor, Bakelite was found to be the ideal material for billiard balls. Bakelite's elasticity was very similar to that of ivory, and Bakelite billiard balls, when they collided, made the same agreeable clicking sound as did ivory billiard balls, an important factor that was lacking in celluloid versions. By 1912 almost all nonivory billiard balls were made of Bakelite. Numerous other applications followed, and within the space of a few years Bakelite was everywhere. Telephones, bowls, washing machine agitators, pipe stems, furniture, car parts, fountain pens, plates, glasses, radios, cameras, kitchen equipment, handles for knives, brushes, drawers, bathroom fittings, and even artworks and decorative items were all being made of Bakelite. Bakelite became known as the "material for a thousand uses"—though nowadays other phenolic resins have superseded the original brown material. Later resins were colorless and could be easily tinted.

A Phenol for Flavor

The creation of Bakelite is not the only example where a phenol molecule was the basis of the development of an artificial substance to take the place of a natural substance for which demand had exceeded supply. The market for vanillin has long surpassed the supply available from the vanilla orchid. So synthetic vanillin is manufactured and comes from a surprising source: the waste pulp liquor from the sulfite treatment of wood pulp in the making of paper. Waste liquor consists mainly of lignin, a substance found in and between the cell walls of land plants. Lignin contributes to the rigidity of plants and makes up about 25 percent of the dry weight of wood. It is not one compound but a variable cross-bonded polymer of different phenolic units.

There is a difference in the lignin composition between softwoods and hardwoods as shown by the structures of the building blocks of their respective lignins. In lignin, as in Bakelite, the rigidity of wood depends on the degree of cross-linking between phenolic molecules. The triple-substituted phenols, found only in hardwoods, allow for more cross-linkages and thus explain why hardwoods are "harder" than softwoods.

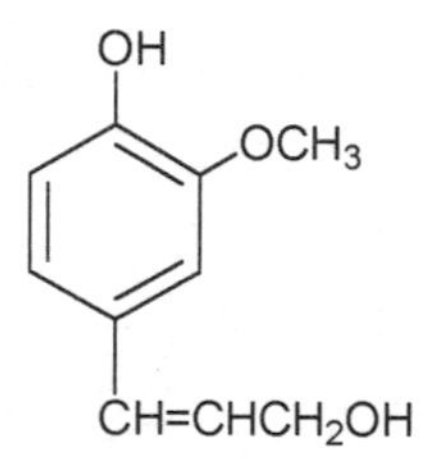

Softwood and hardwood building block (double-substituted phenol)

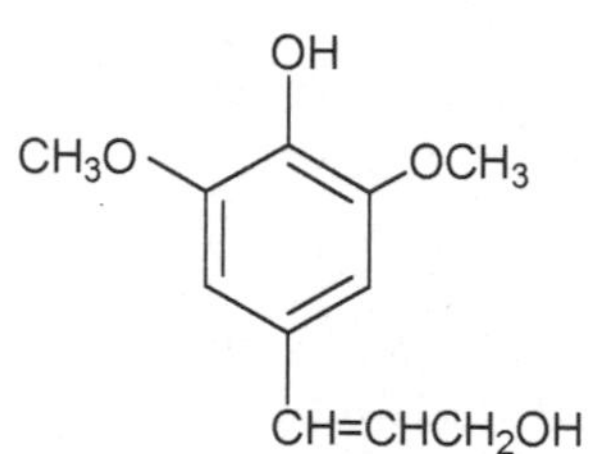

Hardwood building block (triple-substituted phenol)

A representative structure for lignin showing some of the cross-links between these building-block units is illustrated below. It has definite similarities to Baekeland's Bakelite.

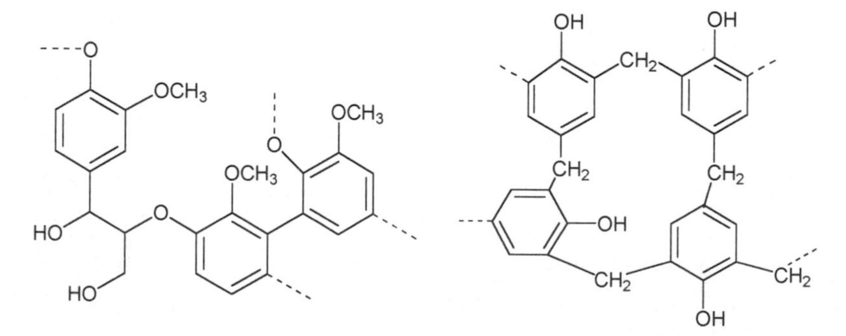

Part of the structure of lignin (left). The dotted lines indicate connection to the rest of the molecule. The structure of Bakelite (right) also has cross-links between phenol units.

The circled part in the drawing of lignin (below) highlights part of the structure that is very similar to the vanillin molecule. When a lignin molecule is broken up under controlled conditions, vanillin can be produced.

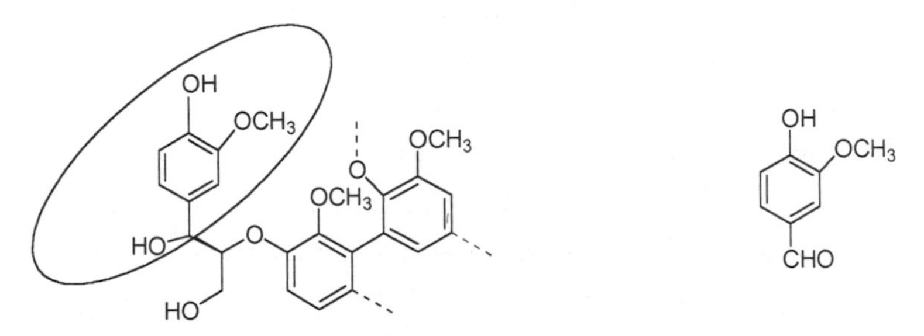

Lignin (left), with the circled part of its structure very similar to the vanillin molecule (right)

Synthetic vanillin is not just a chemical imitation of the real thing; rather, it is pure vanillin molecules made from a natural source and is chemically absolutely identical to vanillin from the vanilla bean. Vanilla flavoring obtained from using the whole vanilla bean does, however, contain trace amounts of other compounds that, together with the vanillin molecule, give the overall flavor of true vanilla. Artificial vanilla flavoring contains synthetic vanillin molecules in a solution containing caramel as a coloring agent.

Odd as it may sound, there is a chemical connection between vanilla and the phenol molecule, found as carbolic acid. Under great pressures and moderate temperatures over a long time, coal forms from decomposing plant materials, including, of course, lignin from woody tissues, as well as cellulose, another major component of vegetation. In the process of heating coal to obtain the important coal gas fuel for homes and industries, a black viscous liquid with an acrid smell is obtained. This is coal tar, Lister's source of carbolic acid. His antiseptic phenol was ultimately derived from lignin.

It was phenol that first permitted antiseptic surgery, allowing operations to be performed without risk of life-threatening infection. Phenol changed the prospects of survival for thousands injured in accidents or wars. Without phenol and later antiseptics the amazing surgical feats of today—hip replacements, open-heart surgery, organ transplants, neurosurgery, and microsurgical repairs—would never have been possible.

By investing in Baekeland's invention for photographic paper, George Eastman was able to offer a better film that, together with the introduction in 1900 of a very inexpensive camera—the Kodak Brownie, which sold for one dollar—changed photography from a pursuit of the wealthy to a hobby available to everyone. Eastman's investment financed the development—with phenol as a starting material—of the first truly synthetic material of the Age of Plastics, Bakelite, used to make the insulators necessary for the widespread use of electrical energy, a major factor in the modern industrial world.

The phenols we have discussed have changed our lives in many large ways (antiseptic surgery, the development of plastics, explosive phenols) and in many small ways (potential health factors, spicy foods, natural dyes, affordable vanilla). With such a wide variety of structures it is likely that phenols will continue to shape history.

8. ISOPRENE

CAN YOU IMAGINE what the world would be like without tires for our cars and trucks and planes? Without gaskets and fan belts for our engines, elastic in our clothes, waterproof soles for our shoes? Where would we be without such mundane but useful items as rubber bands?

Rubber and rubber products are so common that we probably never think about what rubber is and how it has changed our lives. Although mankind has been aware of its existence for hundreds of years, only in the last century and a half has rubber become an essential component of civilization. Its chemical structure gives rubber its unique properties, and the chemical manipulation of this structure produced a molecule from which fortunes have been made, lives have been lost, and countries have been changed forever.

Origins of Rubber

Some form of rubber was long known throughout most of Central and South America. The first use of rubber, for both decorative and practical purposes, is often attributed to Indian tribes of the Amazon basin. Rubber balls from a Mesoamerican archaeological site near Veracruz, Mexico, date back to between 1600 and 1200 B.C. On his second voyage to the New World, in 1495, Christopher Columbus saw Indians on the island of Hispaniola playing with heavy balls made of a plant gum that bounced surprisingly high. "Better than those fill'd with Wind in Spain," he reported, presumably referring to the inflated animal bladders used by the Spanish for ball games. Columbus brought some of this new material back to Europe, as did other travelers to the New World after him. The samples of rubber latex remained mainly a novelty item, however; they became sticky and smelly during hot weather and hard and brittle in European winters.

A Frenchman named Charles-Marie de La Condamine was the first to investigate whether there was a serious use for this strange substance. La Condamine—variously described as a mathematician, a geographer, and an astronomer, as well as a playboy and an adventurer—had been sent by the French Academy of Sciences to measure a meridian through Peru to help in determining whether the Earth was actually slightly flattened at the poles. After completing his work for the academy, La Condamine took the opportunity to explore the South American jungle, arriving back in Paris in 1735 with a number of balls of the coagulated gum from the *caoutchouc* tree (the "tree that weeps"). He had observed the Omegus Indians of Ecuador collecting the sticky white caoutchouc sap, then holding it over a smoky fire and molding it into a variety of shapes to make containers, balls, hats, and boots. Unfortunately, La Condamine's samples of raw sap, which remained as latex when not preserved through smoking, fermented during shipping and arrived in Europe as a useless smelly mass.

Latex is a colloidal emulsion, a suspension of natural rubber particles in water. Many tropical trees and shrubs produce latex, including *Ficus elastica,* the houseplant usually referred to as the "rubber plant." In parts of Mexico latex is still harvested in the traditional manner from wild rubber trees, *Castilla elastica.* All members of the widely distributed *Euphorbia* (milkweed or spurge) family are latex producers, including the familiar Christmas poinsettia, the cactuslike succulent *Euphorbias* from desert regions, deciduous and evergreen shrubby *Euphorbias,* and "Snow-on-the-Mountain," an annual, fast-growing North American *Euphorbia. Parthenium argentatum,* or guayule, a shrub that grows in the southern United States and northern Mexico, also produces much natural rubber. Though it is neither tropical nor a *Euphorbia,* the humble dandelion is yet another latex producer. The single greatest producer of natural rubber is a tree that originated in the Amazon region of Brazil, *Hevea brasiliensis.*

CIS AND TRANS

Natural rubber is a polymer of the molecule isoprene. Isoprene, with only five carbon atoms, is the smallest repeating unit of any natural polymer, making rubber the simplest natural polymer. The first chemical experiments on the structure of rubber were conducted by the great English scientist Michael Faraday. Nowadays more often considered a physicist than a chemist, Faraday thought of himself a "natural philosopher," the boundaries between chemistry and physics being less distinct during his time. Though he is mainly remembered for his physics discoveries in electricity, magnetism, and optics, his contributions to the field of chemistry were substantial and included establishing the chemical formula of rubber as a multiple of C_5H_8 in 1826.

By 1835 it had been shown that isoprene could be distilled from rubber, suggesting that it was a polymer of repeating C_5H_8 or isoprene units. Some years later this was confirmed when isoprene was polymerized to a rubberlike mass. The structure of the isoprene molecule is usually written as

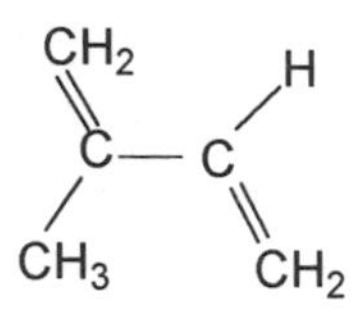

with two double bonds on adjacent carbon atoms. But rotation occurs freely around any single bond between two carbon atoms, as shown.

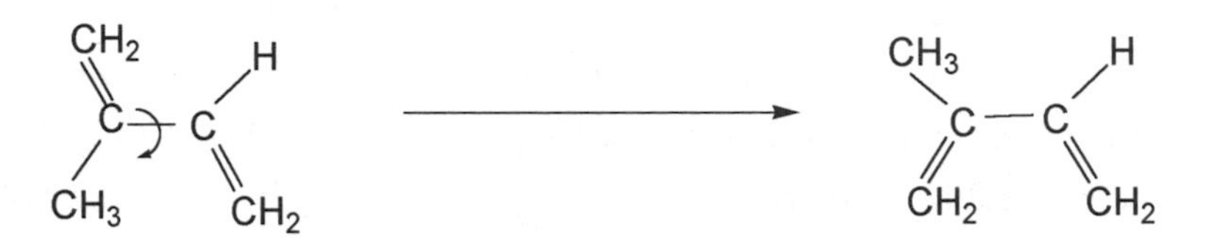

So these two structures—and all the other possible twistings around this single bond—are still the same compound. Natural rubber is formed when isoprene molecules add to one another in an end-to-end fashion. This *polymerization* in rubber produces what are called *cis* double bonds. A double bond supplies rigidity to a molecule by preventing rotation. The result of this is that the structure on the left, below, known as the *cis* form, is not the same as the structure on the right, known as the *trans* form.

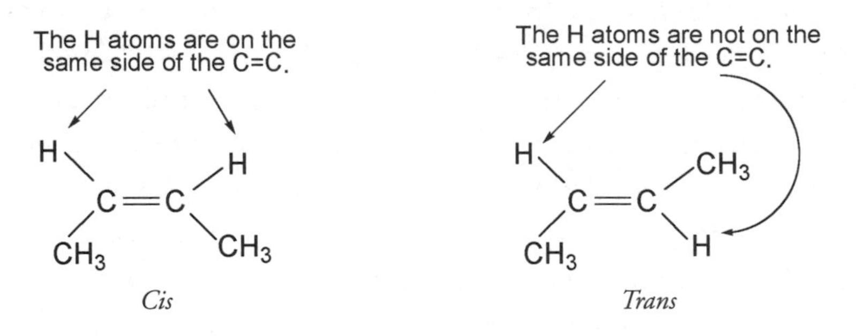

Cis *Trans*

For the cis structure, the two H atoms (and also the two CH_3 groups) are both on the same side of the double bond, whereas in the trans structure the two H atoms (and also the two CH_3 groups) are on different sides of the double bond. This seemingly small difference in the way various groups and atoms are arranged around the double bond has enormous consequences for the properties of the different polymers from the isoprene molecule. Isoprene is only one of many organic compounds with cis and trans forms; they often have quite different properties.

Below four isoprene molecules are shown ready to link up end to end, as indicated with the double-headed arrows, to form the natural rubber molecule.

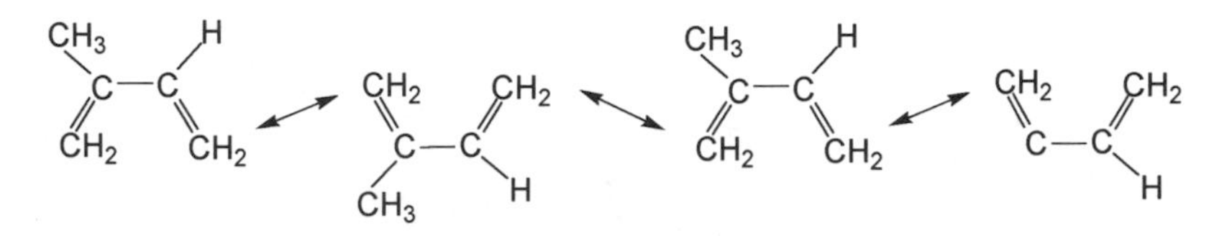

In the next drawing, the dashed lines indicate where the chain continues with the polymerization of further isoprene molecules.

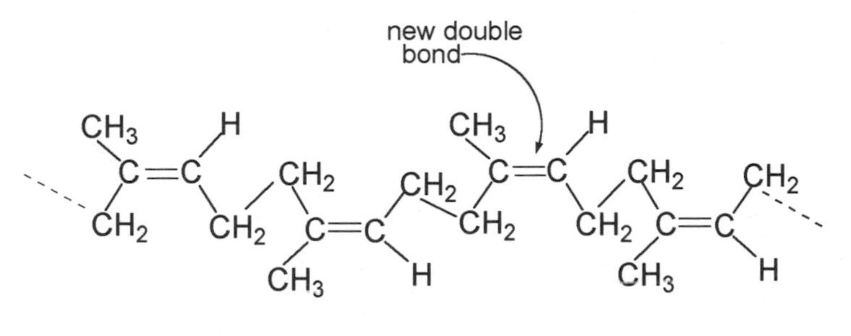

Natural rubber

New double bonds form when isoprene molecules combine; they are all cis with respect to the polymer chain, that is, the continuous chain of carbon atoms that makes the rubber molecule is on the same side of each double bond.

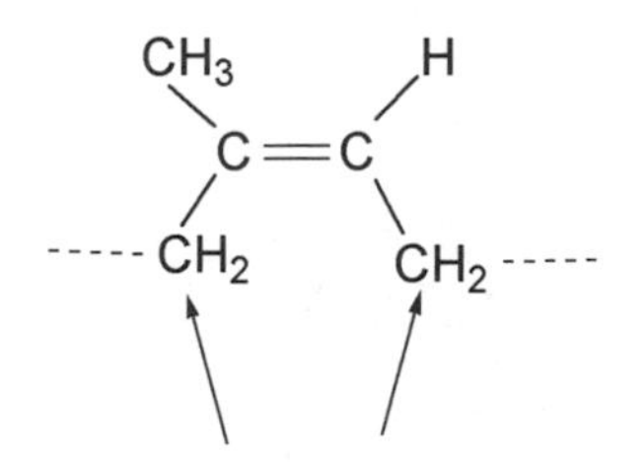

The carbons of the continuous chain are on the same side of this double bond, so this is a cis structure.

This cis arrangement is essential for the elasticity of rubber. But natural polymerization of the isoprene molecule is not always cis. When the arrangement around the double bond in the polymer is trans, another natural polymer with very different properties from those of rubber is produced. If we use the same isoprene molecule but twisted to the position shown,

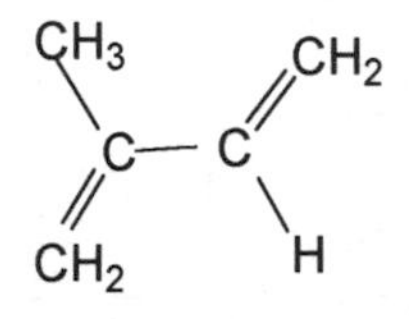

and then have four molecules like this add in an end-to-end fashion, joining as indicated again by double-headed arrows;

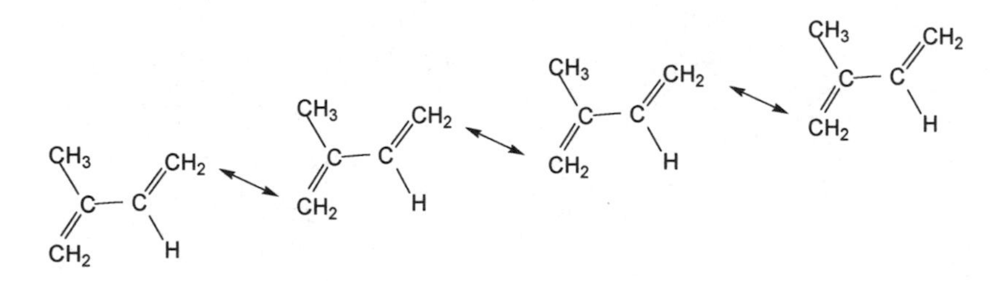

the result is the trans product.

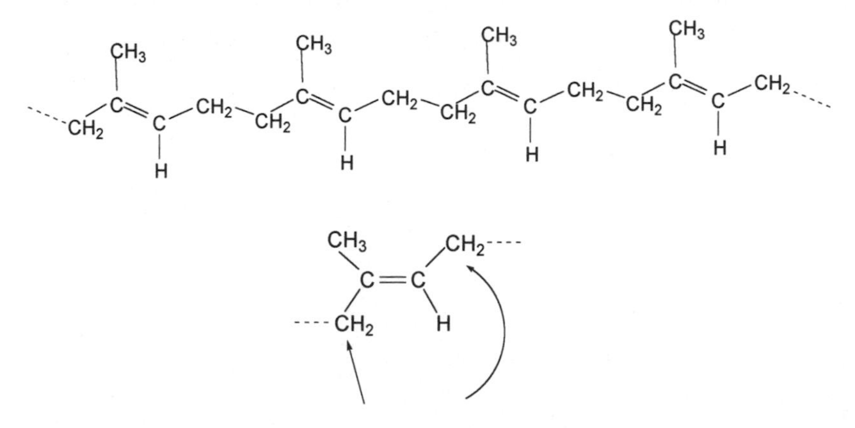

The continuous carbon chain crosses from one side of this double bond to the other, so this is a trans structure.

This trans isoprene polymer occurs naturally in two substances, gutta-percha and balata. Gutta-percha is obtained from latex of various members of the *Sapotaceae* family, particularly the *Palaquium* tree native to the Malay peninsula. About 80 percent of gutta-percha is the trans polymer of isoprene. Balata, made of similar latex from the *Mimusops globosa,* native to Panama and the northern parts of South America, contains the identical trans polymer. Both gutta-percha and balata can be melted and molded, but after exposure to the air for some time, they become hard and hornlike. As this change does not occur when these substances are kept under water, gutta-percha was used extensively as underwater cable coating during the late nineteenth and early twentieth centuries. Gutta-percha was also used by the medical and dental professions in splints, catheters, and forceps, as a poultice for skin eruptions, and as a filling for cavities in teeth and gums.

The peculiar properties of gutta-percha and balata are probably most appreciated by golfers. The original golf ball was wooden, usually made out of elm or beech. But by sometime around the early part of the eighteenth century, the Scots had invented the "feathery," a leather outer casing stuffed with goose feathers. A feathery could be hit about twice as far as a wooden ball, but it would become soggy and performed poorly in wet weather. Featheries also had a tendency to split and were more than ten times as expensive as wooden balls.

In 1848 the *gutty* was introduced. Made from gutta-percha that had been boiled in water, molded into a sphere by hand (or later in metal molds), and then allowed to harden, the gutty quickly became popular. But it too had disadvantages. The trans isomer of isoprene tends to become hard and brittle with time, so an older gutta-percha golf ball was likely to break up in midair. The rules of golf were changed to allow play to continue if this happened by substituting a new ball at the position where the largest piece of the old ball had fallen. Balls that became scuffed or scored were observed to travel farther, so factories started to prescore new balls, eventually leading to the dimpled ball of today. At the end of the nineteenth century, isoprene's cis isomer also invaded golf when a ball with rubber wound around a gutta-percha core was introduced; the cover was still made of gutta-percha. A variety of materials are used in modern golf balls; even now many include rubber in their construction. The trans isoprene polymer, often from balata rather than gutta-percha, may still be found in the covers.

RUBBER'S PROMOTERS

Michael Faraday was not alone in experimenting with rubber. In 1823, Charles Macintosh, a Glasgow chemist, used naphtha (a waste product from the local gas works) as a solvent to convert rubber into a pliable coating for fabric. Waterproofed coats made from this treated fabric were known as "macintoshes," and raincoats are still called this (or "macs") in Britain today. Macintosh's discovery led to an increased use of rubber in engines, hoses, boots, and overshoes as well as hats and coats.

A period of rubber fever hit the United States in the early 1830s. But despite waterproof qualities, the popularity of these early rubberized garments declined as people realized they became iron-hard and brittle in the winter and melted to a smelly gluelike mess in the summer. Rubber fever was over almost as soon as it began, and it seemed that rubber would remain a curiosity, its only practical use as an eraser. The word *rubber* had been coined, in 1770, by the English chemist Joseph Priestley, who found that a small piece of caoutchouc rubbed out pencil marks more effectively than the moistened bread method then in use. Erasers were marketed as India Rubbers in Britain, furthering the mistaken perception that rubber came from India.

Just as the first round of rubber fever waned, around 1834, American inventor and entrepreneur Charles Goodyear began a series of experiments that launched a much more prolonged period of worldwide rubber fever. Goodyear was a better inventor than entrepreneur. He was in and out of debt all his life, went bankrupt a number of times, and was known to refer to debtors' prisons as his "hotels." He had the idea that mixing a dry powder with rubber could absorb the excess moisture that made the substance so sticky during hot weather. Following this line of reasoning, Goodyear tried mixing various substances with natural rubber. Nothing worked. Every time the right formulation seemed to be achieved, summertime proved him wrong; rubber-impregnated boots and clothes sagged into an odorous mess whenever the temperature soared. Neighbors complained about the smell from his workshop and his financial backers retreated, but Goodyear still persisted.

One line of experimentation did seem to offer hope. When treated with nitric acid, rubber would turn into a seemingly dry, smooth material that Goodyear hoped would stay that way even when the temperature fluctuated. He once again found financial backers, who managed to get a government contract for nitric acid–treated rubberized mailbags. This time Goodyear was certain that success was finally his. Storing the finished mailbags in a locked room, he took his family on a summer holiday. But on his return he found that his mailbags had melted into the familiar shapeless mess.

Goodyear's great discovery occurred in the winter of 1839, when he had been experimenting with powdered sulfur as his drying agent. He accidentally dropped some rubber mixed with sulfur on the top of a hot stove. Somehow he recognized potential in the charred glutinous mass that formed. He was now certain that sulfur and heat changed rubber in a way that he had been hoping to find, but he did not know how much sulfur or how much heat was necessary. With the family kitchen serving as his laboratory, Goodyear continued his experiments. Sulfur-impregnated rubber samples were pressed between hot irons, roasted in the oven, toasted over the fire, steamed over the kettle, and buried in heated sand.

Goodyear's perseverance finally paid off. After five years he had hit on a process that produced uniform results: a rubber that was consistently tough, elastic, and stable in hot and cold weather. But having shown his ability as an inventor with the successful formulation of rubber, Goodyear proceeded to demonstrate his inability as a businessman. The royalties he gained from his many rubber patents were minimal. Those to whom he sold the rights, however, made fortunes from them. Despite his taking at least thirty-two cases all the way to the U.S. Supreme Court and winning, Goodyear continued to experience patent infringement throughout his life. His heart was not in the business end of rubber. He was still infatuated by what he saw as the substance's endless possibilities: rubber banknotes, jewelry, sails, paint, car springs, ships, musical instruments, floors, wetsuits, life rafts—many of which later appeared.

He was equally inept with foreign patents. He sent a sample of his newly formulated rubber to Britain and prudently did not reveal any details of the vulcanization process. But Thomas Hancock, an English rubber expert, noticed traces of powdered sulfur on one of the samples. When Goodyear finally applied for a British patent, he found that Hancock had filed for a patent on the almost identical vulcanization process just weeks before. Goodyear, declining an offer of a half-share in Hancock's patent if he would drop his claim, sued and lost. In the 1850s, at a World's Fair in London and another in Paris, pavilions built completely of rubber showcased the new material. But Goodyear, unable to pay his bills when his French patent and royalties were canceled on a technicality, once again spent time in debtors' prison. Bizarrely, while he was incarcerated in a French jail, Goodyear was awarded the French Cross of the Legion of Honor. Presumably Emperor Napoleon III was recognizing the inventor and not the entrepreneur when he bestowed this medal.

What Makes It Stretch?

Goodyear, who wasn't a chemist, had no idea why sulfur and heat worked so well on natural rubber. He was unaware of the structure of isoprene, unaware that natural rubber was its polymer and that, with sulfur, he had achieved the all-important cross-linking between rubber molecules. When heat was supplied, sulfur atoms formed cross-links that held the long chains of rubber molecules in position. It was more than seventy years after Goodyear's fortuitous discovery—named vulcanization after the Roman god of fire, Vulcan—before the English chemist Samuel Shrowder Pickles suggested that rubber was a linear polymer of isoprene and the vulcanization process was finally explained.

The elastic properties of rubber are a direct result of its chemical structure. Randomly coiled chains of the isoprene polymer, on being stretched, straighten out and align themselves in the direction of the stretch. Once the stretching force is removed, the molecules reform coils. The long flexible chains of the all-cis configuration of the natural rubber molecule do not lie close enough together to produce very many effective cross-links between the chains, and the aligned molecules can slip past one another when the substance is under tension. Contrast this with the highly regular zigzags of the all-trans isomer. These molecules can fit closely together, forming effective cross-links that prevent the long chains from slipping past one another—stretching is not possible. Thus gutta-percha and balata, trans isoprenes, are hard, inflexible masses, while rubber, the cis isoprene, is a flexible elastomer.

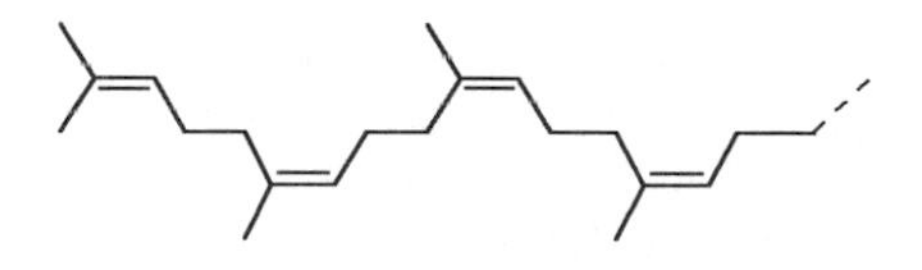

The extended cis isomer chain of the rubber molecule cannot fit closely to another rubber molecule, and so few cross-links occur. On stretching, the molecules slip past one another.

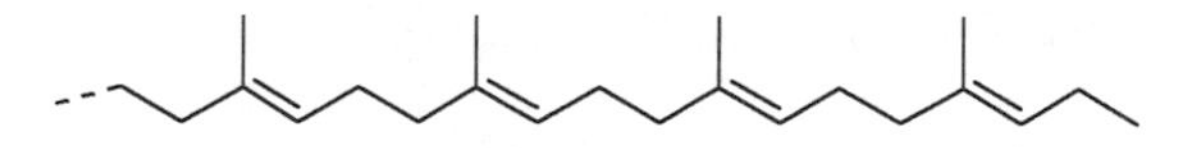

The zigzag trans isomer chains can fit closely together, allowing for many cross-links between adjacent molecules. This prevents them from slipping; gutta-percha and balata don't stretch.

By adding sulfur to natural rubber and heating, Goodyear created cross-links formed through sulfur to sulfur bonds; heating was necessary to help the formation of these new bonds. Creating enough of these disulfur bonds allows rubber molecules to remain flexible but hinders them slipping past one another.

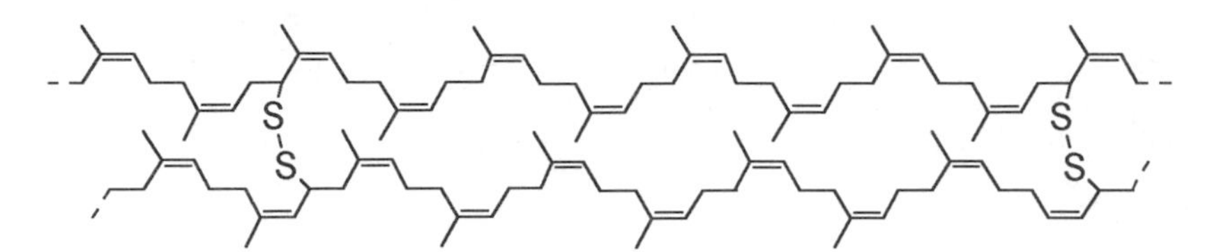

Molecules of rubber with disulfur (-S-S-) cross-links that hinder slippage

After Goodyear's discovery vulcanized rubber became one of the important commodities of the world and a vital material in wartime. As little as 0.3 percent of sulfur changes the limited temperature range of the elasticity of natural rubber so that it is no longer sticky when warm and brittle when cold. Soft rubber, used to make rubber bands, contains about 1 to 3 percent sulfur; rubber made with 3 to 10 percent sulfur has more cross-links, is more inflexible, and is used for vehicle tires. With even more cross-links, rubber becomes too rigid to be used in applications where flexibility is required, although ebonite—developed by Goodyear's brother, Nelson—a very hard, black material used as an insulator, is vulcanized rubber containing 23 to 35 percent sulfur.

RUBBER AFFECTS HISTORY

Once the possibilities of vulcanized rubber were recognized, the demand for it began in earnest. Although many tropical trees yield rubberlike latex products, the Amazon rainforests had a monopoly on the *Hevea* species. Within a very few years so-called rubber barons became extremely wealthy from the work of indentured laborers, mainly natives of the Amazon basin region. While it has not generally been recognized as such, this system of indebted bondage must be considered close to slavery. Once workers were signed on, they were advanced credit to buy equipment and supplies from their employer. Their debt increased as wages never quite covered costs. Rubber harvesters worked from sunrise to sunset tapping rubber trees, collecting latex, curing the coagulating mass over dense smoky fires, and hauling solid balls of blackened latex to waterways for shipping. During the December-to-June rainy season, when latex would not congeal, the workers remained in their dismal camps, guarded by brutal overseers who would not hesitate to shoot would-be escapers.

Less than 1 percent of the trees in the Amazon basin were rubber trees. The best trees gave only about three pounds of rubber a year. A good tapper could manage to produce about twenty-five pounds of smoke-cured rubber a day. Balls of cured latex were taken downstream by canoe to trading posts and eventually reached the city of Manaus, nine hundred miles inland from the Atlantic Ocean, on the Negro River, eleven miles above its confluence with the Amazon River. Manaus grew from a small tropical river town to a boom city on the basis of rubber. The huge profits made by the hundred or so rubber barons—mainly Europeans—and the disparity between their luxurious lifestyle and the wretched conditions of the indentured workers laboring upstream were most obvious in Manaus. Enormous mansions, fancy carriages, luxury stores carrying every manner of exotic goods, manicured gardens, and every other indication of wealth and prosperity could be found in Manaus at the height of the Amazon rubber monopoly between 1890 and 1920. A great opera house featured top stars from the theaters of Europe and America. At one time Manaus even had the distinction of having the greatest number of diamond purchases in the world.

But the rubber bubble was about to burst. As early as the 1870s Britain began to worry about the continual felling of wild rubber trees in tropical forests. Increased yields of latex could be drained from each fallen tree, up to one hundred pounds compared with the three pounds per year from tapping. The *Castilla* tree, a species that produced an inferior grade of natural rubber known as Peruvian slab, which was used for household goods and children's toys, faced extinction from this practice. In 1876 an Englishman, Henry Alexander Wickham, left the Amazon on a chartered ship carrying seventy thousand seeds from *Hevea brasiliensis,* which later proved to be the most prolific source of rubber latex. Amazonian forests had seventeen different species of *Hevea* trees, and it is not clear whether Wickham knew that the oily seeds he collected were from the most promising species or whether luck played a part in his collecting choice. Nor is it clear why his chartered ship was not searched by Brazilian officials except, possibly, that the authorities thought the rubber tree could not grow anywhere other than the Amazon basin.

Wickham took great care in transporting his cargo, packing his oily seeds carefully to prevent them from becoming rancid or germinating. Early one morning in June 1876 he arrived at the home of the eminent botanist Joseph Hooker, curator of the Royal Botanical Gardens at Kew, outside London. A propagating house was established, and the rubber tree seeds were planted. A few days later some of these seeds began to germinate, the forerunners of more than nineteen hundred rubber tree seedlings that were to be sent to Asia—the start of another great rubber dynasty. The first seedlings, sealed in miniature greenhouses and carefully tended, were shipped to Colombo in Ceylon (now Sri Lanka).

Very little was known at the time about the growth habits of the rubber tree or how growing conditions in Asia would affect the production of latex. Kew Gardens established a program of intensive scientific study of all aspects of cultivation of *Hevea brasiliensis* and found that, contrary to current belief, well-cared-for trees could be tapped daily for latex. Cultivated trees would begin producing after four years, whereas the age for starting to tap wild rubber trees had always been assumed to be around twenty-five years.

The first two rubber plantations were established in Selangor, in what is now western Malaysia. In 1896 clear, amber-colored Malayan rubber first arrived in London. The Dutch soon established plantations in Java and Sumatra, and by 1907 the British had some ten million rubber trees, planted in orderly rows, spread over 300,000 acres in Malaya and Ceylon. Thousands of immigrant workers were imported, Chinese into Malaya and Tamils into Ceylon, to supply the labor force necessary for the cultivation of natural rubber.

Africa was also affected by the demand for rubber, in particular the central African region of the Congo. During the 1880s, Leopold II of Belgium, finding that the British, French, Germans, Portuguese, and Italians had already partitioned much of the western, southern, and eastern African continent, colonized areas of less coveted central Africa, where for centuries the slave trade had reduced the region's population. The nineteenth-century ivory trade had an equally devastating effect, disrupting traditional ways of life. A favorite method of ivory traders was to capture local people, demand ivory for their release, and force whole villages into dangerous elephant-trapping expeditions to save their families. As ivory became scarce and the worldwide price of rubber increased, traders switched to requiring, as ransom, red rubber from the wild rubber vine that grew in the forests of the Congo basin.

Leopold used the rubber trade to finance the first formal colonial rule in central Africa. He leased enormous tracts of land to commercial companies such as the Anglo-Belgian India Rubber Company and the Antwerp Company. Profit from rubber depended on volume. Sap collecting became compulsory for the Congolese, and military force was used to convince them to abandon their agricultural livelihoods to harvest rubber. Whole villages would hide from the Belgians to avoid being enslaved. Barbaric punishments were common; those who didn't gather enough rubber could have their hands cut off by machete. Despite some humanitarian protest against Leopold's regime, other colonizing nations allowed companies leasing their rubber concessions to use forced labor on a large scale.

HISTORY AFFECTS RUBBER

Unlike other molecules, rubber was as much changed by history as history changed it. The word *rubber* now applies to a variety of polymer structures whose development was hastened by events of the twentieth century. The supply of plantation-grown natural rubber quickly surpassed the supply of natural rubber from the Amazon rain forests, and by 1932, 98 percent of rubber came from plantations in Southeast Asia. Its dependence on this source was of great concern to the government of the United States, as—despite a program of stockpiling rubber—much more rubber was needed for the nation's growing industrialization and transportation sector. After the December 1941 Japanese attack on Pearl Harbor brought the United States into World War II, President Franklin Delano Roosevelt appointed a special commission to investigate proposed solutions to looming wartime rubber shortages. The commission concluded that "if we fail to secure quickly a large rubber supply, our war effort and our domestic economy both will fail." They dismissed the idea of extracting natural rubbers from a variety of different plants grown in different states: rabbit brush in California, dandelions in Minnesota. Though Russia actually did use its native dandelions as an emergency source of rubber during the war, Roosevelt's commission felt that latex yields from such sources would be low and the quality of the latex questionable. The only reasonable solution, they thought, was the manufacture of synthetic rubber.

Attempts to make synthetic rubber by the polymerization of isoprene had hitherto been without success. The problem was rubber's cis double bonds. When natural rubber is produced, enzymes control the polymerization process so that the double bonds are cis. No such control was available for the synthetic process, and the result was a product in which the double bonds were a randomly arranged mix of both the cis and trans form.

A similar variable isoprene polymer was already known to occur naturally in latex from the South American sapodilla tree, *Achras sapota.* It was known as "chicle" and had long been used for making chewing gum. The chewing of gum appears to have been an ancient practice; pieces of chewed tree resins have been unearthed with prehistoric artifacts. Ancient Greeks would chew the resin of the mastic tree, a shrub found in parts of the Middle East, Turkey, and Greece, where it is still chewed today. In New England local Indians chewed the hardened sap of the spruce tree, a habit adopted by European settlers. Spruce gum had a distinctive and overpowering flavor. But it often contained difficult-to-remove impurities, so gum made of paraffin wax became more popular among colonists.

Chicle, chewed by the Mayan people of Mexico, Guatemala, and Belize for at least a thousand years, was introduced to the United States by General Antonio López de Santa Anna, conqueror of the Alamo. As president of Mexico, around 1855, Santa Anna agreed to land deals in which Mexico relinquished all territories north of the Rio Grande; as a result he was deposed and exiled from his homeland. He hoped the sale of chicle—as a substitute for rubber latex—to American rubber interests would allow him to raise a militia and reclaim the presidency of Mexico. But he did not count on the random cis and trans double bonds of chicle. Despite numerous efforts by Santa Anna and his partner Thomas Adams, a photographer and an inventor, gum chicle could not be vulcanized to an acceptable rubber substitute, nor could it be effectively blended with rubber. Chicle appeared to have no commercial value until Adams saw a child buying a penny's worth of paraffin chewing gum in a drugstore and remembered that the natives of Mexico had chewed chicle for years. He decided that this might be the solution for the supply of chicle stored in his warehouse. Chicle-based gum, sweetened with powdered sugar and variously flavored, soon became the basis of a growing chewing gum industry.

Though it was issued to troops during World War II to keep the men alert, chewing gum could hardly be considered a strategic material in wartime. Experimental procedures trying to make rubber from isoprene produced only chicle-like polymers, so an artificial rubber made from materials other than isoprene still had to be developed. Ironically, the technology for the process that made this possible came from Germany. During World War I, German supplies of natural rubber from Southeast Asia had been subjected to an Allied blockade. In response, the large German chemical companies had developed a number of rubberlike products, the best of which was styrene butadiene rubber (SBR), which had properties very similar to those of natural rubber.

Styrene was first isolated in the late eighteenth century, from the balsam of the oriental sweet gum tree, *Liquidamber orientalis,* native to southwestern Turkey. After a few months, it was noted, the extracted styrene would become jellylike, indicating that it was beginning to polymerize.

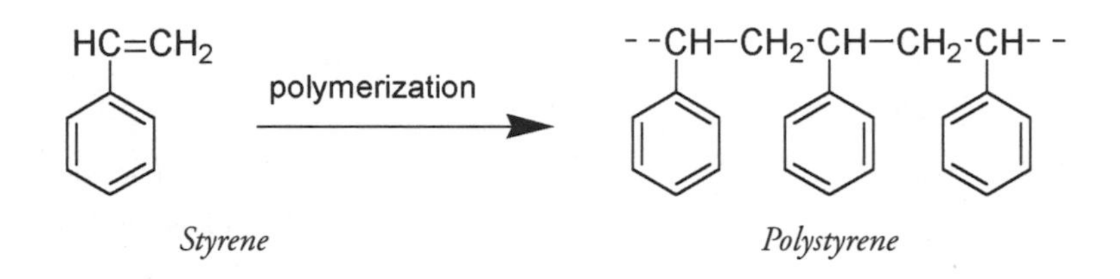

This polymer is today known as *polystyrene* and is used for plastic films, packing materials, and "Styrofoam" coffee cups. Styrene—prepared synthetically as early as 1866—and butadiene were the starting materials used by the German chemical company IG Farben in the manufacture of artificial rubber. The ratio of butadiene (CH_2=CH–CH=CH_2) to styrene is about three to one in SBR; though the exact ratio and structure are variable, it is thought that the double bonds are randomly cis or trans.

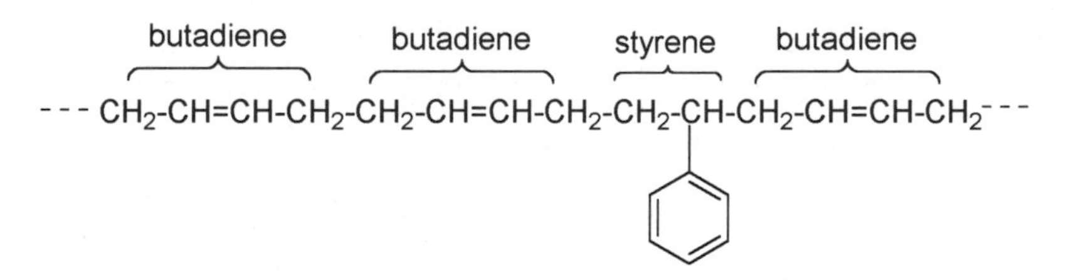

Partial structure of styrene butadiene rubber (SBR), also known as government rubber styrene (GR-S) or Buna-S. SBR can be vulcanized with sulfur.

In 1929 the Standard Oil Company of New Jersey formed a partnership with IG Farben based on shared processes relating to synthetic oil. Part of the agreement specified that Standard Oil would have access to certain of IG Farben's patents, including the SBR process. IG Farben was not obligated to share its technical details, however, and in 1938 the Nazi government informed the company that the United States was to be denied any information on Germany's advanced rubber-manufacturing technology.

IG Farben did eventually release the SBR patent to Standard Oil, certain that it contained insufficient technical information for the Americans to use in making their own rubber. But this judgment proved wrong. The chemical industry in the United States mobilized, and development of an SBR manufacturing process proceeded rapidly. In 1941 American synthetic rubber production was only eight thousand tons, but by 1945 it had expanded to over 800,000 tons, a significant proportion of the country's total rubber consumption. The production of such huge quantities of rubber in such a short period of time has been described as the second greatest feat of engineering (and chemistry) of the twentieth century, after the building of the atomic bomb. Over the following decades other synthetic rubbers (neoprene, butyl rubber, and Buna-N) were created. The meaning of the word *rubber* came to include polymers made from starting materials other than isoprene but with properties closely related to those of natural rubber.

In 1953, Karl Ziegler in Germany and Giulio Natta in Italy further refined the production of synthetic rubber. Ziegler and Natta independently developed systems that produced either cis or trans double bonds depending on the particular catalyst used. It was now possible to make natural rubber synthetically. The so-called Ziegler-Natta catalysts, for which their discoverers received the 1963 Nobel Prize in chemistry, revolutionized the chemical industry by allowing the synthesis of polymers whose properties could be precisely controlled. In this way rubber polymers could be made that were more flexible, stronger, more durable, stiffer, less likely to be affected by solvents or ultraviolet light, and with greater resistance to cracking, heat, and cold.

Our world has been shaped by rubber. The collecting of raw material for rubber products had an enormous effect on society and the environment. The felling of the rubber trees in the Amazon basin, for example, was just one episode in the exploitation of the resources of tropical rain forests and destruction of a unique environment. The shameful treatment of the area's indigenous population has not changed; today prospectors and subsistence farmers continue to invade the traditional lands of the descendants of the native peoples who harvested latex. The brutal colonization of the Belgian Congo left a legacy of instability, violence, and strife that is still very much present in the region today. The mass migrations of workers to the rubber plantations of Asia over a century ago continue to affect the ethnic, cultural, and political face of the countries of Malaysia and Sri Lanka.

Our world is still shaped by rubber. Without rubber the enormous changes brought on by mechanization would not have been possible. Mechanization requires essential natural or man-made rubber components for machines—belts, gaskets, joints, valves, o-rings, washers, tires, seals, and countless others. Mechanized transportation—cars, trucks, ships, trains, planes—has changed the way we move people and goods. Mechanization of industry has changed the jobs we do and the way we do them. Mechanization of agriculture has allowed the growth of cities and changed our society from rural to urban. Rubber has played an essential part in all these events.

Our exploration of future worlds may be shaped by rubber, as this material—an essential part of space stations, space suits, rockets and shuttles—is now enabling us to explore worlds beyond our own. But our failure to consider long-known properties of rubber has already limited our push to the stars. Despite NASA's sophisticated knowledge of polymer technology, rubber's lack of resistance to cold—a trait known to La Condamine, to Macintosh, to Goodyear—doomed the space shuttle *Challenger* on a chilly morning in January 1986. The launch temperature was 36°F, 15° lower than the next-coldest previous launch. On the shuttle system's solid rocket motor aft field joint, the rubber o-ring in the shade on the side away from the sun was possibly as cold as 28°F. At that frigid temperature it would have lost its normal pliability and, by not returning to its proper shape, led to failure of a pressure seal. The resulting combustion gas leak caused an explosion that took the lives of the seven *Challenger* astronauts. This is a very recent example of what we might now call the Napoleon's buttons factor, the neglect of a known molecular property being responsible for a major tragic event: *"And all for the want of an O-ring."*

9. DYES

DYES COLOR OUR clothes, our furnishings, our accessories, and even our hair. Yet even as we ask for a different shade, a brighter hue, a softer tint, or a deeper tone, we rarely give a passing thought to the variety of compounds that allow us to indulge our passion for color. Dyes and dyestuffs are composed of natural or man-made molecules whose origins stretch back thousands of years. The discovery and exploitation of dyes has led to the creation and growth of the biggest chemical companies in the world today.

The extraction and preparation of dyestuffs, mentioned in Chinese literature as long ago as 3000 B.C., may have been man's earliest attempts at the practice of chemistry. Early dyes were obtained mainly from plants: their roots, leaves, bark, or berries. Extraction procedures were well established and often quite complicated. Most substances did not adhere permanently to untreated fibers; fabrics first had to be treated with mordants, compounds that helped fix the color to the fiber. Although early dyes were highly sought after and very valuable, there were numerous problems with their use. They were frequently difficult to obtain, their range was limited, and the colors were not strong or faded quickly to dull, muddy shades in sunlight. Early dyes were rarely colorfast, bleeding out at every wash.

Primary Colors

Blue, in particular, was a much-sought-after color. Compared with red and yellow, blue shades are not common in plants, but one plant, *Indigofera tinctoria,* a member of the legume family, was known to be a major source of the blue dye indigo. Named by the famed Swedish botanist Linnaeus, *Indigofera tinctoria* grows up to six feet tall in both tropical and subtropical climates. Indigo is also produced in more temperate regions from *Isatis tinctoria,* one of the oldest dye plants of Europe and Asia, known as "woad" in Britain and "pastel" in France. On his travels seven hundred years ago, Marco Polo was reputed to have seen indigo being used in the Indus valley; hence the name *indigo.* But indigo was also prevalent in many other parts of the world, including Southeast Asia and Africa, well before the time of Marco Polo.

The fresh leaves of indigo-producing plants do not appear to be blue. But after fermentation under alkaline conditions followed by oxidation, the blue color appears. This process was discovered by numerous cultures around the world, possibly when the plant leaves were accidentally soaked in urine or covered with ashes, then left to ferment. In these circumstances the conditions necessary for production of the intense blue color of indigo would be present.

The indigo precursor compound, found in all indigo-producing plants, is *indican,* a molecule that contains an attached glucose unit. Indican itself is colorless, but fermentation under alkaline conditions splits off the glucose unit to produce the *indoxol* molecule. Indoxol reacts with oxygen from the air to produce blue-colored indigo (or indigotin, as chemists call this molecule).

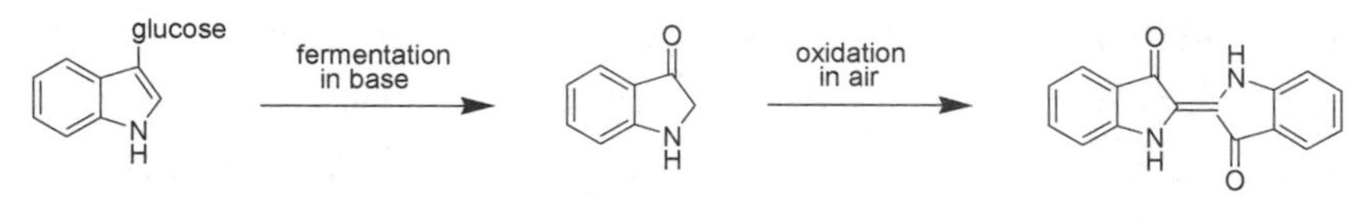

Indican (colorless) *Indoxol (colorless)* *Indigo or indigotin (blue)*

Indigo was a very valuable substance, but the most expensive of the ancient dyes was a very similar molecule known as Tyrian purple. In some cultures the wearing of purple was restricted by law to the king or emperor; hence the other name for this dye—royal purple—and the phrase "born to the purple," implying an aristocratic pedigree. Even today purple is still regarded as an imperial color, an emblem of royalty. Mentioned in writings dating to around 1600 B.C., Tyrian purple is the dibromo derivative of indigo; that is, an indigo molecule that contains two bromine atoms. Tyrian purple was obtained from an opaque mucus secreted by various species of a marine mollusk or snail, most commonly of the genus *Murex.* The compound secreted by the mollusk is, as in the indigo plant, attached to a glucose unit. It is only through oxidation in the air that the brilliant color of Tyrian purple develops.

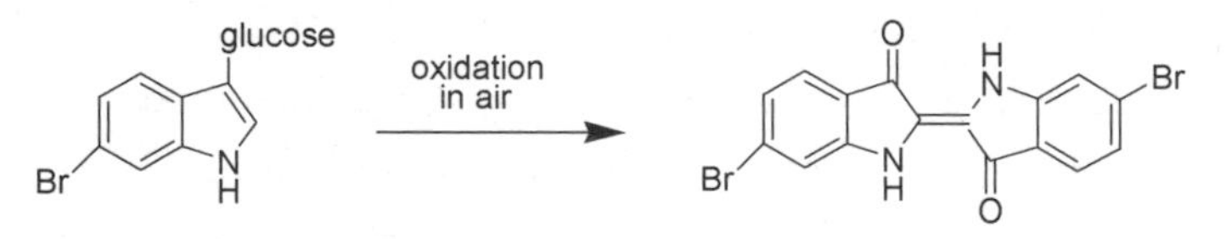

Compound secreted by the mollusk (bromoindican molecule)

Tyrian purple (dibromoindigo molecule)

Bromine is rarely found in terrestrial plants or animals, but as there is a lot of bromine, as well as chlorine and iodine, in seawater, it is not that surprising to find bromine incorporated into compounds from marine sources. What is perhaps surprising is the similarity of these two molecules, given their very different sources—that is, indigo from a plant and Tyrian purple from an animal.

Mythology credits the discovery of Tyrian purple to the Greek hero Hercules, who observed his dog's mouth becoming stained a deep purple color as the animal crunched on some shellfish. The manufacture of the dye is believed to have started in the Mediterranean port city of Tyre in the Phoenician Empire (now part of Lebanon). An estimated nine thousand shellfish were needed to produce one gram of Tyrian purple. Mounds of shells from *Murex brandaris* and *Purpura haemastoma* can still be found on the beaches of Tyre and Sidon, another Phoenician city involved in the ancient dye trade.

To obtain the dye, workers cracked open the shell of these mollusks and, using a sharp stick, extracted a small veinlike gland. Cloth was saturated with a treated solution from this gland, then exposed to the air for the color to develop. Initially the dye would turn cloth a pale yellow-green shade, then gradually blue, and then a deep purple. Tyrian purple colored the robes of Roman senators, Egyptian pharaohs, and European nobility and royalty. It was so sought after that by A.D. 400 the species of shellfish that produced it were in danger of becoming extinct.

Indigo and Tyrian purple were manufactured by these labor-intensive methods for centuries. It was not until the end of the nineteenth century that a synthetic form of indigo became available. In 1865 the German chemist Johann Friedrich Wilhelm Adolf von Baeyer began investigating the structure of indigo. By 1880 he had found a way to make it in the laboratory from easily obtainable starting materials. It took another seventeen years, however, before synthetic indigo, prepared by a different route and marketed by the German chemical company Badische Anilin und Soda Fabrik (BASF), became commercially viable.

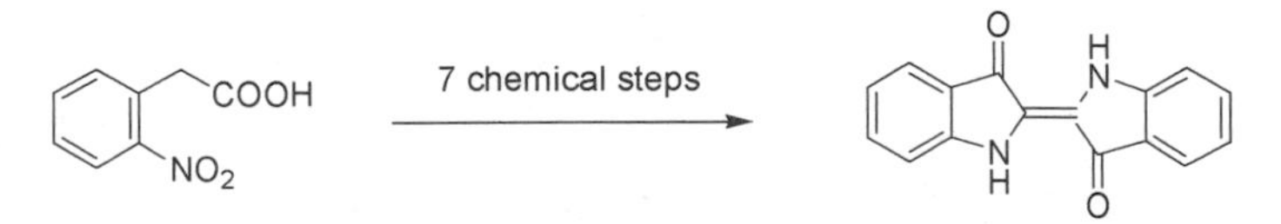

Baeyer's first synthesis of indigo required seven separate chemical reactions.

This was the beginning of the decline of the large natural indigo industry, a change that altered the way of life for thousands whose livelihood depended on the cultivation and extraction of natural indigo.

Today an annual production of over fourteen thousand tons makes synthetic indigo a major industrial dye. Though synthetic indigo (like the natural compound) notoriously lacks colorfastness, it is most often used to dye blue jeans, where this property is considered a fashion advantage. Millions of pairs of jeans are now made from specially prefaded indigo-dyed denim. Tyrian purple, the dibromo derivative of indigo, was also produced synthetically through a process similar to indigo synthesis, although other purple dyes have superseded it.

Dyes are colored organic compounds that are incorporated into the fibers of textiles. The molecular structure of these compounds allows the absorption of certain wavelengths of light from the visible spectrum. The actual color of the dye we see depends on the wavelengths of the visible light that is reflected back rather than absorbed. If all the wavelengths are absorbed, no light is reflected, and the color of the dyed cloth we see is black; if no wavelengths are absorbed, all light is reflected and the color we see is white. If only the wavelengths of red light are absorbed, then the reflected light is the complementary color of green. The relationship between the wavelength absorbed and the chemical structure of the molecule is very similar to the absorption of ultraviolet rays by sunscreen products—that is, it depends on the presence of double bonds alternating with single bonds. But for the absorbed wavelength to be in the visible range rather than the ultraviolet, there must be a greater number of these alternating double and single bonds. This is shown in the molecule β-carotene, below, which is responsible for the orange color of carrots, pumpkin, and squash.

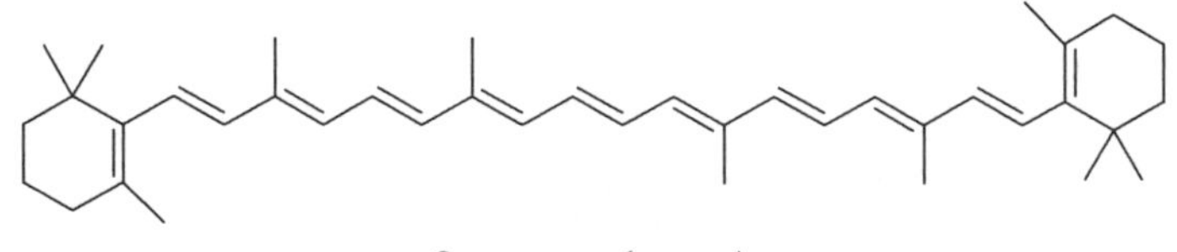

β-*carotene (orange)*

Such alternating double and single bonds, as shown in carotene, are referred to as *conjugated.* β-carotene has eleven of these conjugated double bonds. Conjugation can be extended and the wavelength of absorbed light changed, when atoms such as oxygen, nitrogen, sulfur, bromine, or chlorine are also part of the alternating system.

The indican molecule, from indigo and woad plants, has some conjugation but not enough to appear colored. The indigo molecule, however, has twice indican's number of alternating single and double bonds and also has oxygen atoms as part of the conjugation combination. It thus has enough to absorb light from the visible spectrum, which is why indigo is highly colored.

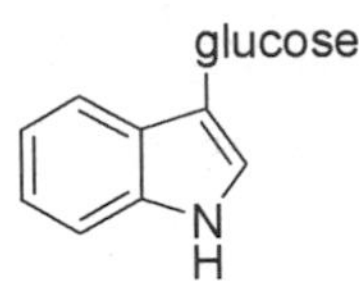

Indican (colorless)

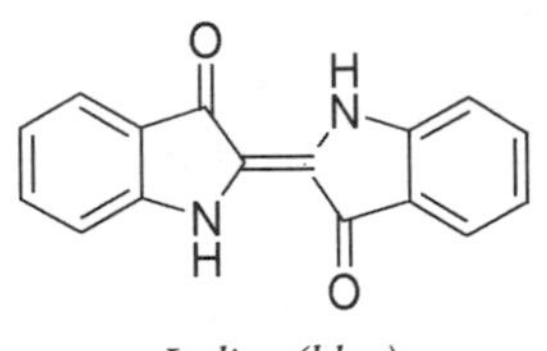

Indigo (blue)

Distinct from organic dyes, finely ground minerals and other inorganic compounds have also been used since antiquity to create color. But while the color of these *pigments*—found in cave drawings, tomb decorations, paintings, wall murals, and frescoes—is also due to absorption of certain wavelengths of visible light, it has nothing to do with conjugated double bonds.

The two common ancient dyes that were used for red shades have very different sources but surprisingly similar chemical structures. The first of these comes from the root of the madder plant. Madder plants, belonging to the *Rubiaceae* family, contain a dye called alizarin. Alizarin was probably first used in India, but it was also known in Persia and Egypt long before its use by the ancient Greeks and Romans. It is a mordant dye, that is, a dye where another chemical—a metal ion—is needed to fix color to a fabric. Different colors can be obtained by first treating the fabric with different metal salt mordant solutions. The aluminum ion as a mordant produces a rose-red color; a magnesium mordant gives a violet color; chromium a brownish-violet; and calcium a reddish-purple. The bright red color obtained when both aluminum and calcium ions are in the mordant would have resulted from using clay with dried, crushed, and powdered madder root in the dye process. This is likely the dye/mordant combination used by Alexander the Great in 320 B.C., in a ruse to lure his enemy into an unnecessary battle. Alexander had his soldiers dye their uniforms with large patches of a blood-red dye. The attacking Persian army, assuming they were assaulting wounded survivors and expecting little resistance, were easily defeated by the smaller number of Alexander's soldiers—and, if the story is true, by the alizarin molecule.

Dyes have long been associated with army uniforms. The blue coats supplied by France to the Americans during the American Revolution were dyed with indigo. The French army used an alizarin dye, known as Turkey red as it had been grown for centuries in the eastern Mediterranean, although it probably originated in India and gradually moved westward through Persia and Syria to Turkey. The madder plant was introduced into France in 1766, and by the end of the eighteenth century it had become one of the country's most important sources of wealth. Government subsidies to industry may have started with the dye industry: Louis Philippe of France decreed that soldiers in the French army were to wear trousers dyed with Turkey red. Well over a hundred years before, James II of England had banned the exportation of undyed cloth to protect English dyers.

The process of dyeing with natural dyes did not always produce consistent results and was often laborious and time consuming. But Turkey red, when it was obtained, was a beautiful bright red and very colorfast. The chemistry of the process was not understood, and today some of the operations involved seem somewhat bizarre and were probably unnecessary. Of the ten individual steps recorded in dyers' handbooks of the time, many are repeated more than once. At various stages the fabric or yarn is boiled in potash and in soap solution; mordanted with olive oil, alum, and a little chalk; treated with sheep dung, with tanning material, and with a tin salt; and rinsed overnight in river water, in addition to being dyed with madder.

We now know the structure of the alizarin molecule that is responsible for the color of Turkey red and other shades from the madder plant. Alizarin is a derivative of anthraquinone, the parent compound of a number of naturally occurring coloring materials. More than fifty anthraquinone-based compounds have been found in insects, plants, fungi, and lichens. As with indigo, the parent anthraquinone is not colored. But the two OH groups on the right-hand ring in alizarin, combined with alternating double and single bonds in the rest of the molecule, provide enough conjugation for alizarin to absorb visible light.

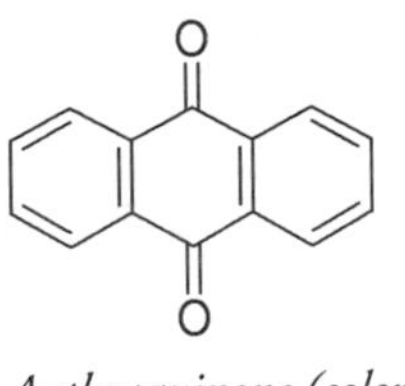

Anthraquinone (colorless)

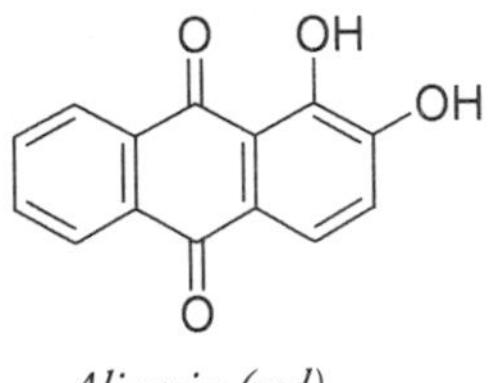

Alizarin (red)

OH groups are more important for producing color in these compounds than the number of rings. This is shown as well in compounds derived from naphthoquinone, a molecule with two rings compared to anthraquinone's three.

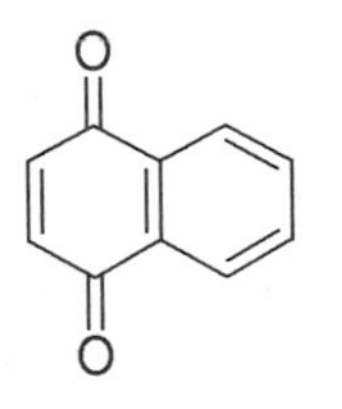

Naphthoquinone (colorless)

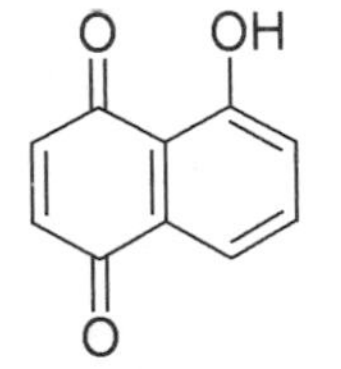

Juglone (walnuts) (brown)

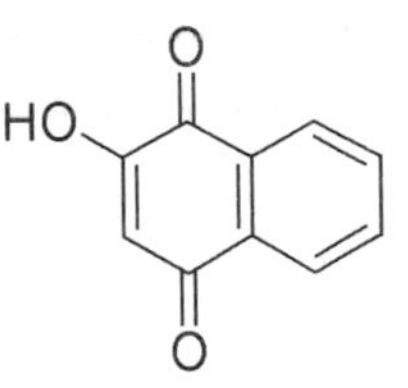

Lawsone (henna) (reddish-orange)

The naphthoquinone molecule is colorless; colored derivatives of naphthoquinone include *juglone,* found in walnuts, and *lawsone,* the coloring matter in Indian henna (used for centuries as a hair and skin dye). Colored naphthoquinones can have more than one OH group, as shown by *echinochrome,* a red pigment found in sand dollars and sea urchins.

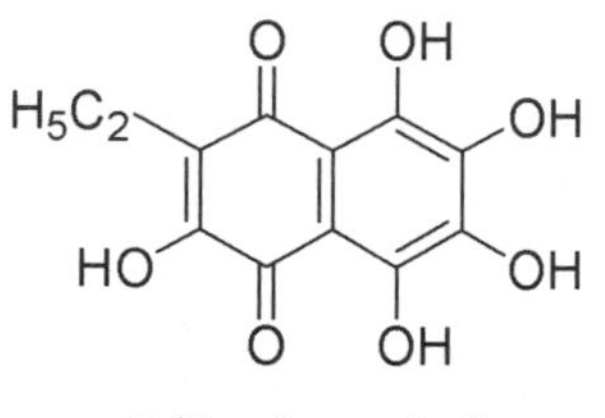

Echinochrome (red)

Another anthraquinone derivative, chemically similar to alizarin, is carminic acid, the principal dye molecule from cochineal, the other red dye from ancient times. Obtained from the crushed bodies of the female cochineal beetle, *Dactylopius coccus,* carminic acid contains numerous OH groups.

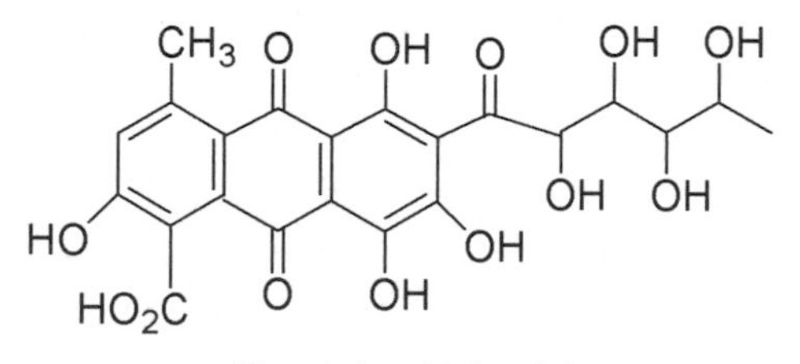

Carminic acid (scarlet)

Cochineal was a dye of the New World, used by the Aztecs long before the arrival of the Spanish conquistador Hernán Cortés in 1519. Cortés introduced cochineal to Europe, but its source was kept secret until the eighteenth century, in order to protect the Spanish monopoly over this precious scarlet dye. Later, British soldiers became known as "redcoats" from their cochineal-dyed jackets. Contracts for English dyers to produce fabric in this distinctive color were still in place at the beginning of the twentieth century. Presumably this was another example of government support of the dye industry, as by then British colonies in the West Indies were major cochineal producers.

Cochineal, also called carmine, was expensive. It took about seventy thousand insect bodies to produce just one pound of the dye. The small dried cochineal beetles looked a little like grain; hence the name "scarlet grain" was often applied to the contents of the bags of raw material that were shipped from cactus plantations in tropical regions of Mexico, and Central and South America for extraction in Spain. Today the major producer of the dye is Peru, which makes about four hundred tons annually, around 85 percent of the world's production.

The Aztecs were not the only people to use insect extractions as dyes. Ancient Egyptians colored their clothes (and the women their lips) with red juice squeezed from the bodies of the kermes insect (*Coccus ilicis*). The red pigment from this beetle is mainly kermesic acid, a molecule extraordinarily similar to its New World counterpart of carminic acid from cochineal. But unlike carminic acid, kermesic acid never went into widespread use.

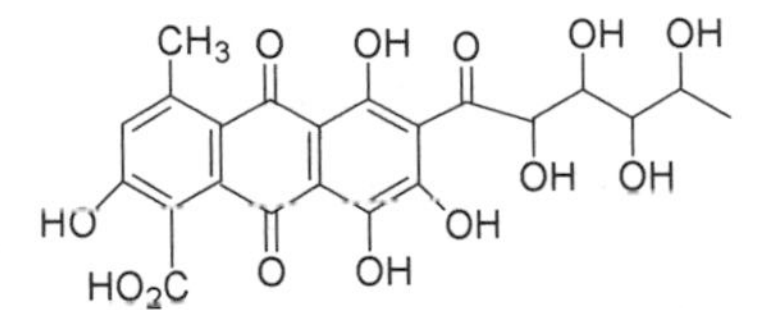

Carminic acid (scarlet)

CH3 O OH O
HO OH
HO2C O OH

Kermesic acid (bright red)

Although kermesic acid, cochineal, and Tyrian purple were derived from animals, plants supplied most of the starting materials for dyers. Blue from indigo and woad, and red from the madder plant were the standards. The third remaining primary color was a bright yellow-orange shade from the saffron crocus, *Crocus sativus.* Saffron is obtained from the stigmas of flowers, the part that catches pollen for the ovary. This crocus was native to the eastern Mediterranean and was used by the ancient Minoan civilization of Crete as early as 1900 B.C. It was also found extensively throughout the Middle East and was used in Roman times as a spice, a medicine, and a perfume as well as a dye.

Once widespread over Europe, saffron growing declined during the Industrial Revolution for two reasons. First, the three stigmas in each hand-picked blossom had to be individually removed. This was a very labor-intensive process, and laborers at this time had largely moved to the cities to work in factories. The second reason was chemical. Although saffron produced a beautiful brilliant shade, especially when applied to wool, the color was not particularly fast. When man-made dyes were developed, the once-large saffron industry faded away.

Saffron is still grown in Spain, where each flower is still hand-picked in the traditional way and at the traditional time, just after sunrise. The majority of the crop is now used for the flavoring and coloring of food in such traditional dishes as Spanish paella and French bouillabaise. Because of the way it is harvested, saffron is the most expensive spice in the world today; thirteen thousand stigmas are required to produce just one ounce.

The molecule responsible for the characteristic yellow-orange of saffron is known as *crocetin,* and its structure is reminiscent of the orange color of β-carotene, each having the same chain of seven alternating double bonds indicated, below, by the brackets.

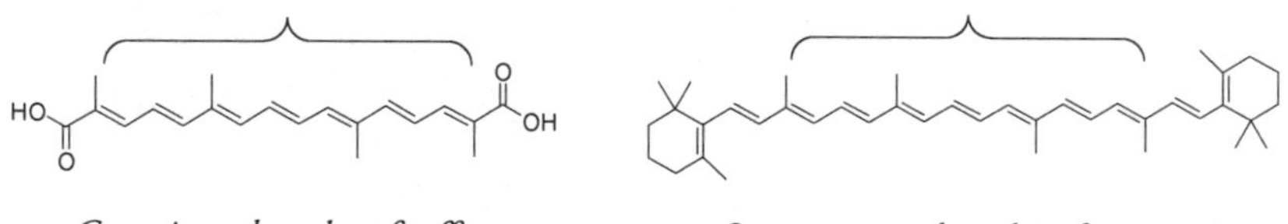

Crocetin—the color of saffron

β-carotene—the color of carrots

Although the art of dyeing no doubt started as a cottage craft and indeed continues in this mode to some extent today, dyeing has been recorded as a commercial enterprise for thousands of years. An Egyptian papyrus from 236 B.C. has a description of dyers—"stinking of fish, with tired eyes and hands working unceasingly." Dyers' guilds were well established in medieval times, and the industry flourished along with the woolen trade of northern Europe and silk production in Italy and France. Indigo, cultivated with slave labor, was an important export crop in parts of the southern United States during the eighteenth century. As cotton became an important commodity in England, so too were dyers' skills in great demand.

Synthetic Dyes

Beginning in the late 1700s synthetic dyes were created that changed the centuries-old practices of these artisans. The first of these man-made dyes was picric acid, the triply nitrated molecule that was used in munitions in World War I.

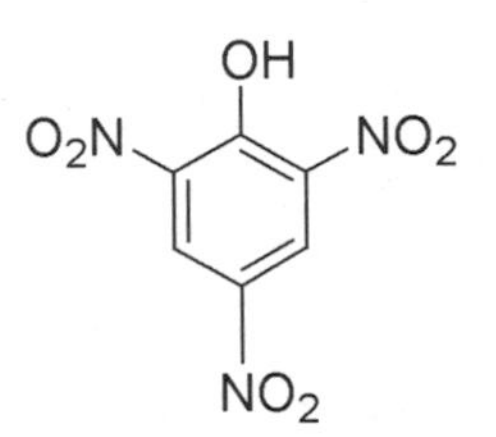

Picric acid (trinitrophenol)

An example of a phenolic compound, it was first synthesized in 1771 and used as a dye for both wool and silk from about 1788. Although picric acid produced a wonderfully strong yellow hue, its drawback, like that of many nitrated compounds, was its explosive potential, something that dyers did not have to worry about with natural yellow dyes. Two other disadvantages were that picric acid had poor light fastness and that it was not easily obtained.

Synthetic alizarin became available in good quantity and quality in 1868; synthetic indigo became available in 1880. In addition, totally new man-made dyes were prepared; dyes that gave bright, clear shades, were colorfast, and produced consistent results. By 1856 eighteen-year-old William Henry Perkin had synthesized an artificial dye that radically changed the dye industry. Perkin was a student at London's Royal College of Chemistry; his father was a builder who had little time for the pursuit of chemistry because he felt it was unlikely to lead to a sound financial future. But Perkin proved his father wrong.

Over his Easter holidays of 1856, Perkin decided to try to synthesize the antimalarial drug quinine, using a tiny laboratory that he had set up in his home. His teacher, one August Hofmann, a German chemistry professor at the Royal College, was convinced that quinine could be synthesized from materials found in coal tar, the same oily residue that was, a few years later, to yield phenol for surgeon Joseph Lister. The structure of quinine was not known, but its antimalarial properties were making it in short supply and great demand. The British Empire and other European nations were expanding their colonies into malaria-ridden areas of tropical India, Africa, and Southeast Asia. The only known cure and preventive for malaria was quinine, obtained from the increasingly scarce bark of the South American cinchona tree.

A chemical synthesis of quinine would be a great achievement, but none of Perkin's experiments were successful. One of his trials did, however, produce a black substance that dissolved in ethanol to give a deep purple solution. When Perkin dropped a few strips of silk into his mixture, the fabric soaked up the color. He tested this dyed silk with hot water and with soap and found that it was colorfast. Perkin exposed the samples to light; the color did not fade—it remained a brilliant lavender purple. Aware that purple was a rare and costly shade in the dye industry and that a purple dye, colorfast on both cotton and silk, could be a commercially viable product, Perkin sent a sample of the dyed cloth to a leading dyeing company in Scotland. Back came a supportive reply: "If your discovery does not make the goods too expensive, it is decidedly one of the most valuable that has come out for a very long time."

This was all the encouragement Perkin needed. He left the Royal College of Chemistry and, with financial help from his father, patented his discovery, set up a small factory to produce his dye in larger quantities and at a reasonable cost, and investigated the problems associated with dyeing wool and cotton as well as silk. By 1859 *mauve,* as Perkin's purple was called, had taken the fashion world by storm. Mauve became the favorite color of Eugénie, empress of France, and the French court. Queen Victoria wore a mauve dress to the wedding of her daughter and to open the London Exhibition of 1862. With royal approval from Britain and France, the popularity of the color soared; the 1860s were often referred to as the mauve decade. Indeed, mauve was used for printing British postal stamps up until the late 1880s.

Perkin's discovery had far-reaching consequences. As the first true multistep synthesis of an organic compound, it was quickly followed by a number of similar processes leading to many different colored dyes from the coal tar residues of the coal gas industry. These are often known collectively as coal tar dyes or aniline dyes. By the end of the nineteenth century dyers had around two thousand synthetic colors in their repertoire. The chemical dye industry had effectively replaced the millennia-old enterprise of extracting dyes from natural sources.

While Perkin did not make money from the quinine molecule, he did make a vast fortune from *mauveine,* the name he gave to the molecule that produced the beautiful deep purple shade of mauve, and from his later discoveries of other dye molecules. He was the first person to show that the study of chemistry could be extremely profitable, no doubt necessitating a retraction of his father's original pessimistic opinion. Perkin's discovery also emphasized the importance of structural organic chemistry, the branch of chemistry that determines exactly how the various atoms in a molecule are connected. The chemical structures of the new dyes needed to be known, as well as the structures of the older natural dyes like alizarin and indigo.

Perkin's original experiment was based on incorrect chemical suppositions. At the time it had been determined that quinine had the chemical formula of $C_{20}H_{24}N_2O_2$, but little was known about the structure of the substance. Perkin also knew that another compound, allyltoluidine, had the chemical formula $C_{10}H_{13}N$, and it seemed to him possible that combining two molecules of allyltoluidine, in the presence of an oxidizing agent like potassium dichromate to supply extra oxygen, might just form quinine.

From a chemical formula perspective, Perkin's idea may not seem unreasonable, but we now know that this reaction would not occur. Without knowledge of the actual structures of allyltoluidine and quinine, it is not possible to devise the series of chemical steps necessary to transform one molecule into another molecule. This is why the molecule Perkin created, mauveine, was chemically very different from quinine, the one he had intended to synthesize.

To this day the structure of mauveine remains a bit of a mystery. Perkin's starting materials, isolated from coal tar, were not pure, and it is now thought that his purple color was obtained from a mixture of very closely related compounds. The following is presumed to be the major structure responsible for the color:

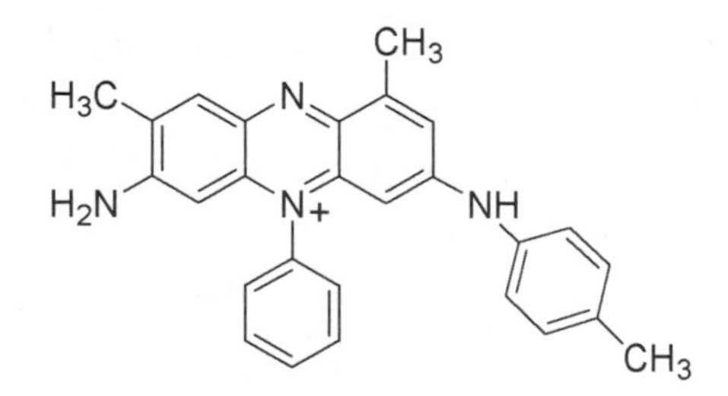

Part of the mauveine molecule, the major contributor to Perkin's mauve color

Perkin's decision to manufacture the mauve dye commercially was undoubtedly a leap of faith. He was young, a novice chemistry student with little knowledge of the dye industry and absolutely no experience of large-scale chemical production. As well, his synthesis had a very low yield, at best maybe 5 percent of the theoretically possible amount, and there were real difficulties associated with obtaining a steady supply of the coal tar starting materials. For a more experienced chemist these problems would have been daunting, and probably we can attribute Perkin's success in part to the fact that he did not let his lack of experience deter him. With no comparable manufacturing process as a guide, Perkin had to devise and test new apparatus and procedures. Solutions to problems associated with scaling up his chemical synthesis were found: large glass vessels were made, as the acid in the process would attack iron containers; cooling devices were used to prevent overheating during the chemical reactions; hazards such as explosions and toxic fume releases were controlled. In 1873 Perkin sold his factory after operating it for fifteen years. He retired a wealthy man and spent the rest of his life studying chemistry in his home laboratory.

THE LEGACY OF DYES

The dye trade, which now mainly produces chemically synthesized artificial dyes, became the forerunner to an organic chemical enterprise that would eventually produce antibiotics, explosives, perfumes, paints, inks, pesticides, and plastics. The fledgling organic chemistry industry developed not in England—the home of mauve—or in France, where dyes and dyeing had been of crucial importance for centuries. Rather, it was Germany that developed a huge organic chemical empire along with the technology and science on which it was based. Britain already had a strong chemical industry, supplying the raw materials needed for bleaching, printing, pottery, porcelain, glassmaking, tanning, brewing, and distilling, but these compounds were mainly inorganic: potash, lime, salt, soda, acid, sulfur, chalk, and clay.

There are several reasons why Germany—and to a lesser extent Switzerland—became major players in synthetic organic chemicals. By the 1870s a number of British and French dye manufacturers had been forced out of business as a result of an endless series of patent disputes over dyes and dye processes. Britain's major entrepreneur, Perkin, had retired, and no one else with the requisite chemical knowledge, manufacturing skills, and business talents replaced him. So Britain, perhaps not realizing it was against the country's interests, became an exporter of the raw materials for the growing synthetic dye industry. Britain had gained industrial supremacy on the basis of importing raw materials and converting them into finished goods for export, so its failure to recognize the usefulness of coal tar and the importance of the synthetic chemical industry was a major mistake that benefited Germany.

Another important reason for the growth of the German dye industry was the collaborative effort between industry and universities. Unlike other countries where chemical research remained the prerogative of universities, German academics tended to work closely with their counterparts in industry. This pattern was vital to the success of the German chemical industry. Without knowledge of the molecular structures of organic compounds and a scientific understanding of the chemical steps in the reactions of organic synthesis, scientists could not have developed the sophisticated technology that eventually led to modern-day pharmaceuticals.

The chemical industry in Germany grew from three companies. In 1861 the first company, Badische Anilin und Soda Fabrik (BASF), was established at Ludwigshafen on the Rhine River. Although it was originally formed to produce inorganic compounds such as soda ash and caustic soda, BASF soon became active in the dye industry. In 1868 two German academics, Carl Graebe and Carl Liebermann, announced the first synthetic alizarin. BASF's head chemist, Heinrich Caro, contacted the chemists in Berlin and collaborated with them to produce a commercially viable synthesis of alizarin. By the beginning of the twentieth century BASF was producing around two thousand tons of this important dye and was well on its way to becoming one of the big-five chemical companies that dominate the world today.

The second big German chemical company, Hoechst, was formed only a year later than BASF. Originally established to produce aniline red, a brilliant red dye also known as magenta or fuchsine, Hoechst chemists patented their own synthesis for alizarin, which proved very profitable. Synthetic indigo, the product of years of research and considerable financial investment, was also very lucrative for both BASF and Hoechst.

The third big German chemical company shared in the synthetic alizarin market as well. Although the name of Bayer is most often associated with aspirin, Bayer and Company, set up in 1861, initially made aniline dyes. Aspirin had been synthesized in 1853, but it was not until around 1900 that profits from synthetic dyes, especially alizarin, allowed Bayer and Company to diversify into pharmaceuticals and to market aspirin commercially.

In the 1860s these three companies produced only a small percentage of the world's synthetic dyes, but by 1881 they accounted for half of the global output. At the turn of the century, despite a huge increase in overall world production of synthetic dyes, Germany had gained almost 90 percent of the dye market. Along with domination of dye manufacturing went a commanding lead in the business of organic chemistry as well as a heavy role in the development of German industry. With the advent of World War I the German government was able to enlist dye companies to become sophisticated producers of explosives, poisonous gases, drugs, fertilizers, and other chemicals necessary to support the war.

After World War I the German economy and the German chemical industry were in trouble. In 1925, in the hopes of alleviating stagnating market conditions, the major German chemical companies consolidated into a giant conglomerate, Interessengemeinschaft Farbenindustrie Aktiengesellschaft (Syndicate of Dyestuff Industry Corporation), generally known as IG Farben. Translated literally, *Interessengemeinschaft* means "community of interest," and this conglomeration was definitely in the interest of the German chemical-manufacturing community. Reorganized and revitalized, IG Farben, by now the world's largest chemical cartel, invested its considerable profits and economic power into research, diversified into new products, and developed new technologies with the aim of achieving a future monopoly of the chemical industry.

With the arrival of World War II IG Farben, already a major contributor to the Nazi Party, became a major player in Adolf Hitler's war machine. As the German army advanced through Europe, IG Farben took over control of chemical plants and manufacturing sites in German-occupied countries. A large chemical plant to make synthetic oil and rubber was built at the Auschwitz concentration camp in Poland. Inmates of the camp labored in the plant and were also subjected to experimentation with new drugs.

After the war nine of IG Farben's executives were tried and found guilty of plunder and property crimes in occupied territories. Four executives were convicted of imposing slave labor and of treating prisoners of war and civilians inhumanely. IG Farben's growth and influence was halted; the giant chemical group was split up, so that the major players again were BASF, Hoechst, and Bayer. These three companies have continued to prosper and expand and today constitute a sizable portion of the organic chemical industry, with interests ranging from plastics and textiles to pharmaceuticals and synthetic oil.

Dye molecules changed history. Sought after from their natural sources for thousands of years, they created some of humankind's first industries. As the demand for color grew, so did guilds and factories, towns and trade. But the appearance of synthetic dyes transformed the world. Traditional means of obtaining natural dyes vanished. In their place, less than a century after Perkin first synthesized mauve, giant chemical conglomerates dominated not only the dye market but also a burgeoning organic chemistry industry. This in turn provided the financial capital and the chemical knowledge for today's huge production of antibiotics, analgesics, and other pharmaceutical compounds.

Perkin's mauve was only one of the synthetic dye compounds involved in this remarkable transformation, but many chemists consider it the molecule that turned organic chemistry from an academic pursuit into a major global industry. From mauve to monopoly, the dye concocted by a British teenager on his vacation had a powerful influence on the course of world events.

10. WONDER DRUGS

IT PROBABLY WOULD not have surprised William Perkin that his synthesis of mauve became the basis for the huge commercial dye enterprise. After all, he had been so sure that the manufacture of mauve would be profitable that he had persuaded his father to finance his dream—and he had been extremely successful in his lifetime. But even he likely could not have predicted that his legacy would include one of the major developments evolving from the dye industry: pharmaceuticals. This aspect of synthetic organic chemistry would far surpass the production of dyes, change the practice of medicine, and save millions of lives.

In 1856, the year when Perkin prepared the mauve molecule, the average life expectancy in Britain was around forty-five years. This number did not change markedly for the rest of the nineteenth century. By 1900 the average life expectancy in the United States had only increased to forty-six years for a male and forty-eight years for a female. A century later, in contrast, these figures have soared to seventy-two for males and seventy-nine for females.

For such a dramatic increase after so many centuries of much lower life expectancies, something amazing had to have happened. One of the major factors in longer lifespans was the introduction, in the twentieth century, of molecules of medicinal chemistry and in particular of the miracle molecules known as antibiotics. Literally thousands of different pharmaceutical compounds have been synthesized over the past century, and hundreds of them were life changing for many people. We will look at the chemistry and development of only two types of pharmaceuticals: the pain-relieving molecule aspirin, and two examples of antibiotics. Profits from aspirin helped convince chemical companies that there was a future in pharmaceuticals; the first antibiotics—sulfa drugs and penicillins—are still prescribed today.

For thousands of years medicinal herbs have been used to heal wounds, cure sickness, and relieve pain. Every human society has developed unique traditional remedies, a number of which have yielded extremely useful compounds or have been chemically modified to produce modern medicines. Quinine, which comes from the South American cinchona tree and originally used by the Indians of Peru to treat fevers, is still today an antimalarial. Foxglove containing digitalis, which is still prescribed as a contemporary heart stimulant, has long been used in western Europe to treat heart ailments. The analgesic properties of sap from the seed capsules of a poppy plant were well known from Europe to Asia and morphine extracted from this source still plays a major role in pain relief.

Historically, however, few if any effective remedies were known for treating bacterial infections. Until relatively recently even a small cut or a tiny puncture wound could, if infected, become life threatening. Fifty percent of soldiers wounded during the American Civil War died of bacterial infections. Thanks to antiseptic procedures and molecules like phenol, introduced by Joseph Lister, this percentage was smaller during the First World War. But although use of antiseptics helped prevent infection from surgery, it did little to stop an infection once it had started. The great influenza pandemic of 1918–1919 killed more than twenty million people worldwide, a far greater toll than that of World War I. The influenza itself was viral; the actual cause of death was usually a secondary infection of bacterial pneumonia. Contracting tetanus, tuberculosis, cholera, typhoid fever, leprosy, gonorrhea, or any of a host of other illnesses was often a death sentence. In 1798 an English doctor, Edward Jenner, successfully demonstrated for the smallpox virus the process of artificially producing immunity to a disease, although the concept of acquiring immunity in this way had been known from earlier times and from other countries. Starting in the last decades of the nineteenth century, similar methods of providing immunity against bacteria were investigated as well, and gradually inoculation became available for a number of bacterial diseases. By the 1940s fear of the dreaded childhood duo of scarlet fever and diphtheria had receded in countries where vaccination programs were available.

Aspirin

In the early twentieth century the German and Swiss chemical industries were prospering from their investment in the manufacture of dyestuffs. But this success was more than just financial. Along with profits from dye sales came a new wealth of chemical knowledge, of experience with large-scale reactions, and of techniques for separation and purification that were vital for expansion into the new chemical business of pharmaceuticals. Bayer and Company, the German firm that got its start from aniline dyes, was one of the first to recognize the commercial possibilities in the chemical production of medicines—in particular aspirin, which has now been used by more people worldwide than any other medication.

In 1893 Felix Hofmann, a chemist working for the Bayer company, decided to investigate the properties of compounds that were related to salicylic acid, a molecule obtained from salicin, a pain-relieving molecule originally isolated from the bark of trees of the willow genus

(*Salix*) in 1827. The curative properties of the willow and related plants such as poplars had been known for centuries. Hippocrates, the famed physician of ancient Greece, had used extracts from willow bark to reduce fevers and relieve pain. Although the bitter-tasting salicin molecule incorporates a glucose ring into its structure, the rest of the molecule overwhelms any sweetness from the sugar part.

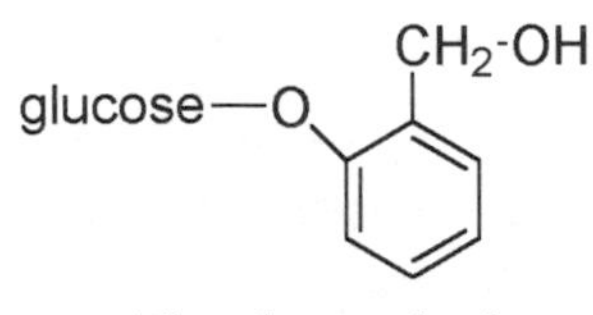

The salicin molecule

Like the glucose-containing indican molecule that produces indigo, salicin breaks into two parts: glucose and salicyl alcohol, which can be oxidized to salicylic acid. Both salicyl alcohol and salicylic acid are classified as phenols because they have an OH group directly attached to the benzene ring.

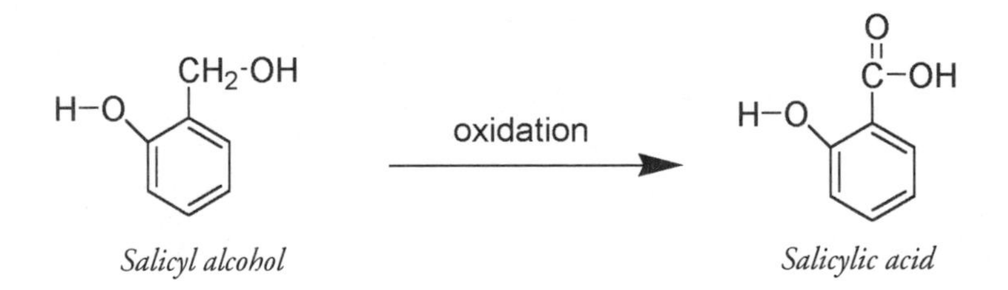

Salicyl alcohol *Salicylic acid*

These molecules are also similar in structure to isoeugenol, eugenol, and zingerone from cloves, nutmeg, and ginger. It is probable that like these molecules, salicin acts as a natural pesticide to protect the willow tree. Salicylic acid is also produced from the flowers of meadowsweet or *Spiraea ulmaria,* a wetlands perennial native to Europe and western Asia.

Salicylic acid, the active portion of the salicin molecule, not only reduces fever and relieves pain but also acts as an anti-inflammatory. It is much more potent than the naturally occurring salicin, but it can be very irritating to the lining of the stomach, reducing its medicinal value. Hofmann's interest in compounds related to salicylic acid arose out of concern for his father, whose rheumatoid arthritis was little relieved by salicin. Hoping that the anti-inflammatory properties of salicylic acid would be retained but its corrosive properties lessened, Hofmann gave his father a derivative of salicylic acid—acetyl salicylic acid, first prepared by another German chemist forty years previously. In ASA, as acetyl salicylic acid has come to be called, the acetyl group (CH_3CO) replaces the H of the phenolic OH group of salicylic acid. The phenol molecule is corrosive; perhaps Hofmann reasoned that converting the OH attached to the aromatic ring into an acetyl group might mask its irritating characteristics.

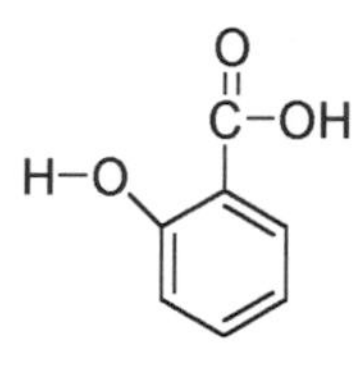

Salicylic acid

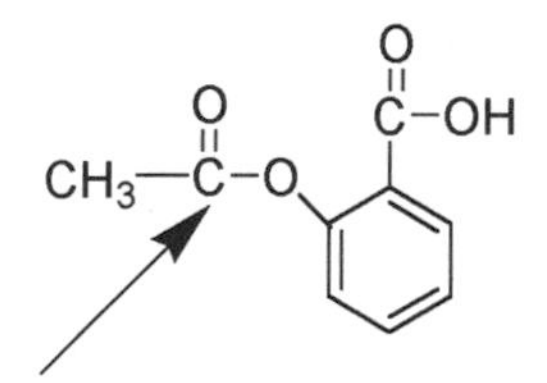

Acetyl salicylic acid. The arrow shows where the acetyl group replaces H of the phenol group.

Hofmann's experiment paid off—for his father and for the Bayer company. The acetylated form of salicylic acid turned out to be effective and well tolerated. Its potent anti-inflammatory and analgesic properties persuaded the Bayer company, in 1899, to begin marketing small packets of powdered "aspirin." The name is a combination of the *a* from *acetyl* and the *spir* from *Spiraea ulmaria,* the meadowsweet plant. The Bayer company name became synonymous with aspirin, marking Bayer's entrance into the world of medicinal chemistry.

As the popularity of aspirin increased, the natural sources from which salicylic acid was produced—meadowsweet and willow—were no longer sufficient to satisfy world demand. A new synthetic method using the phenol molecule as the starting material was introduced. Aspirin sales soared; during World War I the American subsidiary of the original Bayer company purchased as much phenol as possible from both national and international sources in order to guarantee an adequate supply for the manufacture of aspirin. The countries that supplied Bayer with phenol thus had reduced capacity to make picric acid (trinitrophenol), an explosive also prepared from this same starting material (see Chapter 5). What effect this may have had on the course of World War I we can only speculate, but aspirin production may have reduced reliance on picric acid for munitions and hastened the development of TNT-based explosives.

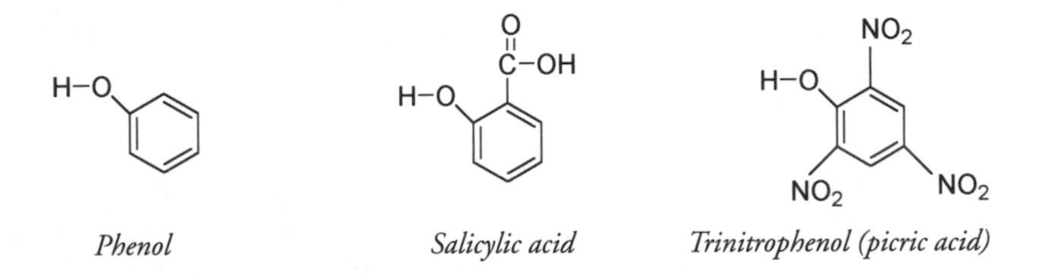

Phenol *Salicylic acid* *Trinitrophenol (picric acid)*

Today aspirin is the most widely used of all drugs for treating illness and injury. There are well over four hundred aspirin-containing preparations, and over forty million pounds of aspirin are produced in the United States annually. As well as relieving pain, lowering body temperature, and reducing inflammation, aspirin also has blood-thinning properties. Small doses of aspirin are being recommended as a preventive against strokes and for deep-vein thrombosis, the condition known as "economy class syndrome" in long-haul airline passengers.

THE SAGA OF SULFA

Around the time of Hofmann's experiment on his father—a drug-trial procedure that is not recommended—the German doctor Paul Ehrlich was conducting experiments of his own. Ehrlich was, by all accounts, a truly eccentric character, said to smoke twenty-five cigars a day and spend many hours in philosophical discussions in beer halls. But along with his eccentricity came the determination and insight that gained him the 1908 Nobel Prize in medicine. Despite having no formal training in experimental chemistry or applied bacteriology, Ehrlich noted that different coal tar dyes would stain some tissues and some microorganisms but not others. He reasoned that if one microorganism absorbed a dyestuff and another did not, this differentiation might allow a toxic dye to kill tissue that absorbed it without damaging nonstaining tissue. Hopefully the infecting microorganism would be eliminated while the host was unharmed. Ehrlich termed this theory the "magic bullet" approach, the magic bullet being the dye molecule targeting the tissue it stained.

Ehrlich's first success was with a dye called trypan red I, which acted very much as he had hoped against trypanosomes—a protozoic parasite—in laboratory mice. Unfortunately it was not effective against the type of trypanosome responsible for the human disease known as African sleeping sickness, which Ehrlich had hoped to cure.

Undeterred, Ehrlich continued. He had shown that his method could work, and he knew it was only a matter of finding a suitable magic bullet for the right disease. He began investigating syphilis, an affliction caused by a corkscrew-shaped bacterium known as a spirochete. Theories of how syphilis came to Europe abound; one of the most widely acknowledged was that it returned from the New World with Columbus's sailors. A form of "leprosy" reported in Europe before Columbus's time, however, was known to be highly contagious and venereally spread. Like syphilis it also sometimes responded to treatment with mercury. None of these observations fit what we know about leprosy, and it is possible that what was described was actually syphilis.

By the time Ehrlich began looking for a magic bullet against this bacterium, mercury cures had been claimed for syphilis for over four hundred years. Yet mercury could hardly be considered a magic bullet for syphilis, as it often killed its patients. Victims died of heart failure, dehydration, and suffocation during the process of being heated in an oven while breathing mercury fumes. If one survived this procedure, typical symptoms of mercury poisoning—loss of hair and teeth, uncontrollable drooling, anemia, depression, and kidney and liver failure—took their toll.

In 1909, after testing 605 different chemicals, Ehrlich finally found a compound that was both reasonably effective and reasonably safe. "Number 606," an arsenic-containing aromatic compound, proved active against the syphilis spirochete. Hoechst Dyeworks—the company Ehrlich collaborated with—marketed this compound in 1910, under the name salvarsan. Compared with the torture of the mercury remedy, the new treatment was a great improvement. Despite some toxic side effects and the fact that it did not always cure syphilitic patients even after a number of treatments, salvarsan greatly reduced the incidence of the disease wherever it was used. For Hoechst Dyeworks it proved extremely profitable, providing the capital to diversify into other pharmaceuticals.

After the achievement of salvarsan, chemists sought further magic bullets by testing tens of thousands of compounds for their effect on microorganisms, then making slight changes to chemical structures and testing again. There were no successes. It seemed as if the promise of what Ehrlich had termed "chemotherapy" would not live up to expectations. But then in the early 1930s Gerhard Dogmak, a doctor working with the IG Farben research group, decided to use a dye called prontosil red to treat his daughter, who was desperately ill with a streptococcal infection contracted from a simple pinprick. He had been experimenting with prontosil red at the IG Farben laboratory, and though it had shown no activity against bacteria grown in laboratory cultures, it did inhibit the growth of streptococci in laboratory mice. No doubt deciding he had nothing to lose, Dogmak gave his daughter an oral dose of the still-experimental dye. Her recovery was fast and complete.

It was at first assumed that the dye action—the actual staining of the cells—was responsible for the antibacterial properties of prontosil red. But researchers soon realized that antibacterial effects had nothing to do with dye action. In the human body the prontosil red molecule breaks down to produce sulfanilamide, and it is the sulfanilamide that has the antibiotic effect.

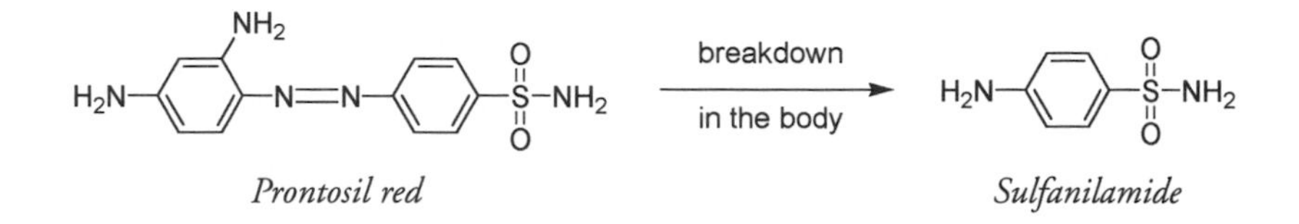

Prontosil red *Sulfanilamide*

This, of course, was why prontosil red had been inactive in test tubes (in vitro) but not in live animals (in vivo). Sulfanilamide was found to be effective against many diseases other than streptococcal infections, including pneumonia, scarlet fever, and gonorrhea. Having recognized sulfanilamide as an antibacterial agent, chemists quickly started to synthesize similar compounds, hoping that slight modifications of the molecular structure would increase effectiveness and lessen any side effects. The knowledge that prontosil red was not the active molecule was extremely important. As can be seen from the structures, prontosil red is a more complicated molecule than sulfanilamide and it is more difficult to synthesize and to modify.

Between 1935 and 1946 more than five thousand variations of the sulfanilamide molecule were made. A number of them proved superior to sulfanilamide, whose side effects can include allergic response—rashes and fever—and kidney damage. The best results from varying the sulfanilamide structure were obtained when one of the hydrogen atoms of the SO_2NH_2 was replaced with another group.

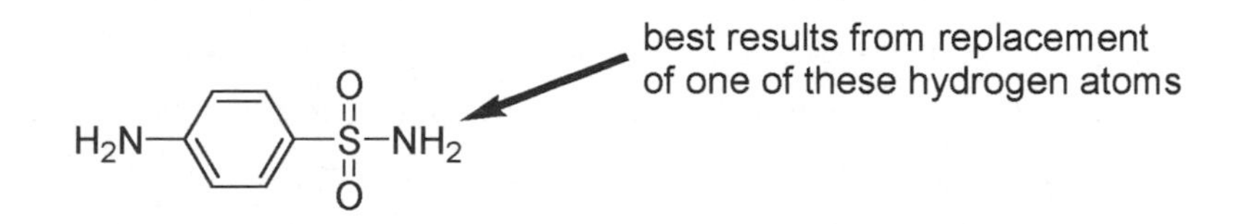

The resulting molecules are all part of the family of antibiotic drugs known collectively as *sulfanilamides* or *sulfa drugs.* A few of the many examples are

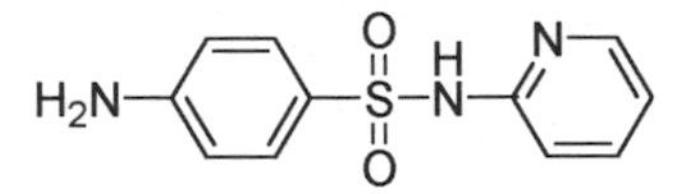

Sulfapyridine—used for pneumonia

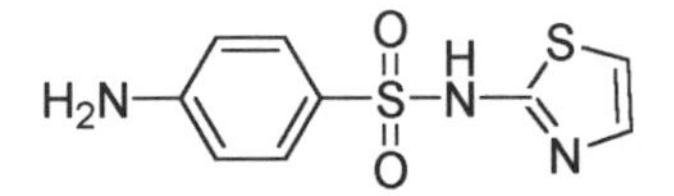

Sulfathiazole—used for gastrointestinal infections

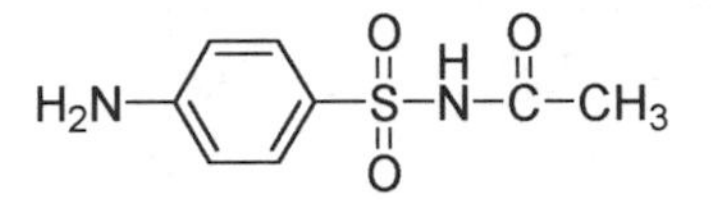

Sulfacetamide—used for urinary tract infections

Sulfa drugs were soon being described as wonder drugs and miracle cures. While such descriptives may seem unduly exaggerated nowadays, when numerous effective treatments against bacteria are available, the results obtained from these compounds in the early decades of the twentieth century appeared to be extraordinary. For example, after the introduction of sulfanilamides, the number of deaths from pneumonia dropped by twenty-five thousand a year in the United States alone.

In World War I, between 1914 and 1918, death from wound infection was as likely as death from injury on the battlefields of Europe. The major problem in the trenches and in any army hospital was a form of gangrene known as gas gangrene. Caused by a very virulent species of the *Clostridium* bacteria, the same genus responsible for the deadly botulism food poisoning, gas gangrene usually developed in deep wounds, typical of injuries from bombs and artillery where tissue was pierced or crushed. In the absence of oxygen, these bacteria multiply quickly. A brown foul-smelling pus is exuded, and gases from bacterial toxins bubble to the skin's surface, causing a distinctive stench.

Before the development of antibiotics there was only one treatment for gas gangrene—amputation of the infected limb above the site of infection, in the hope of removing all the gangrenous tissue. If amputation was not possible, death was inevitable. During World War II, thanks to antibiotics such as sulfapyridine and sulfathiazole—both effective against gangrene—thousands of injured were spared disfiguring amputations, not to mention death.

We now know that the effectiveness of these compounds against bacterial infection has to do with the size and shape of the sulfanilamide molecule preventing bacteria from making an essential nutrient, folic acid. Folic acid, one of the B vitamins, is required for human cell growth. It is widely distributed in foods, such as leafy vegetables (hence the word *folic* from *foliage*), liver, cauliflower, yeast, wheat, and beef. Our bodies do not manufacture folic acid, so it is essential that we take it in with what we eat. Some bacteria, on the other hand, do not require supplemental folic acid, as they are able to make their own.

The folic acid molecule is fairly large and looks complicated:

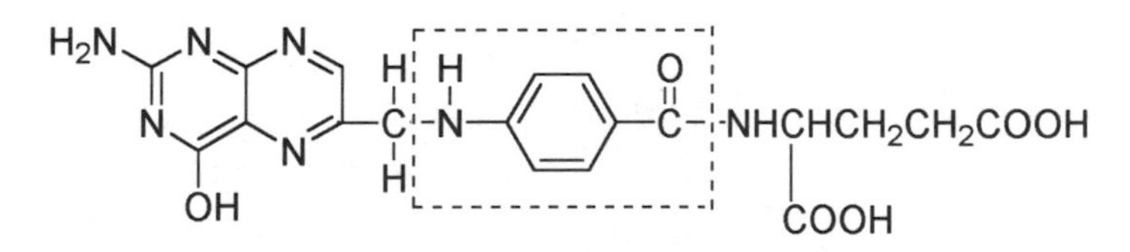

Folic acid with the middle portion from the p-*aminobenzoic acid molecule outlined*

Consider just the part of its structure shown inside the outlined box in the structure above. This middle portion of the folic acid molecule is derived (in bacteria that make their own folic acid) from a smaller molecule, *p*-aminobenzoic acid. *p*-Aminobenzoic acid is thus an essential nutrient for these microorganisms.

The chemical structures of *p*-aminobenzoic acid and sulfanilamide are remarkably similar in shape and size, and it is this similarity that accounts for the antimicrobial activity of sulfanilamide. The lengths (as indicated by the square brackets) of each of these molecules measured from the hydrogen of the NH_2 group to the doubly bonded oxygen atom are within 3 percent of each other. As well they have almost the same width.

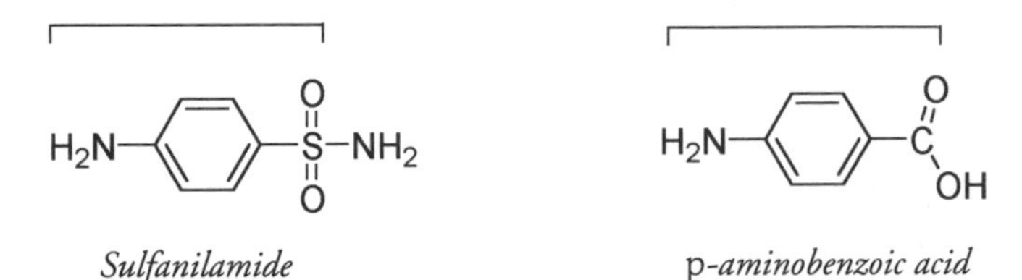

Sulfanilamide

p-*aminobenzoic acid*

The bacterial enzymes involved in synthesizing folic acid appear to be unable to distinguish between the molecules of *p*-aminobenzoic acid that they need and the look-alike sulfanilamide molecules. Bacteria will thus unsuccessfully attempt to use sulfanilamide instead of *p*-aminobenzoic acid—and ultimately die because they are unable to make enough folic acid. We, relying on folic acid absorbed from our food, are not negatively affected by the action of sulfanilamide.

Technically, sulfanilamide-based sulfa drugs are not true antibiotics. Antibiotics are properly defined as "substances of microbial origin that in very small amounts have antimicrobial activity." Sulfanilamide is not derived from a living cell. It is man-made and is properly classified as an antimetabolite, a chemical that inhibits the growth of microbes. But the term *antibiotic* is now commonly used for all substances, natural or artificial, that kill bacteria.

Although sulfa drugs were not the very first synthetic antibiotic—that honor belongs to Ehrlich's syphilis-fighting molecule salvarsan—they were the first group of compounds that had widespread use in the fight against bacterial infection. Not only did they save the lives of hundreds of thousands of wounded soldiers and pneumonia victims, they were also responsible for an astounding drop in deaths of women in childbirth, because the streptococcus bacteria that cause puerperal or childbed fever also proved susceptible to sulfa drugs. More recently, however, the use of sulfa drugs has decreased worldwide, for a number of reasons: concern over their long-term side effects, the evolution of sulfanilamide-resistant bacteria, and the development of newer and more powerful antibiotics.

PENICILLINS

The earliest true antibiotics, from the penicillin family, are still in widespread use today. In 1877, Louis Pasteur was the first to demonstrate that one microorganism could be used to kill another. Pasteur showed that the growth of a strain of anthrax in urine could be prevented by the addition of some common bacteria. Subsequently Joseph Lister, having convinced the world of medicine of the value of phenol as an antiseptic, investigated the properties of molds, supposedly curing a persistent abscess in one of his patients with a compress soaked in a *Penicillium*-mold extract.

Despite these positive results, further investigation of the curative properties of molds was sporadic until 1928, when a Scottish physician named Alexander Fleming, working at St. Mary's Hospital Medical School of London University, discovered that a mold of the *Penicillium* family had contaminated cultures of the staphylococci bacteria he was studying. He noted that a colony of the mold became transparent and disintegrated (undergoing what is called *lysis*). Unlike others before him Fleming was intrigued enough to follow through with further experimentation. He assumed that some compound produced by the mold was responsible for the antibiotic effect on the staphylococcus bacteria, and his tests confirmed this. A filtered broth, made from cultured samples of what we now know was *Penicillium notatum,* proved remarkably effective in laboratory tests against staphylococci grown in glass dishes. Even if the mold extract was diluted eight hundred times, it was still active against the bacterial cells. Moreover mice, injected with the substance that Fleming was now calling *penicillin,* showed no toxic effects. Unlike phenol, penicillin was nonirritating and could be applied directly to infected tissues. It also seemed to be a more powerful bacterial inhibitor than phenol. It was active against many bacteria species, including those causing meningitis, gonorrhea, and streptococcal infections like strep throat.

Although Fleming published his results in a medical journal, they aroused little interest. His penicillin broth was very dilute, and his attempts to isolate the active ingredient were not successful; we now know that penicillin is easily inactivated by many common laboratory chemicals, and by solvents and heat.

Penicillin did not undergo clinical trials for more than a decade, during which time sulfanilamides became the major weapon against bacterial infections. In 1939 the success of sulfa drugs encouraged a group of chemists, microbiologists, and physicians at Oxford University to start working on a method to produce and isolate penicillin. The first clinical trial with crude penicillin was in 1941. Sadly, the results were much like the old punch line "The treatment was a success, but the patient died." Intravenous penicillin treatment was given to one patient, a policeman suffering from both severe staphylococcal and streptococcal infections. After twenty-four hours an improvement was noted; five days later his fever was gone and his infection was clearing. But by then all of the penicillin available—about a teaspoon of the unrefined extract—had been used up. The man's infection was still virulent. It expanded unchecked, and he soon died. A second patient also died. In a third trial, however, enough penicillin had been produced to completely eliminate a streptococcal infection in a fifteen-year-old boy. After that success penicillin cured staphylococcal blood poisoning in another child, and the Oxford group knew they had a winner. Penicillin proved active against a range of bacteria, and it had no harsh side effects, such as the kidney toxicity that had been reported with sulfanilamides. Later studies indicated that some penicillins inhibit the growth of streptococci at a dilution of one to fifty million, an amazingly small concentration.

At this time the chemical structure of penicillin was not yet known, and so it was not possible to make it synthetically. Penicillin still had to be extracted from molds, and the production of large amounts was a challenge for microbiologists and bacteriologists rather than chemists. The U.S. Department of Agriculture laboratory in Peoria, Illinois, had expertise in growing microorganisms and became the center of a massive research program. By July 1943 American pharmaceutical companies were producing 800 million units of the new antibiotic. One year later the monthly production topped 130 billion units.

It has been estimated that during World War II a thousand chemists in thirty-nine laboratories in the United States and in Britain worked on the problems associated with establishing the chemical structure of and finding a way to synthesize penicillin. Finally, in 1946, the structure was determined, although it was successfully synthesized only in 1957.

The structure of penicillin may not be as large or look as complicated a molecule as others we have discussed, but for chemists it is a most unusual molecule in that it contains a four-membered ring, known in this case as the β-lactam ring.

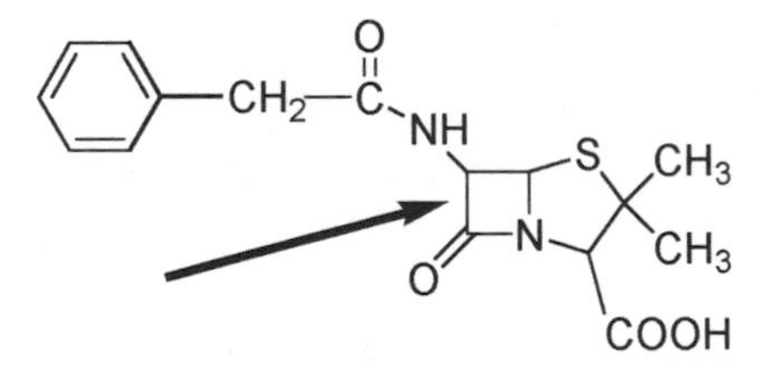

The structure of the penicillin G molecule. The arrow indicates the four-membered β-lactam ring.

Molecules with four-membered rings do exist in nature, but they are not common. Chemists can make such compounds, but it can be quite difficult. The reason is that the angles in a four-membered ring—a square—are 90 degrees, while normally the preferred bond angles for single-bonded carbon and nitrogen atoms are near 109 degrees. For a double-bonded carbon atom, the preferred bond angle is around 120 degrees.

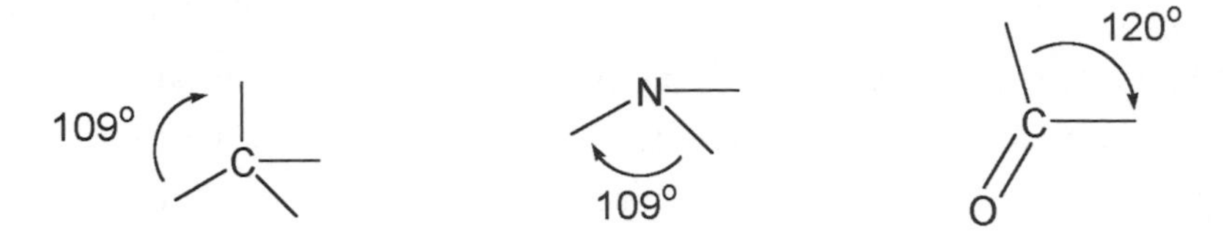

The single-bonded carbon and nitrogen atoms are three-dimensionally arranged in space, while the carbon double-bonded to an oxygen atom is in the same plane.

In organic compounds a four-membered ring is not flat; it buckles slightly, but even this cannot reduce what chemists call *ring strain,* an instability that results mainly from atoms being forced to have bond angles too different from the preferred bond angle. But it is precisely this instability of the four-membered ring that accounts for the antibiotic activity of penicillin molecules. Bacteria have cell walls and produce an enzyme that is essential for cell wall formation. In the presence of this enzyme, the β-lactam ring of the penicillin molecule splits open, relieving ring strain. In the process an OH group on the bacterial enzyme is *acylated* (the same type of reaction that converted salicylic acid into aspirin). In this acylation reaction penicillin attaches the ring-opened molecule to the bacterial enzyme. Note that the five-membered ring is still intact, but the four-membered ring has opened up.

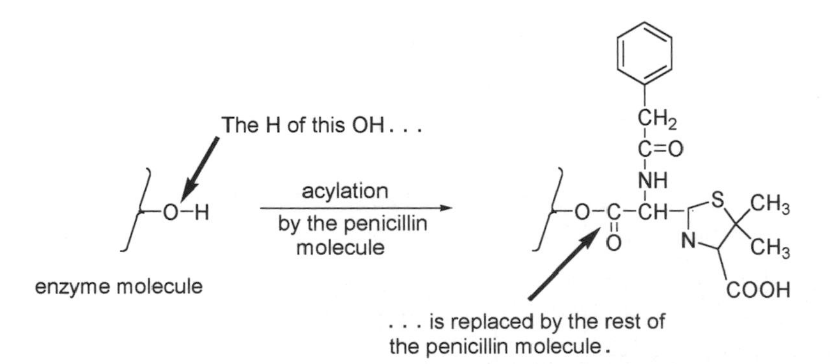

The penicillin molecule attaches to the bacterial enzyme in this acylation reaction.

This acylation deactivates the cell wall–forming enzyme. Without the ability to build cell walls, the growth of new bacteria in an organism is inhibited. Animal cells have a cell membrane rather than a cell wall and so do not have the same wall-forming enzyme as these bacteria. We are therefore not affected by the acylation reaction with the penicillin molecule.

The instability of the four-membered β-lactam ring of penicillin is also the reason that penicillins, unlike sulfa drugs, need to be stored at low temperatures. Once the ring opens—a process accelerated by heat—the molecule is no longer an effective antibiotic. Bacteria themselves seem to have discovered the secret of ring opening. Penicillin-resistant strains have developed a further enzyme that breaks open the β-lactam ring of penicillin before it has a chance to deactivate the enzyme responsible for cell wall formation.

The structure of the penicillin molecule shown below is that of penicillin G, first produced from mold in 1940 and still widely used. Many other penicillin molecules have been isolated from molds, and a number have been synthesized chemically from the naturally occurring versions of this antibiotic. The structures of different penicillins vary only in the part of the molecule circled below.

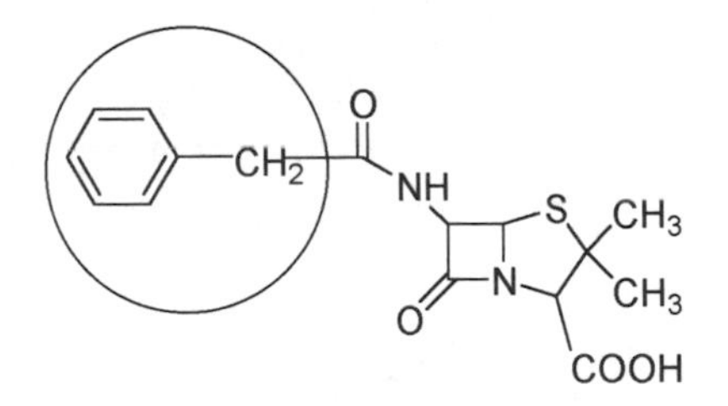

Penicillin G. The variable part of the molecule is circled.

Ampicillin, a synthetic penicillin effective against bacteria that are resistant to penicillin G, is only slightly different. It has an extra NH_2 group attached.

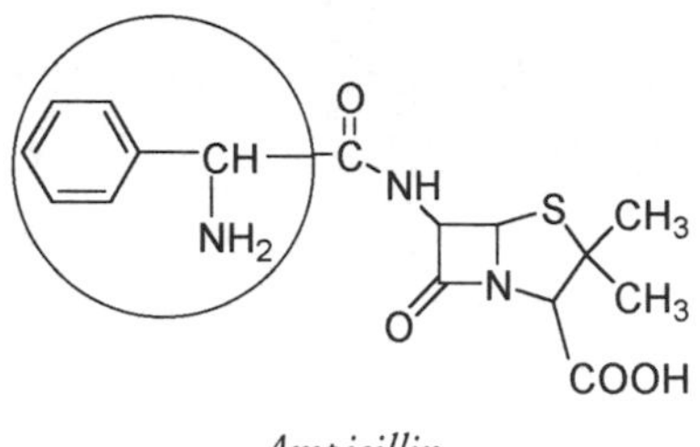

Ampicillin

The side group in amoxicillin, today one of the most widely prescribed drugs in the United States, is very similar to ampicillin but with an extra OH. The side group can be very simple, as in penicillin O, or more complicated, as in cloxacillin.

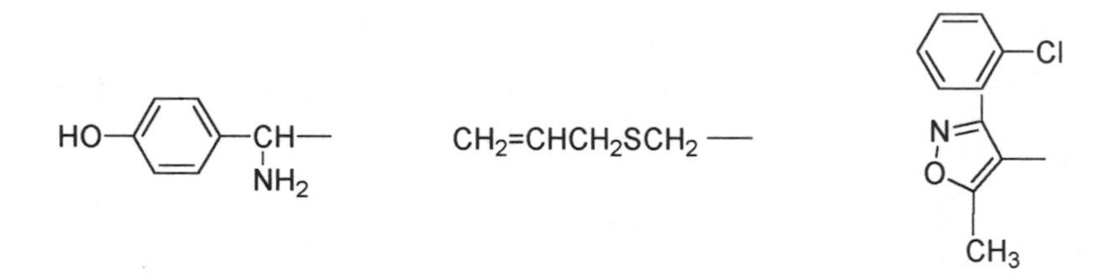

The structure of the side groups in the circled portion of the molecule for amoxicillin (left), penicillin O (center), and cloxacillin (right)

These are only four of the ten or so different penicillins still in use today. (Many more exist that are no longer used clinically.) The structural modifications, at the same (circled) site on the molecule, can be very variable, but the four-membered β-lactam ring is always present. It is this piece of the molecular structure that may have saved your life, if you have had the occasion to need a penicillin antibiotic.

Although it is impossible to obtain accurate statistics of mortality during previous centuries, demographers have estimated average lifespans in some societies. From 3500 B.C. through to about A.D. 1750, a period of over five thousand years, life expectancy among European societies hovered between thirty and forty years; in classical Greece, around 680 B.C., it rose as high as forty-one years; in Turkey of A.D. 1400 it was just thirty-one years. These numbers are similar to those in the underdeveloped countries of the world today. The three main reasons for these high mortality rates—inadequate food supplies, poor sanitation, and epidemic disease—are closely interrelated. Poor nutrition leads to increased susceptibility to infection; poor sanitation produces conditions conducive to disease.

In those parts of the world with efficient agriculture and a good transportation system, food supply has increased. At the same time vastly improved personal hygiene and public health measures—clean water supply, sewage treatment systems, refuse collection and vermin control, and wholesale immunization and vaccination programs—have led to fewer epidemics and a healthier population more able to resist disease. Because of these improvements, death rates in the developed world have been dropping steadily since the 1860s. But the final onslaught against those bacteria that have for generations caused untold misery and death has been by antibiotics.

From the 1930s the effect of these molecules on mortality rates from infectious diseases has been marked. After the introduction of sulfa drugs to treat pneumonia, a common complication with the measles virus, the death rate from measles declined rapidly. Pneumonia, tuberculosis, gastritis, and diphtheria, all among leading causes of death in the United States in 1900, do not make the list today. Where isolated incidents of bacterial diseases—bubonic plague, cholera, typhus, and anthrax—have occurred, antibiotics contained what might otherwise have become a widespread outbreak. Today's acts of bioterrorism have focused public concern on the possibility of a major bacterial epidemic. Our present array of antibiotics would normally be able to cope with such an attack.

Another form of bioterrorism, that waged by bacteria themselves as they adapt to our increasing use and even overuse of antibiotics, is worrying. Antibiotic-resistant strains of some common but potentially lethal bacteria are becoming widespread. But as biochemists learn more about the metabolic pathways of bacteria—and of humans—and how the older antibiotics worked, it should become possible to synthesize new antibiotics able to target specific bacterial reactions. Understanding chemical structures and how they interact with living cells is essential to maintaining an edge in the never-ending struggle with disease-causing bacteria.

11. THE PILL

BY THE MIDDLE of the twentieth century antibiotics and antiseptics were in common use and had dramatically lowered mortality rates, particularly among women and children. Families no longer needed a multitude of children to ensure that some reached maturity. As the specter of losing children to infectious diseases diminished, a demand for ways of limiting family size by preventing conception arose. In 1960 a contraceptive molecule emerged that played a major role in shaping contemporary society.

We are, of course, referring to *norethindrone,* the first oral contraceptive, usually known as "the pill." The molecule has been credited with—or blamed for (depending on your point of view)—the sexual revolution of the 1960s, the women's liberation movement, the rise of feminism, the increased percentage of women in the workplace, and even the breakdown of the family. Despite the varying opinions on the benefits or disadvantages of this molecule, it has played an important role in the enormous changes in society in the forty or so years since the pill was introduced.

Struggles for legal access to birth control information and supplies, fought in the early part of the last century by such notable reformers as Margaret Sanger in the United States and Marie Stopes in Britain, seem remote to us now. Young people today are often incredulous on hearing that just to give information on contraception was, in many countries during the initial decades of the twentieth century, a crime. But the need was clearly present: high rates of infant mortality and maternal death found in poor areas of cities often correlated with large families. Middle-class families were already using the contraceptive methods then available, and working-class women were desperate for the same information and access. Letters written to birth control advocates by mothers of large families detailed the despair they felt while facing yet another unwanted pregnancy. By the 1930s public acceptance of birth control, often couched in the more acceptable term *family planning,* was increasing; health clinics and medical personnel were involved in prescribing contraceptive devices, and laws, at least in some places, were being changed. Where restrictive statutes did remain on the books, prosecutions became less common, especially if contraception matters were handled discreetly.

Early Attempts at Oral Contraception

Over the centuries and in every culture women have swallowed many substances in the hope of preventing conception. None of these substances would have accomplished this goal except, perhaps, by making the woman so ill she would be unable to conceive. Some of the remedies were fairly straightforward: brewed teas of parsley and mint, of leaves or bark of hawthorn, ivy, willow, wallflower, myrtle, or poplar. Mixtures containing spiders' eggs or snake were also suggested. Fruits, flowers, kidney beans, apricot kernels, and mixed herbal potions were other recommendations. At one time the mule featured prominently in contraception, supposedly because a mule is the sterile offspring of a female horse and a male donkey. Sterility was allegedly assured if a woman ate the kidney or uterus of a mule. For male sterility the animal's contribution was no less tasty; a man was to eat the burned testicles of a castrated mule. Mercury poisoning might have been an effective means of attaining sterility for a woman swallowing a seventh-century Chinese remedy of quicksilver (an old name for mercury) fried in oil—that is, if it did not kill her first. Solutions of different copper salts were drunk as contraceptives in ancient Greece and in parts of Europe in the 1800s. A bizarre method from the Middle Ages required a woman to spit three times into the mouth of a frog. It was the woman who would become sterile, not the frog!

Steroids

Though some of the substances smeared on various parts of the body to prevent pregnancy could possibly have had spermicidal properties, the advent of oral contraceptives in the middle of the twentieth century marked the first truly safe and effective chemical means of birth control. Norethindrone is one of a group of compounds known as *steroids,* a perfectly good chemical name that now is often applied to performance-enhancing drugs illegally used by some athletes. Such drugs are definitely steroids, but so are many other compounds that have nothing to do with athletic prowess; we will be using the term *steroid* in the wider chemical sense.

In many molecules very small changes in structure can have very large changes in effect. This is nowhere more pronounced than in the structures of the sex hormones: the male sex hormones (androgens), the female sex hormones (estrogens), and the hormones of pregnancy (progestins).

All compounds that are classified as steroids have the same basic molecular pattern, a series of four rings fused in the same way. Three of the rings have six carbon atoms each, and the fourth has five. These rings are referred to as the A, B, C, and D rings—the D ring always being five-membered.

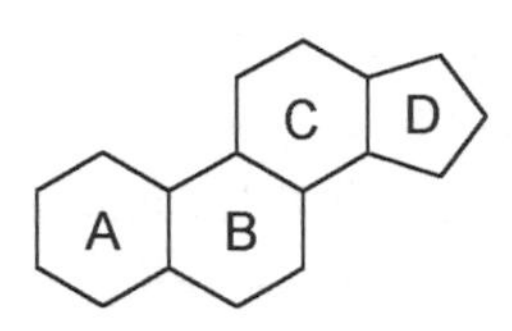

The four basic rings of the steroid structure, showing the A, B, C, and D designations

Cholesterol, the most widespread of all animal steroids, is found in most animal tissues, with especially high levels in egg yolks and human gallstones. It is a molecule with an undeservedly bad reputation. We need cholesterol in our system; it plays a vital role as the precursor molecule of all our other steroids, including bile acids (compounds that enable us to digest fats and oils) and the sex hormones. What we do not need is a lot of extra cholesterol in our diet, as we synthesize enough of our own. The molecular structure of cholesterol shows the four basic fused rings as well as the side groups, including a number of methyl groups (CH_3, sometimes written as H_3C, just to fit more easily on the drawing).

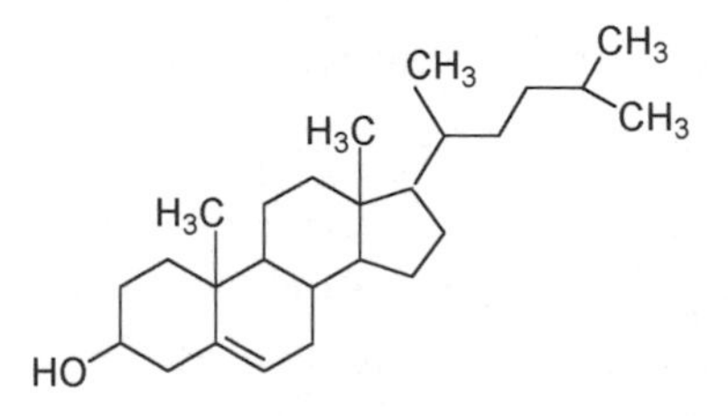

Cholesterol, the most widespread animal steroid

Testosterone, the principal male sex hormone, was first isolated from ground-up bull testes in 1935, but it was not the first male sex hormone to be isolated. That hormone was androsterone, a metabolized and less potent variation of testosterone that is excreted in the urine. As you can see from a comparison of the two structures, there is very little difference between them, androsterone being an oxidized version where a double-bonded oxygen atom has replaced the OH of testosterone.

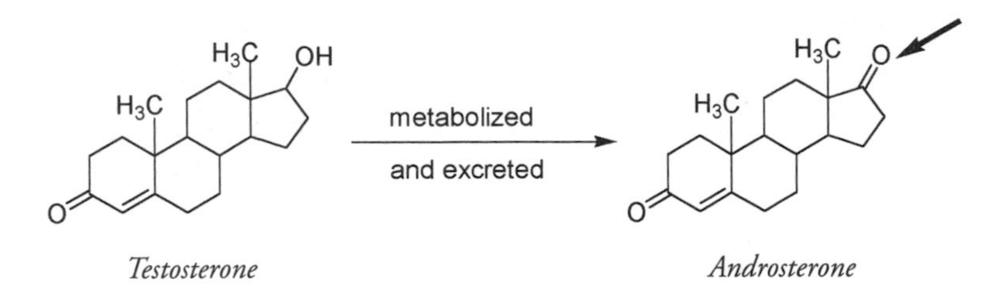

Androsterone varies from testosterone at only one position (arrowed).

The first isolation of a male hormone was in 1931 when fifteen milligrams of androsterone was obtained from fifteen thousand liters of urine collected from the Belgian police force, presumably an all-male group in those days.

The first sex hormone ever isolated was the female sex hormone estrone, obtained in 1929 from urine of pregnant women. As with androsterone and testosterone, estrone is a metabolized variation of the principal and more potent female sex hormone, estradiol. A similar oxidation process changes an OH on estradiol into a double-bonded oxygen.

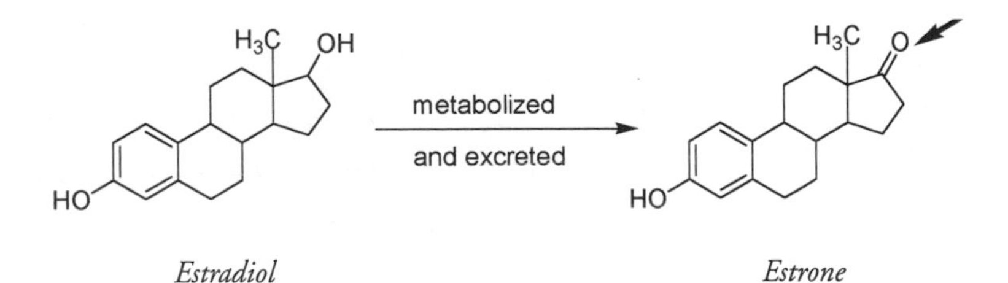

Estrone varies from estradiol at only one position (arrowed).

These molecules are present in our bodies in very small amounts: four tons of pig ovaries were used to extract only twelve milligrams of the first estradiol that was isolated.

It is interesting to consider how structurally similar the male hormone testosterone and the female hormone estradiol are. Just a few changes in molecular structure make an enormous difference.

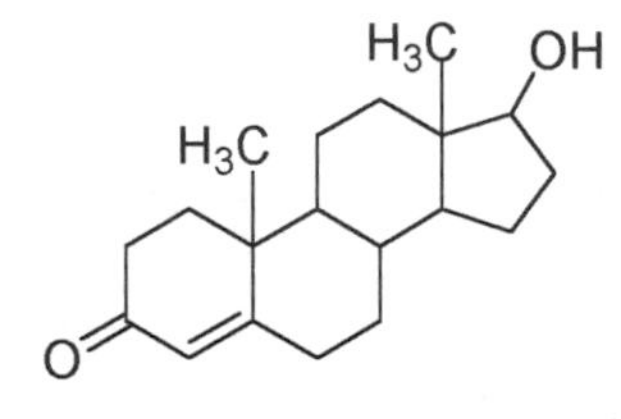

Testosterone

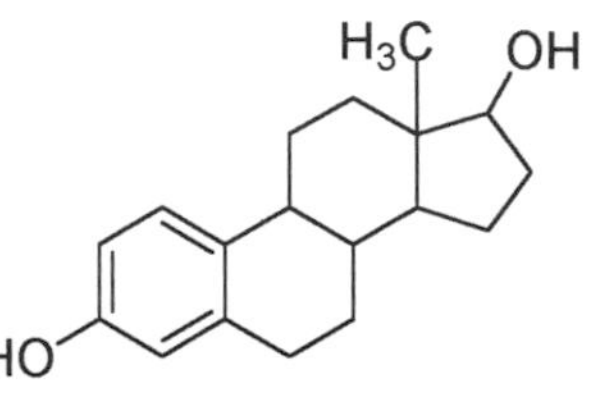

Estradiol

If you have one less CH_3, an OH instead of a double-bonded O, and a few more C=C bonds, then at puberty instead of developing male secondary sex characteristics (facial and body hair, deep voice, heavier muscles), you will grow breasts and wider hips and start menstruating.

Testosterone is an anabolic steroid, meaning that it is a steroid that promotes muscle growth. Artificial testosterones—manufactured compounds that also stimulate the growth of muscle tissue—have similar structures to testosterone. They were developed for use with injuries or diseases that cause debilitating muscle deterioration. At prescription-level doses these drugs help rehabilitate with minimal masculinizing effect, but when these synthetic steroids, like Dianabol and Stanozolol, are used at a ten or twenty times normal rate by athletes wanting to "bulk up" the side effects can be devastating.

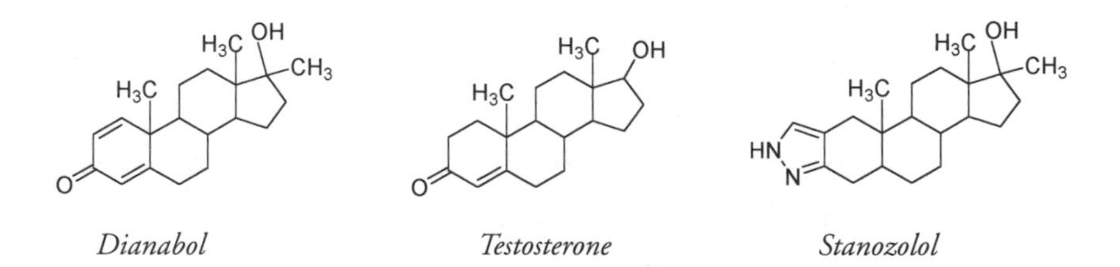

Dianabol *Testosterone* *Stanozolol*

The synthetic anabolic steroids Dianabol and Stanozolol compared with natural testosterone

Increased risk of liver cancer and heart disease, heightened levels of aggression, severe acne, sterility, and shriveled testicles are just a few of the dangers from the misuse of these molecules. It may seem a bit odd that a synthetic androgenic steroid, one that promotes male secondary characteristics, causes the testes to shrink, but when artificial testosterones are supplied from a source outside the body, the testes—no longer needing to function—atrophy.

Just because a molecule has a similar structure to testosterone does not necessarily mean it acts like a male hormone. Progesterone, the main pregnancy hormone, not only has a structure closer to that of testosterone and androsterone than Stanozolol but is also more like the male sex hormones than the estrogens. In progesterone a CH_3CO group (circled in the diagram) replaces the OH of testosterone.

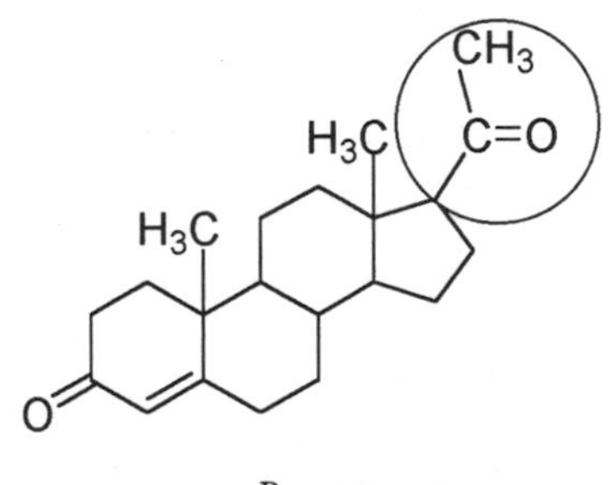

Progesterone

This is the only variation in the chemical structure between progesterone and testosterone, but it makes a vast difference in what the molecule does. Progesterone signals the lining of the uterus to prepare for the implantation of a fertilized egg. A pregnant woman does not conceive again during her pregnancy because a continuous supply of progesterone suppresses further ovulation. This is the biological basis of chemical contraception: an outside source of progesterone, or a progesteronelike substance, is able to suppress ovulation.

There are major problems with the use of the progesterone molecule as a contraceptive. Progesterone has to be injected; its effectiveness is very much reduced when taken orally, presumably because it reacts with stomach acids or other digestive chemicals. Another problem (as we saw with the isolation of milligrams of estradiol from tons of pig ovaries) is that natural steroids occur in very minute quantities in animals. Extraction from these sources is just not practical.

The solution to these problems is the synthesis of an artificial progesterone that retains its activity when taken orally. For such a synthesis to be possible on a large scale, one needs a starting material where the four-ring steroid system, with CH_3 groups at set positions, is already in place. In other words, synthesis of a molecule that mimics the role of progesterone requires a convenient source of large amounts of another steroid whose structure can be altered in the laboratory with the right reactions.

THE AMAZING ADVENTURES OF RUSSELL MARKER

We have stated the chemical problem here, but it should be emphasized that we are seeing it from the perspective of hindsight. Synthesis of the first birth control pill was the result of an attempt to solve quite a different set of puzzles. The chemists involved had no idea that they would eventually produce a molecule that would promote social change, give women control over their lives, and alter traditional gender roles. Russell Marker, the American chemist whose work was crucial to the development of the pill, was no exception; the goal of his chemical experimentation was not to produce a contraceptive molecule but to find an affordable route to produce another steroid molecule—cortisone.

Marker's life was a continual conflict with tradition and authority, perhaps fittingly for the man whose chemical achievements helped identify the molecule that would also have to battle tradition and authority. He attended high school and then college against the wishes of his sharecropper father, and by 1923 he had obtained a bachelor's degree in chemistry from the University of Maryland. Although he maintained that he continued his education to "get out of farm work," Marker's ability and interest in chemistry must also have been factors in his decision to pursue graduate degrees.

With his doctoral thesis completed and already published in the *Journal of the American Chemical Society,* Marker was told that he needed to take another course, one in physical chemistry, to fulfill the requirements for a Ph.D. Marker felt this would be a waste of valuable time that could be more profitably spent in the laboratory. Despite repeated warnings by his professors about the lack of opportunity for a career in chemical research without a doctoral degree, he left the university. Three years later he joined the research faculty of the prestigious Rockefeller Institute in Manhattan, his talents obviously having overcome the handicap of not finishing his doctorate.

There Marker became interested in steroids, particularly in developing a method to produce large enough quantities to allow chemists to experiment with ways to change the structure of the various side groups on the four steroid rings. At this time the cost of progesterone isolated from the urine of pregnant mares—over $1,000 a gram—was beyond the means of research chemists. The small amounts extracted from this source were mainly used by wealthy racehorse owners to prevent miscarriages in their valuable breeding stock.

Marker knew that steroid-containing compounds existed in a number of plants, including foxglove, lily of the valley, sarsaparilla, and oleander. Though isolating just the four-ring steroid system had not been possible so far, the quantity of these compounds found in plants was much greater than in animals. To Marker this was obviously the route to follow, but once again he was up against tradition and authority. The tradition at the Rockefeller Institute was that plant chemistry belonged in the department of pharmacology, not in Marker's department. Authority, in the person of the president of Rockefeller, forbade Marker to work on plant steroids.

Marker left the Rockefeller Institute. His next position was a fellowship at Pennsylvania State College, where he continued to work on steroids, eventually collaborating with the Parke-Davis drug company. It was from the plant world that Marker was eventually to produce the large quantities of steroids that he needed for his work. He began with roots of the sarsaparilla vine (used in flavoring root beer and similar drinks), which were known to contain compounds called *saponins,* so named because of their ability to make soapy or foamy solutions in water. Saponins are complex molecules, though nowhere near as large as polymer molecules like cellulose or lignin. Sarsasaponin, the saponin from the sarsaparilla plant, consists of three sugar units attached to a steroid ring system, which in turn is fused at the D ring to two other rings.

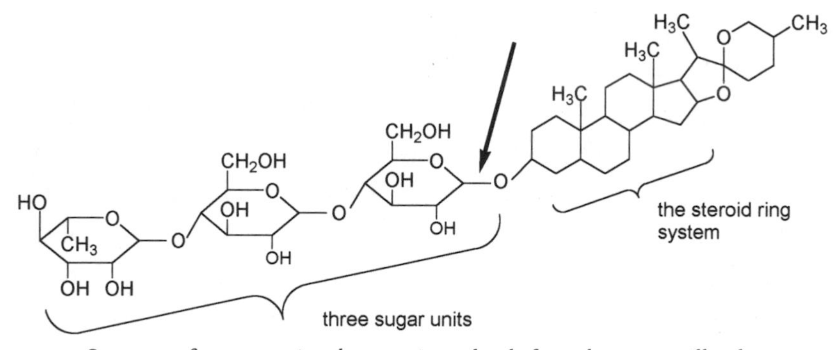

Structure of sarsasaponin, the saponin molecule from the sarsaparilla plant

It was known that removing the three sugars—two glucose units and a different sugar unit called rhamnose—was simple. With acid the sugar units split off at the point indicated by the arrow in the structure above.

sarsasaponin —*reaction with acid or enzymes*→ sarsasapogenin + 2 glucose + rhamnose

It was the remaining portion of the molecule, a *sapogenin,* that presented problems. To obtain the steroid ring system from sarsasapogenin, it was necessary to remove the side grouping circled in the next diagram. The prevailing chemical wisdom of the day was that this could not be done, not without destroying other parts of the steroid structure.

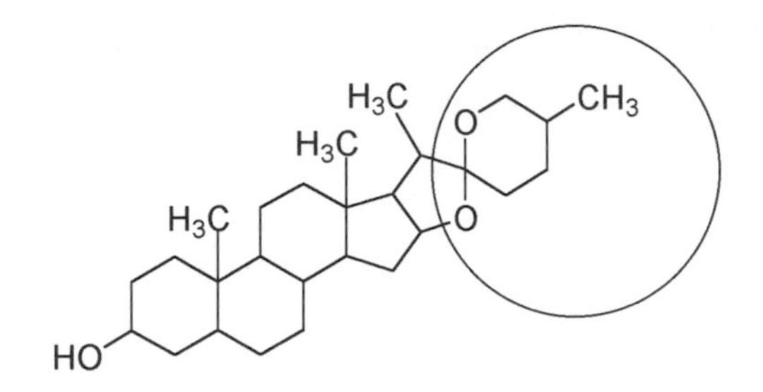

Sarsasapogenin, the sapogenin from the sarsaparilla plant

Marker was sure it could be done, and he was right. The process he developed produced the basic four-ring steroid system that, with only a few more steps, gave pure synthetic progesterone, chemically identical to that produced in the female body. And once the side group was removed, synthesis of many other steroidal compounds became possible. This procedure—the removal of the sapogenin side grouping from the steroid system—is still used today in the multibillion-dollar synthetic hormone industry. It is known as the "Marker degradation."

Marker's next challenge was to find a plant that contained more of the starting material than sarsaparilla. Steroid sapogenins, derived by removing sugar units from the parent saponins, can be found in numerous plants other than the sarsaparillas, including trillium, yucca, foxglove, agave cactus, and asparagus. His search, involving hundreds of tropical and subtropical plants, eventually led to a species of *Dioscorea,* a wild yam found in the mountains of the Mexican province of Veracruz. By now it was early 1942, and the United States was involved in World War II. Mexican authorities were not issuing plant-collecting permits, and Marker was advised not to venture into the area to collect the yam. Such advice had not stopped Marker before, and he did not let it get in his way now. Traveling by local bus, he eventually reached the area where he had been told the plant grew. There he collected two bags of the black foot-long roots from *cabeza de negro* (black head), as the yams were known locally.

Back in Pennsylvania he extracted a very similar sapogenin to the sarsasapogenin from the sarsaparilla plant. The only difference was an extra double bond (arrowed) found in diosgenin, the sapogenin from the wild yam.

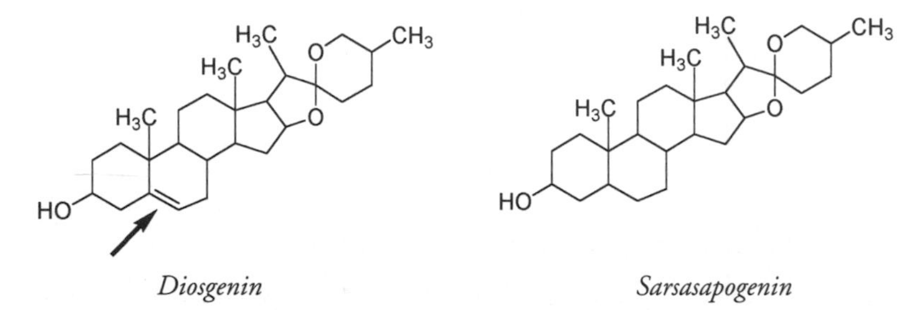

Diosgenin, from the Mexican yam, differs from the sarsaparilla sapogenin, sarsasapogenin, only by an extra double bond (arrowed).

The Marker degradation removed the unwanted side group, and further chemical reactions produced a generous amount of progesterone. Marker was convinced that the way to obtain sizable amounts of steroidal hormones at a reasonable cost would be to set up a laboratory in Mexico and use the abundant source in the Mexican yam.

But if this solution seemed practical and sensible to Marker, it did not appear that way to the major pharmaceutical companies that he tried to interest in his scheme. Tradition and authority once more blocked his way. Mexico had no history of carrying out such complicated chemical syntheses, the drug company authorities told him. Unable to get financial backing from established companies, Marker resolved to enter the hormone-production business himself. He resigned from Pennsylvania State College and eventually moved to Mexico City, where, in 1944 and in partnership with others, he established Syntex (for Synthesis and Mexico), the pharmaceutical company that was to become a world leader in steroidal products.

But Marker's relationship with Syntex was not to last. Arguments over payments, profits, and patents led to his departure. Another company he established, Botanica-Mex, was ultimately bought out by Eu-

Russell Marker, whose development of the series of chemical steps known as the Marker degradation allowed chemists access to abundant plant steroid molecules. (Photo courtesy of Pennsylvania State University)

ropean drug companies. By this time Marker had discovered other species of *Dioscorea* that were even richer in the steroid-containing diosgenin molecule. The cost of synthetic progesterone steadily decreased. These yams, once an obscure root used only as a fish poison by local farmers—the fish became dazed but were still edible—are now grown as a commercial crop in Mexico.

Marker had always been reluctant to patent his procedures, feeling his discoveries should be available to everyone. By 1949 he was so disgusted and disappointed both with his fellow chemists and with the profit motive that he now saw as driving chemical research that he destroyed all his laboratory notes and experiment records in an attempt to remove himself totally from the field of chemistry. Despite these efforts the chemical reactions Marker pioneered are today acknowledged as the work that made possible the birth control pill.

SYNTHESIS OF OTHER STEROIDS

In 1949 a young Austrian immigrant to the United States joined the Syntex research facilities in Mexico City. Carl Djerassi had just finished his Ph.D. at the University of Wisconsin, where his thesis work involved the chemical conversion of testosterone to estradiol. Syntex wanted to find a way to convert the now relatively abundant progesterone from wild yams to the cortisone molecule. Cortisone is one of at least twenty-eight different hormones isolated from the adrenal cortex (the outer part of the adrenal glands adjacent to the kidneys). It is a potent anti-inflammatory agent that is especially effective in treating rheumatoid arthritis. Like other steroids, cortisone is present in minute quantities in animal tissues. Although it could be made in the laboratory, such methods were very expensive. Synthesis required thirty-two steps, and its starting material, desoxycholic acid, had to be isolated from ox bile—which was not at all abundant.

Using the Marker degradation, Djerassi showed how cortisone could be produced at a much lower cost from a plant source like diosgenin. One of the major stumbling blocks in making cortisone is attaching the double-bonded oxygen at carbon number 11 on the C ring, a position that is not substituted in the bile acids or sex hormones.

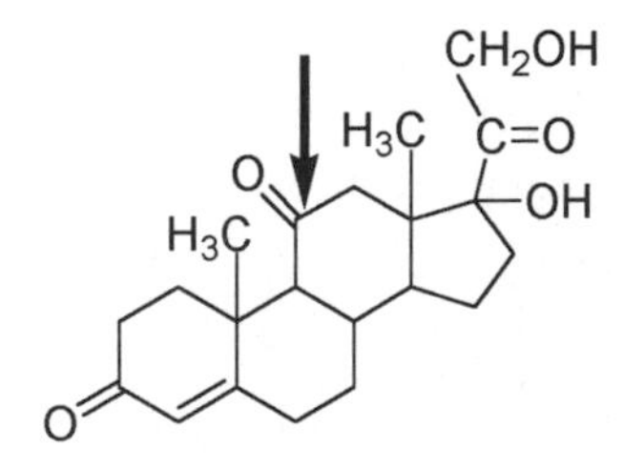

Cortisone. The C=O at C#11 is arrowed.

A novel method of attaching oxygen at this position was later discovered using the mold *Rhizopus nigricans.* The effect of this combination of fungi and chemist was to produce cortisone from progesterone in a total of only eight steps—one microbiological and seven chemical.

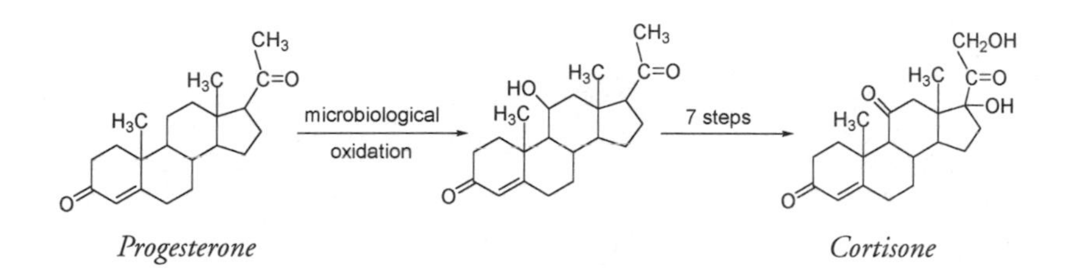

Progesterone *Cortisone*

After his success in making cortisone, Djerassi synthesized both estrone and estradiol from diosgenin, giving Syntex a preeminent position as a major world supplier of hormones and steroids. His next project was to make an artificial progestin, a compound that would have progesteronelike properties but could be taken orally. The aim was not to create a contraceptive pill. Progesterone, now available at a reasonable cost—less than a dollar a gram—was being used to treat women who had a history of miscarriage. It had to be injected and in fairly large doses. Djerassi's reading of the scientific literature led him to suspect that substituting, on the D ring, a group with a carbon-to-carbon triple bond (≡) might allow the molecule to retain its effectiveness when swallowed. Another report had mentioned that removal of a CH_3 group—the carbon designated as number 19—seemed to increase potency in other progesteronelike molecules. The molecule Djerassi and his team produced and patented in November 1951 was eight times more powerful than progesterone and could be taken orally. It was named norethindrone—the *nor* indicating a missing CH_3 group.

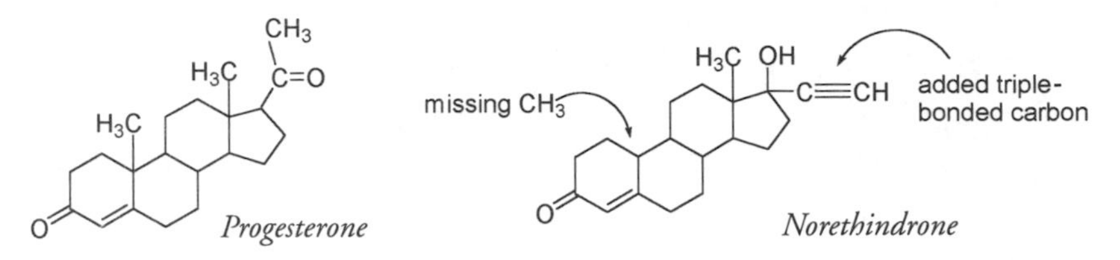

The structure of natural progesterone compared with the artificial progestin norethindrone

Critics of the birth control pill have pointed out that it was developed by men to be taken by women. Indeed, the chemists involved in the synthesis of the molecule that became the pill were men, but as Djerassi, who is now sometimes known as the "Father of the Pill," was to say years later, "Not in our wildest dreams did we imagine that this substance would eventually become the active ingredient of nearly half the oral contraceptives used worldwide." Norethindrone was designed as a hormone treatment to support pregnancy or to relieve menstrual irregularity, especially where severe blood loss was involved. Then in the early 1950s two women became the driving force responsible for changing the role of this molecule from a limited infertility treatment to an everyday factor in the lives of countless millions of women.

THE MOTHERS OF THE PILL

Margaret Sanger, the founder of International Planned Parenthood, was jailed in 1917 for giving contraceptives to immigrant women at a Brooklyn clinic. Throughout her life she was an impassioned believer in a woman's right to control her own body and fertility. Katherine McCormick was one of the first women to receive a degree in biology from the Massachusetts Institute of Technology. She was also, after the death of her husband, immensely wealthy. She had known Margaret Sanger for over thirty years, had even helped her smuggle illegal contraceptive diaphragms into the United States, and had supplied financial help for the birth control cause. Both women were in their seventies when they journeyed to Shrewsbury, Massachusetts, for a meeting with Gregory Pincus, a specialist in female fertility and one of the founders of a small nonprofit organization called the Worcester Foundation for Experimental Biology. Sanger challenged Dr. Pincus to produce a safe, cheap, reliable "perfect contraceptive" that could be "swallowed like an aspirin." McCormick supported her friend's venture with financial backing and over the next fifteen years contributed more than three million dollars to the cause.

Pincus and his colleagues at the Worcester Foundation first verified that progesterone did inhibit ovulation. Their work was done with rabbits; it was not until Pincus met another reproduction researcher, Dr. John Rock of Harvard University, that he realized similar results were available from human subjects. Rock was a gynecologist working to overcome fertility problems in his patients. His rationale for using progesterone to treat infertility assumed that blocking fertility by inhibiting ovulation for a few months would promote a "rebound effect" once the progesterone injections stopped.

In 1952 the state of Massachusetts had some of the most restrictive birth control laws in the United States. It was not illegal to use birth control, but exhibiting, selling, prescribing, and providing contraceptives and even information about contraception were all felonies. This law was not repealed until March 1972. Given these legal constraints, Rock was understandably cautious in explaining his progesterone-injection treatment to his patients. As the procedure was still experimental, informed consent was especially necessary. So the repression of ovulation was explained but was stressed as a temporary side effect to the real aim of increased fertility.

Neither Rock nor Pincus felt that injections of fairly large doses of progesterone would make a feasible long-term contraceptive. Pincus began contacting drug companies to find out if any of the artificial progesterones developed so far might be more potent in smaller doses and also effective orally. The answer came back: there were two synthetic progestins that fitted the requirements. The Chicago-based pharmaceutical company G. D. Searle had patented a molecule very similar to that synthesized by Djerassi at Syntex. Their norethynodrel differed from norethindrone only in the position of a double bond. The effective molecule is assumed to be norethindrone; stomach acids supposedly flipped the position of the double bond of norethynodrel to that of its structural isomer—same formula, different arrangement—norethindrone.

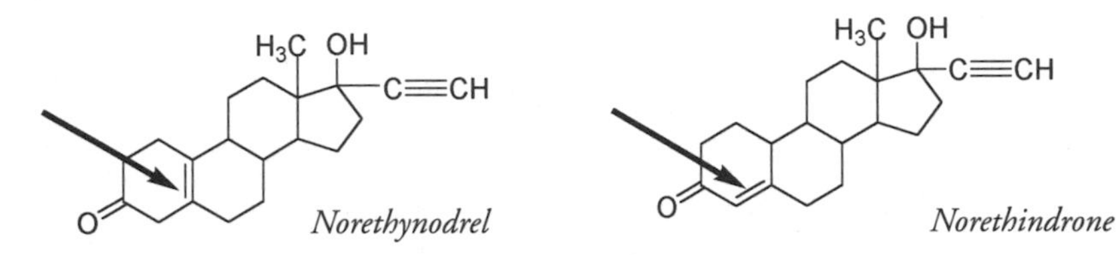

The arrows point to the positions of the double bond, the only difference between Searle's norethynodrel and Syntex's norethindrone.

A patent for each compound was granted. The legal question of whether one molecule that the body changes to another constituted an infringement of patent law was never examined.

Pincus tried both molecules for the suppression of ovulation in rabbits at the Worcester Foundation. The only side effect was no baby rabbits. Rock then started cautious testing of norethynodrel, now given the name Enovid, with his patients. The fiction that he was still investigating infertility and menstrual irregularities was maintained, not without some degree of truth. His patients were still seeking help for this problem, and he was, for all intents and purposes, doing the same experiments as before—blocking ovulation for a few months to make use of the increase in fertility that seemed to occur, at least for some women, after this treatment. He was, however, using artificial progestins, administered orally and at lower doses than synthetic progesterone. The rebound effect seemed to be no different. Careful monitoring of his patients showed that Enovid was 100 percent effective in preventing ovulation.

What was needed now was field trials, which took place in Puerto Rico. In recent years critics have denounced the "Puerto Rico experiment" for its supposed exploitation of poor, uneducated, and uninformed women. But Puerto Rico was well ahead of Massachusetts in terms of enlightenment on birth control. Although it had a predominantly Catholic population, in 1937—thirty-five years before Massachusetts—Puerto Rico amended its laws so that the distribution of birth control supplies was no longer illegal. Family planning clinics, known as "pre-maternity" clinics, existed, and doctors at Puerto Rico's medical school as well as public health officials and nurses supported the idea of field testing an oral contraceptive.

The women selected for the study were carefully screened and meticulously monitored throughout. They may have been poor and uneducated, but they were also pragmatic and practical. These women may not have understood the intricacies of the female hormonal cycle, but they did comprehend the perils of having more children. For a thirty-six-year-old mother of thirteen, eking out a living in a two-room shack on a subsistence farm, the possible side effects of a birth control pill would seem a lot safer than a further unwanted pregnancy. There was no shortage of volunteers in Puerto Rico in 1956; nor would there be for further studies done in Haiti and Mexico City.

More than two thousand women participated in the trials in these three countries. Among them the failure-to-prevent-pregnancy rate was around 1 percent, compared with a failure rate among other forms of contraception of anywhere from 30 to 40 percent. The clinical trials of oral contraception were a success; the concept, proposed by two older women who had seen much of the hardship and misery of unfettered fertility, was workable. Ironically, if these trials had occurred in Massachusetts, even informing the subjects about the aim of the tests would have been illegal.

In 1957 the drug Enovid was given limited approval by the Food and Drug Administration as a treatment for menstrual irregularities. The forces of tradition and authority still prevailed; though the pill's contraceptive properties were definitely known, it was believed that women would be unlikely to take a daily contraceptive pill and that the relatively high cost (about ten dollars monthly) would be a deterrent. Yet two years after the FDA's approval, half a million women were taking Enovid for their "menstrual irregularities."

G. D. Searle finally applied for the approval of Enovid as an oral contraceptive and formally obtained it in May 1960. By 1965 nearly four million American women were "on the pill," and twenty years later it was estimated that as many as eighty million women worldwide were taking advantage of the molecule made possible by Marker's experiments with a Mexican yam.

The ten-milligram dose used during field trials (another point of present-day criticism of the Puerto Rican tests) was soon reduced to five milligrams, then to two milligrams and later to even less. Combining the synthetic progestin with a small percentage of estrogen was found to decrease side effects (weight gain, nausea, breakthrough bleeding, and mood swings). By 1965 Syntex's molecule, norethindrone, through its licensees Parke-Davis and Ortho, a division of Johnson & Johnson, had the major share of the contraceptive market.

Why was a birth control pill not developed for men? Both Margaret Sanger—whose mother died of consumption at fifty after having eleven children and a number of miscarriages—and Katherine McCormick played crucial roles in the development of the pill. Both believed that women should have contraceptive control. It is doubtful they would have supported research for a male pill. If the early pioneers of oral contraceptives had synthesized a molecule to be taken by men, would the criticism now be that "male chemists developed a method allowing men to have contraceptive control"? Probably.

The difficulty with oral contraception for men lies in biology. Norethindrone (and the other artificial progestins) only mimic what natural progesterone tells the body to do—that is, to stop ovulating. Men do not have a hormonal cycle. Preventing, on a temporary basis, the daily production of millions of sperm is much more difficult than preventing the development of a once-a-month egg.

Still, a number of different molecules are being investigated for possible birth control pills for men, in response to a perceived need to share the responsibility for contraception more equally between genders. One nonhormonal approach involves the molecule gossypol, the toxic polyphenol extracted from cottonseed oil that we mentioned in Chapter 7.

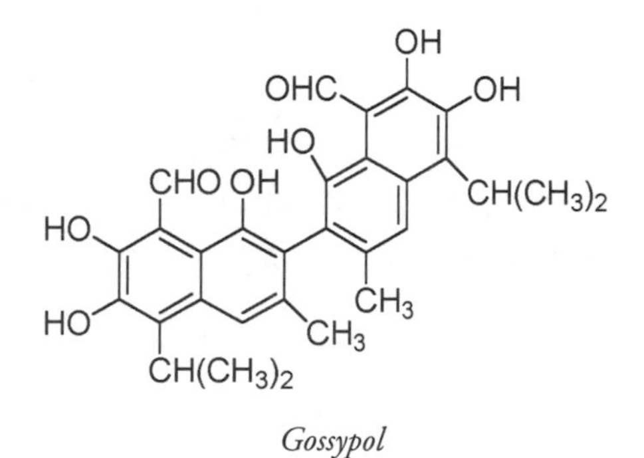

Gossypol

In the 1970s tests in China showed gossypol to be effective in suppressing sperm production, but uncertainty about the reversibility of the process and depleted levels of potassium leading to heartbeat irregularity were problems. Recent tests in both China and Brazil, using lower doses of gossypol (10 to 12.5 mg daily), have indicated that these side effects can be controlled. Wider testing is planned for this molecule.

Whatever happens in the future with new and better methods of birth control, it seems unlikely that another contraceptive molecule could change society to quite the same extent as the pill. This molecule has not gained universal acceptance; issues of morality, family values, possible health problems, long-term effects, and other related concerns are still matters for debate. But there can be little doubt that the major change brought about by the pill—a woman's control of her own fertility—led to a social revolution. In the last forty years, in countries where norethindrone and similar molecules became widely available, the birth rate has dropped, and women have gained more education and have entered the workforce in unprecedented numbers: in politics, in business, and in trade women are no longer an exception.

Norethindrone was more than just a fertility-controlling medication. Its introduction signaled the beginning of an awareness, not only of fertility and contraception, but of openness and opportunity, allowing women to speak out (and do something about) subjects that had been taboo for centuries—breast cancer, family violence, incest. The changes in attitudes in just forty years are astounding. With the option of having babies and raising families, women now govern countries, fly jet fighters, perform heart surgery, run marathons, become astronauts, direct companies, and sail the world.

12. MOLECULES OF WITCHCRAFT

FROM THE MIDDLE of the fourteenth until the late eighteenth century, a group of molecules contributed to the doom of hundreds of thousands of people. It can never be known exactly how many, in almost every country in Europe during these centuries, were burned at the stake, hanged, or tortured as witches. Estimates range from forty thousand to millions. Though accused witches included men, women, and children, aristocrats, peasants, and clergy, mostly the fingers were pointed at women—often poor and elderly women. Many reasons have been advanced for why women became the main victims of the waves of hysteria and delusion that threatened whole populations for hundreds of years. We speculate that certain molecules, while not wholly responsible for the centuries of persecution, played a substantial role in this discrimination.

Belief in sorcery and magic has always been part of human society, long before witch-hunts began at the end of the Middle Ages. Stone

Age carvings of female figures were supposedly venerated for their magical powers of fertility. Legends of all ancient civilizations abound in the supernatural: deities that take on animal form, monsters, goddesses who cast spells, enchanters, specters, goblins, ghosts, fearsome creatures who were half animal and half man, spirit-beings, and gods who lived in the sky, in forests, in lakes, in the ocean, and underground. Pre-Christian Europe, a world full of magic and superstition, was no exception.

As Christianity spread throughout Europe, many old pagan symbols and festivals were incorporated into the rituals and celebrations of the Church. We still celebrate as Halloween the great Celtic festival of the dead, marking the beginning of winter on October 31, although November 1, All Saints' Day, was the Church's attempt to divert attention away from the pagan festivities. Christmas Eve was originally the Roman feast day of Saturnalia. Christmas trees and many other symbols (holly, ivy, candles) that we now associate with Christmas are pagan in origin.

TOIL AND TROUBLE

Before 1350 witchcraft was regarded as the practice of sorcery, a method of trying to control nature in one's own interest. Using charms in the belief that they could protect crops or people, casting spells to influence or provide, and invoking spirits were commonplace. In most parts of Europe sorcery was an accepted part of life, and witchcraft was regarded as a crime only if harm resulted. Victims of *maleficium,* or evildoing by means of the occult, could seek legal recourse from a witch, but if they were unable to prove their case, they themselves became liable for a penalty and trial costs. By this method idle accusations were discouraged. Rarely were witches put to death. Witchcraft was neither an organized religion nor an organized opposition to religion. It was not even organized. It was just part of folklore.

But around the middle of the fourteenth century a new attitude toward witchcraft became apparent. Christianity was not opposed to magic, provided it was sanctioned by the Church and known as a miracle. But magic conducted outside the Church was considered the work of Satan. Witches were in league with the devil. The Inquisition, a court of the Roman Catholic Church originally established around 1233 to deal with heretics—mainly in southern France—expanded its mandate to deal with witchcraft. Some authorities have suggested that once heretics had been virtually eliminated, the Inquisition, needing new victims, set its sights on sorcery. The number of potential witches throughout Europe was large; the potential source of income for the inquisitors, who shared with local authorities the confiscated properties and assets of the condemned, would also have been great. Soon witches were being convicted not for performing evil deeds but for supposedly entering into a pact with the devil.

This crime was considered so horrendous that, by the mid–fifteenth century, ordinary rules of law no longer applied to trials of witches. An accusation alone was treated as evidence. Torture was not only allowed, it was used routinely; a confession without torture was seen as unreliable—a view that seems strange today.

The deeds attributed to witches—orgiastic rituals, sex with demons, flying on broomsticks, child murdering, baby eating—were, for the most part, beyond rationality but were still fervently believed. About 90 percent of accused witches were women, and their accusers were just as likely also to be women as men. Whether so-called witch-hunts revealed an underlying paranoia aimed at women and female sexuality is still being argued. Wherever a natural disaster struck—a flood, a drought, a crop failure—no lack of witnesses would attest that some poor woman, or more likely women, had been seen cavorting with demons at a sabbat (or witches' gathering) or flying around the countryside with a familiar (a malevolent spirit in animal form, such as a cat) at their side.

The mania affected Catholic and Protestant countries alike. At the height of witch-hunt paranoia, from about 1500 to 1650, there were almost no women left alive in some Swiss villages. In regions of Germany there were some small villages where the whole population was burned at the stake. But in England and in Holland the witch craze never became as entrenched as in other parts of Europe. Torture was not allowed under English law, although suspected witches were subjected to the water test. Trussed and thrown into a pond, a true witch floated, to be retrieved and properly punished—by hanging. If the accused sank and drowned, she was considered to have been innocent of the charge of witchcraft—a comfort to the family but little use to the victim herself.

The witch-hunt terror faded only slowly. But with so many people

A Delft tile from the Netherlands (first part of the eighteenth century) showing a witch trial. The accused on the right, only her legs visible above the water, is sinking and would be proclaimed innocent. Satan's hand can be seen supporting the accused woman floating to the left, who—her guilt now proven—would be pulled from the water to be burned alive at the stake. (Courtesy of the Horvath Collection, Vancouver)

accused, economic well-being was threatened. As feudalism retreated and the Age of Enlightenment dawned, as the voices of brave men and women who themselves risked the gallows and the stake to oppose the madness became louder, the mania that had swept Europe for centuries gradually abated. In the Netherlands the last execution of a witch took place in 1610 and in England in 1685. The last witches executed in Scandinavia—eighty-five elderly women burned at the stake in 1699—were convicted solely on the basis of statements from young children who claimed to have flown with the women to sabbats.

By the eighteenth century, execution for witchcraft officially ceased: for Scotland in 1727, France in 1745, Germany in 1775, Switzerland in 1782, and Poland in 1793. But although the Church and the state no longer executed witches, the court of public opinion was less ready to give up the fear and loathing of witchcraft acquired by centuries of persecution. In more remote rural communities old beliefs still held sway, and many a suspected witch met a nasty, if unofficial, fate.

Many of the women accused of witchcraft were herbalists, skilled in the use of local plants to cure disease and provide relief from pain. Often they could also be relied upon to supply love potions, to cast spells, and to remove hexes. That some of their herbs did have healing powers would have seemed as magical as the incantations and rituals surrounding the rest of the ceremonies they would perform.

Using and prescribing herbal medicines would have been then—as it is now—a risky business. Different parts of a plant contain varying levels of effective compounds; plants gathered from different locations can vary in their ability to cure; and different times of the year can change the amount of a plant needed to produce an appropriate dose. Many plants in an elixir might be of little benefit, while others might contain medications that would be extremely effective but also deadly poisonous. The molecules in these plants could enhance the reputation of an herbalist as a sorcerer, but the very success of these molecules might eventually prove deadly for these women. Those herbalists whose healing skills were the greatest might be first to be branded a witch.

HEALING HERBS, HARMFUL HERBS

Salicylic acid, from the willow tree and the meadowsweet plant common throughout Europe, was known centuries before Bayer and Company began marketing aspirin in 1899 (see Chapter 10). The root of wild celery was prescribed to prevent muscle cramps, parsley was believed to induce a miscarriage, and ivy was used to relieve symptoms of asthma. Digitalis, an extract from the common foxglove *Digitalis purpurea,* contains molecules that have long been known to have a powerful effect on the heart—the *cardiac glycosides.* These molecules reduce the heart rate, regularize heart rhythm, and strengthen heartbeat, a potent combination in inexperienced hands. (They are also saponins, very similar to those found in sarsaparilla plants and wild Mexican yams from which the birth control pill norethindrone was synthesized; see Chapter 11.) An example of a cardiac glycoside is the digoxin molecule, one of the most widely prescribed drugs in the United States and a good example of a pharmaceutical based on folk medicine.

In 1795 a British physician named William Withering used extracts of foxglove to treat congestive heart failure after hearing rumors of the plant's curative abilities. But it was well over a century before chemists were able to isolate the molecules responsible.

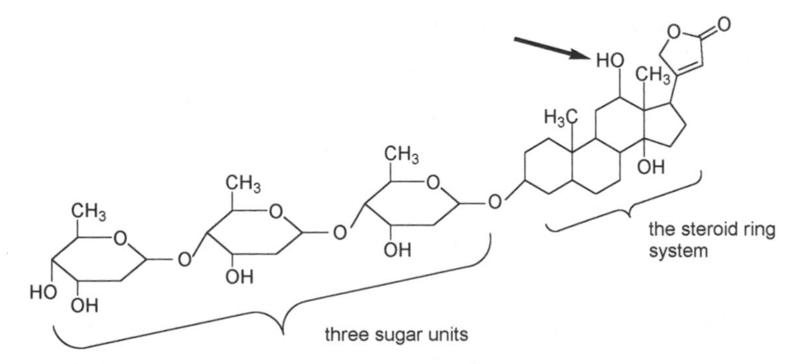

The structure of the digoxin molecule. The three sugar units are different from those in the sarsaparilla or Mexican yam plants. The digitoxin molecule lacks the arrowed OH group on the steroid ring system.

In the *Digitalis* extract there are other very similar molecules to digoxin; for example the digitoxin molecule, which lacks only the OH, as indicated in the structure drawing. Similar cardiac glycoside molecules are found in other plants, usually members of the lily and ranunculus families, but foxglove is still the main source for today's drug. Herbalists have had little difficulty finding heart tonic plants in their own gardens and in local meadows. Ancient Egyptians and Romans used an extract from the sea onion, a member of the hyacinth family, as a heart tonic and (in larger doses) as a rat poison. We now know that this sea onion also contains a different cardiac glycoside molecule.

These molecules all have the same structural feature, which is therefore likely to be responsible for the cardiac effect. All have a five-membered lactone ring attached to the end of the steroid system and an extra OH between the C and D rings of the steroid system, as shown here:

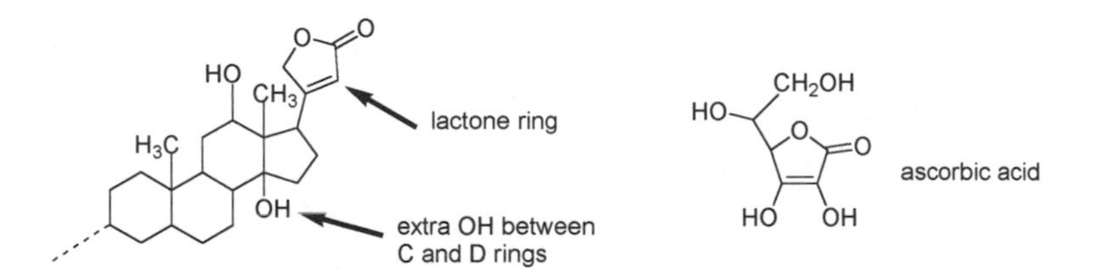

The nonsugar portion of the digoxin molecule with the heart-affecting extra OH and the lactone ring arrowed. This lactone ring is also found in the ascorbic acid molecule (vitamin C).

Molecules that affect the heart are not found only in plants. Toxic compounds that are similar in structure to the cardiac glycosides are found in animals. These molecules do not contain sugars, nor are they used as heart stimulants. Rather, they are convulsive poisons and of little medical value. The source of these venoms is amphibians; extracts from toads and frogs have been used as arrow poisons in many parts of the world. Interestingly, the toad is, after the cat, the most common animal attributed in folklore as a familiar to a witch. Many potions prepared by so-called witches were said to contain parts of toads. The molecule *bufotoxin* is the active component of the venom of the common European toad, *Bufo vulgaris,* and is one of the most toxic molecules known. Its structure shows a striking similarity at the steroid ring system to the digitoxin molecule, with the same extra OH between the C and D rings and a six-membered, instead of five-membered, lactone ring.

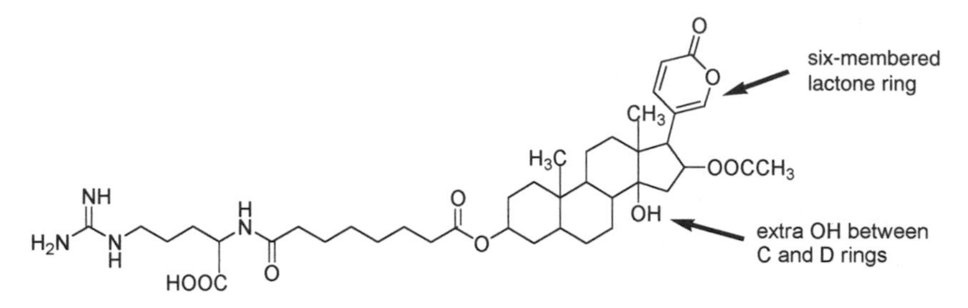

The bufotoxin from the common toad is structurally similar to digitoxin from the foxglove around the steroid portion of the molecule.

Bufotoxin, however, is a cardiac poison rather than a cardiac restorative. Between the cardiac glycosides of foxglove and venoms from toad, supposed witches had access to a potent arsenal of toxic compounds.

In addition to their penchant for toads, one of the most abiding myths about witches is that they were able to fly, often on broomsticks, to attend a sabbat—a midnight tryst, supposedly an orgiastic parody of the Christian mass. Many accused witches confessed, under torture, to flying to such sabbats. This is not surprising—we too would probably make such a confession if we were subject to the same horrific agonies perpetrated in the search for truth. The surprising thing is that a number of accused witches confessed, *before* torture, to the impossible feat of flying to a sabbat on a broomstick. As such a confession would not likely have helped these victims escape torture, it is quite possible that these women truly believed they had flown up the chimney on a broomstick and indulged in all sorts of sexual perversions. There may be a very good chemical explanation for their belief—a group of compounds known as alkaloids.

Alkaloids are plant compounds that have one or more nitrogen atoms, usually as part of a ring of carbon atoms. We have already met a few alkaloid molecules—piperine in pepper, capsaicin in chili peppers, indigo, penicillin, and folic acid. It can be argued that, as a group, alkaloids have had a larger effect on the course of human history than have any other family of chemicals. Alkaloids are often physiologically active in humans, usually affecting the central nervous system, and are generally highly toxic. Some of these naturally occurring compounds have been used as medicines for thousands of years. Derivatives made from alkaloids form the basis of a number of our modern pharmaceuticals, such as the pain-relieving molecule codeine, the local anesthetic benzocaine, and chloroquine, an antimalarial agent

We have already mentioned the role that chemical substances play in protecting plants. Plants cannot run away from danger and cannot hide at the first sign of a predator; physical means of protection such as thorns do not always stop determined grazers. Chemicals are a passive but very effective form of protection from animals as well as fungi, bacteria, and viruses. Alkaloids are natural fungicides, insecticides, and pesticides. It has been estimated that, on average, each of us ingests about a gram and a half of natural pesticide every day, from the plants and plant products in our diet. The estimate for residues from synthetic pesticides is around 0.15 milligrams daily—about ten thousand times less than the natural dose!

In small amounts the physiological effects of alkaloids are often welcomed by humans. Many have been used medicinally for centuries. Acrecaidine, an alkaloid found in betel nuts from the betel palm, *Areca catechu,* has a long history of use in Africa and the East as a stimulant. Crushed betel nuts are wrapped in the betel palm leaves and chewed. Betel users are easily recognized by their characteristic dark-stained teeth and by their habit of spitting copious amounts of dark red saliva. Ephedrine, from *Ephedra sinica* or the ma huang plant, has been used in Chinese herbal medicine for thousands of years and is now used in the West as a decongestant and bronchodilator. Members of the vitamin B family, such as thiamine (B_1), riboflavin (B_2), and niacin (B_4), are all classed as alkaloids. Reserpine, used in the treatment of high blood pressure and as a tranquilizer, is isolated from the Indian snakeroot plant, *Rauwolfia serpentina.*

Toxicity alone has been enough to ensure fame for some alkaloids. The poisonous component of the hemlock plant, *Conium maculatum,* responsible for the death of the philosopher Socrates in 399 B.C., is the alkaloid coniine. Socrates, convicted on charges of irreligion and the corruption of the young men of Athens, was sentenced to death by drinking a potion made from the fruit and seeds of hemlock. Coniine has one of the simplest structures of all the alkaloids, but it can be just as lethal a poison as more complicated alkaloid structures such as that of strychnine, from the seeds of the Asiatic tree *Strychnos nux-vomica.*

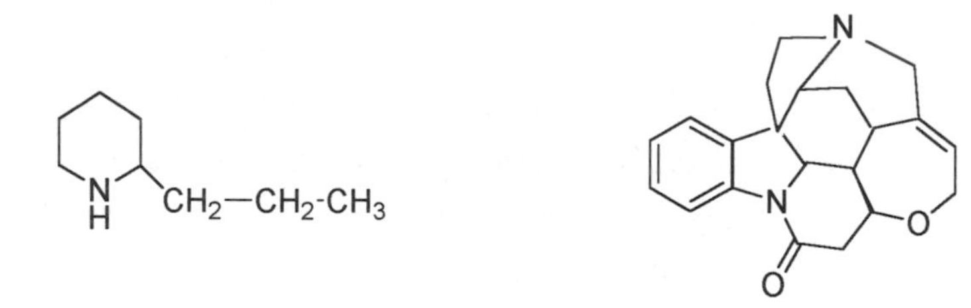

The structures of coniine (left) and strychnine (right)

In their "flying salves"—greases and ointments that supposedly promoted flight—witches often included extracts from mandrake, belladonna, and henbane. These plants all belong to the *Solanaceae* or nightshade family. The mandrake plant, *Mandragora officinarum,* with its branched root said to resemble the human form, is native to the Mediterranean region. It has been used since ancient times as a means of restoring sexual vitality and as a soporific. There are a number of curious legends surrounding the mandrake plant. When pulled from the ground, it was said to emit piercing screams. Whoever was in the vicinity was in danger from both the associated smell and the unearthly cry. That such a characteristic was common knowledge is reflected in Shakespeare's *Romeo and Juliet,* where Juliet says: ". . . with loathsome smells, and shrieks like mandrakes' torn out of the earth, / That living mortals, hearing them, run mad." The mandrake plant was rumored to grow beneath a gallows, springing to life from the released semen of the condemned men hanged there.

The second plant used in flying ointments was belladonna or deadly nightshade (*Atropa belladonna*). The name comes from the practice, common among women in Italy, of dropping juice squeezed from the black berries of this plant into their eyes. The resulting dilation of the pupil was thought to increase their beauty; hence *belladonna,* Italian for "beautiful lady." Greater amounts of deadly nightshade taken internally would eventually induce a deathlike slumber. It was probable that this was also commonly known and possible that this was the potion drunk by Juliet. Shakespeare wrote (in *Romeo and Juliet*) "through all thy veins shall run/A cold and drowsy humour, for no pulse shall keep," but eventually "in this borrow'd likeness of shrunk death/Thou shalt continue two and forty hours,/And then awake as from a pleasant sleep."

The third member of the nightshade family, henbane, was probably *Hyoscyamus niger,* though other species might also have been used in witch's potions. It has a long history as a soporific, a pain reliever (particularly for toothache), an anesthetic, and possibly a poison. The properties of henbane also seem to have been well known: Shakespeare was again only reflecting the common knowledge of his time when Hamlet was told by his father's ghost that "thy uncle stole, / With juice of cursed hebona in a vial, / And in the porches of mine ears did pour / The leperous distilment." The word *hebona* has been ascribed to both the yew and ebony trees as well as henbane, but from a chemical viewpoint we think henbane makes more sense.

Mandrake, deadly nightshade, and henbane all contain a number of very similar alkaloids. The two main ones, hyoscyamine and hyoscine, are found in all three plants in varying proportions. One form of hyoscyamine is known as atropine and is still valued today, in very dilute solutions, to dilate the pupil of the eye for ophthalmic examinations. Large concentrations produce blurry vision, agitation, and even delirium. One of the first symptoms of atropine poisoning is the drying up of bodily fluids. This property is taken advantage of in prescribing atropine, where excess saliva or mucus secretion may interfere with surgery. Hyoscine, also known as scopolamine, has gained a probably undeserved reputation as a truth serum.

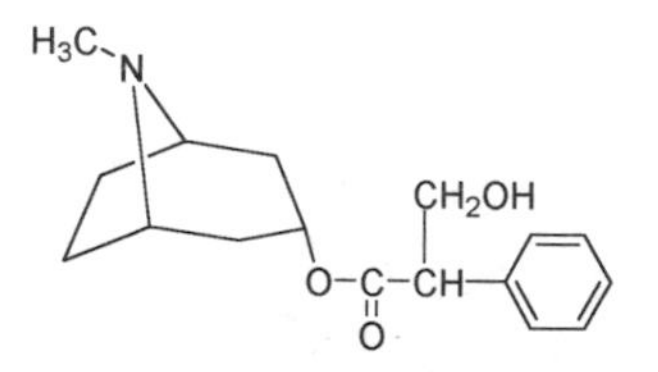

Atropine from hyoscyamine

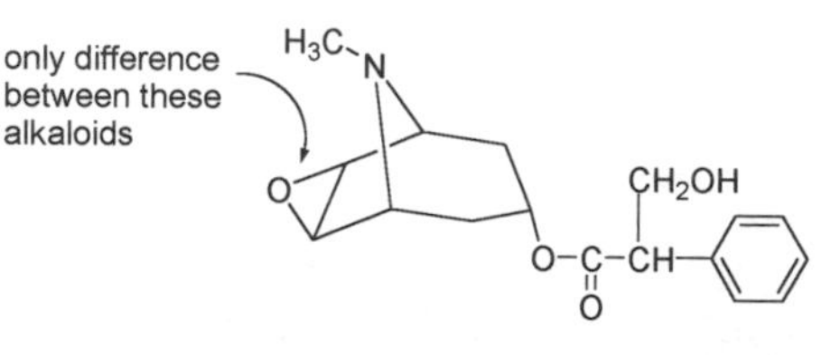

Scopolamine (hyoscine)

Combined with morphine, scopolamine is used as the anesthetic known as "twilight sleep," but whether one babbles the truth under its effect or just babbles is not clear. Still, writers of detective novels have always liked the thought of a truth serum, and it will probably continue to be quoted as such. Scopolamine, like atropine, has antisecretory and euphoric properties. In small amounts it combats travel sickness. U.S. astronauts use scopolamine as a treatment for motion sickness in space.

As bizarre as it might seem, the poisonous compound atropine acts as an antidote for groups of even more toxic compounds. Nerve gases such as sarin—released by terrorists in the Tokyo subway in April of 1995—and organophosphate insecticides, such as parathion, act by preventing the normal removal of a messenger molecule that transmits a signal across a nerve junction. When this messenger molecule is not removed, nerve endings are continuously stimulated, which leads to convulsions and, if the heart or lungs are affected, to death. Atropine blocks the production of this messenger molecule, so provided the right dosage is given, it is an effective remedy for sarin or parathion.

What is now known about the two alkaloids atropine and scopolamine, and was obviously known by the witches of Europe, is that neither is particularly soluble in water. As well, they would have recognized that swallowing these compounds might lead to death rather than the euphoric and intoxicating sensations they wanted. Hence extracts of mandrake, belladonna, and henbane were dissolved in fats or oils, and these greases were applied to the skin. Absorption through the skin—transdermal delivery—is a standard method of taking certain medications today. The nicotine patch for those trying to quit smoking and some travel sickness remedies and hormonal replacement therapies use this route.

As the records of witches' flying salves show, the technique was known hundreds of years ago as well. Today we know that the most efficient absorption is where the skin is the thinnest and blood vessels lie just under the surface; thus vaginal and rectal suppositories are used to ensure rapid absorption of medications. Witches must also have known this fact of anatomy, as flying ointments were said to be smeared all over the body or rubbed under the arms and, coyly, "in other hairy places." Some reports said witches applied the grease to the long handle of a broom and, sitting astride, rubbed the atropine-and-scopolamine-containing mixture onto the genital membranes. The sexual connotations of these accounts are obvious, as are the early engravings of naked or partially clothed witches astride broomsticks, applying salves and dancing around cauldrons.

The chemical explanation is, of course, that the supposed witches did not fly on broomsticks to sabbats. The flights were ones of fancy, illusions brought on by the hallucinatory alkaloids. Modern accounts of hallucinogenic states from scopolamine and atropine sound remarkably like the midnight adventures of witches: the sensation of flying or falling, distorted vision, euphoria, hysteria, a feeling of leaving the body, swirling surroundings, and encounters with beasts. The final stage of the process is a deep, almost comalike sleep.

It is not difficult to imagine how, in a time steeped in sorcery and superstition, users of flying ointments really did believe they had flown through the night sky and taken part in wild dancing and wilder revelries. The hallucinations from atropine and scopolamine have been described as particularly vivid. A witch would have no reason to believe the effects of her flying ointment were solely in her mind. It is also not difficult to imagine how the knowledge of this wonderful secret was passed on—and it would have been considered a wonderful secret. Life for most women in these times was hard. Work was never ending, disease and poverty ever present, and a woman's control over her own destiny unheard of. A few hours of freedom, riding the skies to a gathering where one's sexual fantasies were played out, then waking up safely in one's own bed must have been a great temptation. But unfortunately the temporary escape from reality created by the molecules of atropine and scopolamine often proved fatal, as women accused of witchcraft who confessed to such imagined midnight exploits were burned at the stake.

Along with mandrake, deadly nightshade, and henbane, other plants were included in flying ointments: foxglove, parsley, monkshood, hemlock, and thorn apple are listed in historical accounts. There are toxic alkaloids in monkshood and hemlock, toxic glycosides in foxglove, hallucinogenic myristicin in parsley, and atropine and scopolamine in thorn apple. Thorn apple is a *Datura;* devil's apple, angel's trumpet, stinkweed, and jimsonweed are some of the other plants in this genus. Now widely distributed in the warmer parts of the world, *Datura* furnished alkaloids for witches in Europe as well as for initiation rites and other ceremonial occasions in Asia and the Americas. Folklore associated with *Datura* usage in these countries reveals hallucinations involving animals, a very common aspect of witches' flights. In parts of Asia and Africa *Datura* seeds are included in mixtures to be smoked. Absorption into the bloodstream through the lungs is a very rapid method of obtaining a "hit" from an alkaloid, as European tobacco smokers later discovered in the sixteenth century. Cases of atropine poisoning are still reported today, with thrill-seekers using flowers, leaves, or seeds of *Datura* to pursue a high.

A number of plants from the nightshade family were introduced into Europe from the New World soon after the journeys of Columbus. Some that contained alkaloids—tobacco (*Nicotiana*) and peppers (*Cap-*

sicum)—gained immediate acceptance, but surprisingly, other members of this family—tomatoes and potatoes—were initially regarded with great suspicion.

Other alkaloids that are chemically similar to atropine are found in the leaves of several species of *Erythroxylon,* the coca tree, native to parts of South America. The coca tree is not a member of the nightshade family—an unusual situation, as related chemicals are normally found in related species. But historically plants were classified on morphological features. Revisions now consider chemical components and DNA evidence.

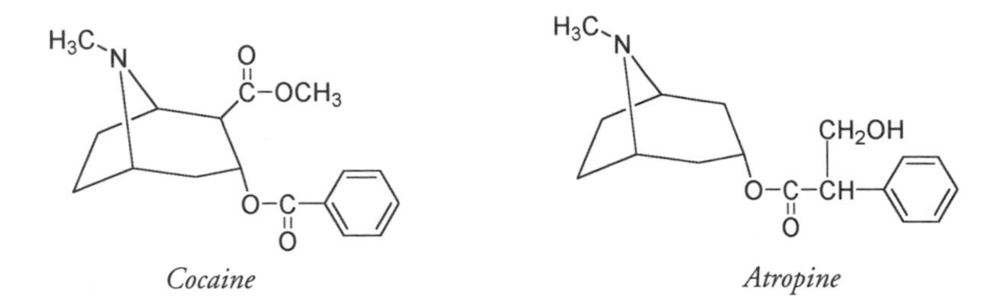

Cocaine *Atropine*

The main alkaloid in the coca tree is cocaine. Coca leaves have been used as a stimulant for hundreds of years in the highland areas of Peru, Ecuador, and Bolivia. The leaves are mixed with a paste of lime, then tucked between the gum and the cheek, where the alkaloids, released slowly, help counter fatigue, hunger, and thirst. It has been estimated that the amount of cocaine absorbed this way is less than half a gram daily, which is not addictive. This traditional method of coca alkaloid use is similar to our use of the alkaloid caffeine in coffee and tea. But cocaine, extracted and purified, is a different matter.

Isolated in the 1880s, cocaine was considered to be a wonder drug. It had amazingly effective local anesthetic properties. Psychiatrist Sigmund Freud considered cocaine a medical panacea and prescribed it for its stimulating properties. He also used it to treat morphine addiction. But it soon became obvious that cocaine itself was extremely addictive, as addictive as any other known substance. It produces a rapid and extreme euphoria, followed by an equally extreme depression, leaving the user craving another euphoric high. The disastrous consequences on human health and modern society of abuse of cocaine are well known. The cocaine structure is, however, the basis for a number of extremely useful molecules developed as topical and local anesthetics. Benzocaine, novocaine, and lidocaine are compounds that mimic the pain-destroying action of cocaine by blocking transmission of nerve impulses, but they lack cocaine's ability to stimulate the nervous system or disrupt heart rhythm. Many of us have thankfully experienced the numbing effect of these compounds in the dentist's chair or the hospital emergency room.

THE ERGOT ALKALOIDS

Another group of alkaloids of quite different structure was probably, although indirectly, responsible for thousands of witch burnings in Europe. But these compounds were not used in hallucinogenic ointments. The effects of some of the alkaloid molecules from this group can be so devastating that whole communities, afflicted with horrendous suffering, assumed that the catastrophe was the result of an evil spell cast by local witches. This group of alkaloids is found in the ergot fungus, *Claviceps purpurea,* that infects many cereal grains but especially rye. Ergotism or ergot poisoning was until fairly recently the next-largest microbial killer after bacteria and viruses. One of these alkaloids, ergotamine, causes blood vessels to constrict; another, ergonovine, induces spontaneous abortions in humans and livestock; while others cause neurological disturbances. Symptoms of ergotism vary depending on the amount of the different ergot alkaloids present but can include convulsions, seizures, diarrhea, lethargy, manic behavior, hallucinations, distortion of the limbs, vomiting, twitching, a crawling sensation on the skin, numbness in the hands and feet, and a burning sensation becoming excruciatingly painful as gangrene from decreased circulation eventually sets in. In medieval times this disease was known by various names: holy fire, Saint Anthony's fire, occult fire, and Saint Vitus' dance. The reference to fire relates to the terrible searing pain and blackened extremities caused by the progression of gangrene. Often there was loss of hands, feet, or genitals. Saint Anthony was considered to have special powers against fire, infection, and epilepsy, making him the saint to appeal to for relief from ergotism. The "dance" of Saint Vitus' dance refers to twitching and convulsive contortions due to the neurological effects of some of the ergot alkaloids.

It is not hard to envisage a situation where a large number of villagers or townsfolk were struck by ergotism. A particularly rainy period just before harvest would encourage fungus growth on rye; poor storage of the cereal in damp conditions would promote further growth. Only a small percent of ergot in flour is needed to cause ergot poisoning. As more and more of a town's inhabitants displayed the dreaded symptoms, people might start to wonder why their community had been singled out for disaster, especially as adjacent towns had no sign of the disease. It could have seemed quite plausible that their village had been bewitched. As in many natural disasters, the blame was often placed on the innocent head of an elderly woman, someone who was no longer useful for childbearing and who may have had no family support. Such women often lived on the outskirts of the community, perhaps surviving on their skills as herbalists and unable to afford even the modest sum required to purchase flour from the miller in town. This level of poverty would have saved a woman from ergotism but ironically, as maybe the only person untouched by the ergot poisons, she became even more vulnerable to the accusation of witchcraft.

Ergotism has been known for a long time. Its cause was hinted at in reports from as early as 600 B.C., when the Assyrians noted "a noxious pustule in the ear of grain." That ergot alkaloids from "noxious grasses" could cause miscarriages in cattle was recorded in Persia around 400 B.C. In Europe the knowledge that fungus or mold on grains was the cause of the problem seems to have been lost—if it was ever known—

during the Middle Ages. With damp winters and improper storage, mold and fungus flourished. In the face of famine, infected grain would have been used rather than discarded.

The first recorded occurrence of ergotism in Europe, in A.D. 857, is from Germany's Rhine valley. Documented reports of forty thousand deaths in France in the year 994 are now attributed to ergotism, as are another twelve thousand in 1129. Periodic outbreaks occurred throughout the centuries and continued into the twentieth century. In 1926–1927 more than eleven thousand people were afflicted with ergotism in an area of Russia near the Ural Mountains. Two hundred cases were reported in England in 1927. In Provence, France, in 1951, four died and hundreds more became ill from ergotism after ergot-infected rye was milled and the flour sold to a baker, although the farmer, miller, and baker were supposedly all aware of the problem.

There are at least four occasions when ergot alkaloids are claimed to have played a role in history. During a campaign in Gaul, in the first century B.C., an epidemic of ergotism among Julius Caesar's legions caused great suffering, reduced the effectiveness of his army, and possibly curtailed Caesar's ambitions to enlarge the Roman Empire. In the summer of 1722 Peter the Great's Cossacks camped at Astrakhan, at the mouth of the Volga River on the Caspian Sea. Both the soldiers and their horses ate contaminated rye. The resulting ergotism supposedly killed twenty thousand troops and so crippled the tsar's army that his planned campaign against the Turks was aborted. Thus Russia's goal of a southern port on the Black Sea was stopped by ergot alkaloids.

In France, in July 1789, thousands of peasants rioted against wealthy landowners. There is evidence that this episode, termed *La Grande Peur* (the Great Fear), was more than just civil unrest associated with the French Revolution. Records attribute the destructive spree to a bout of insanity in the peasant population and cite "bad flour" as a possible cause. The spring and summer of 1789 in northern France had been abnormally wet and warm—perfect conditions for the growth of the ergot fungus. Was ergotism, much more prevalent among the poor, who ate moldy bread out of necessity, a key factor in the French Revolution? Ergotism was also reported to be rife in Napoleon's army during its journey across the Russian plains in the fall of 1812. So maybe the ergot alkaloids, along with the tin in uniform buttons, share some responsibility for the Grande Armée's collapse on the retreat from Moscow.

A number of experts have concluded that ergot poisoning was ultimately responsible for the accusations of witchcraft against some 250 people (mainly women) during 1692 in Salem, Massachusetts. The evidence does indicate an involvement of the ergot alkaloids. Rye was grown in the area in the late seventeenth century; records show warm, rainy weather during the spring and summer of 1691; and the village of Salem was located close to swampy meadows. All of these facts point to the possibility of fungal infestation of the grain used for the community's flour. The symptoms displayed by the victims were consistent with ergotism, particularly convulsive ergotism: diarrhea, vomiting, convulsions, hallucinations, seizures, babbling, bizarre distortions of the limbs, tingling sensations, and acute sensory disturbances.

It seems probable that, at least initially, ergotism may have been the cause of the Salem witch-hunts; almost all the thirty victims who claimed to be bewitched were girls or young women, and young people are known to be more susceptible to the effects of the ergot alkaloids. Later events, however, including the trials of the alleged witches and an increasing number of accusations, often of people outside the community, point more to hysteria or just plain malice.

Symptoms of ergot poisoning cannot be turned off and on. The common phenomenon in the trials—victims throwing a convulsive fit when confronted by the accused witch—is not consistent with ergotism. No doubt relishing the attention and realizing the power they wielded, the so-called victims would denounce neighbors they knew and townsfolk they had scarcely heard of. The suffering of the real victims of the Salem witch-hunts—the nineteen hanged (and one pressed to death by a pile of rocks), those tortured and imprisoned, the families destroyed—may be traced to ergot molecules, but human frailty must bear ultimate responsibility.

Like cocaine, ergot alkaloids, although toxic and dangerous, have had a long history of therapeutic use, and ergot derivatives still play a role in medicine. For centuries herbalists, midwives, and doctors used extracts of ergot to hasten childbirth or produce abortions. Today ergot alkaloids or chemical modifications of these compounds are used as vasoconstrictors for migraine headaches, to treat postpartum bleeding, and as stimulants for uterine contractions in childbirth.

The alkaloids of ergot all have the same common chemical feature; they are derivatives of a molecule known as lysergic acid. The OH group (indicated with an arrow, below) of lysergic acid is replaced by a larger side group, as shown in the ergotamine molecule (used to treat severe migraine headaches) and the ergovine molecule (used to treat post-partum hemorrhages). In these two molecules the lysergic acid portion is circled.

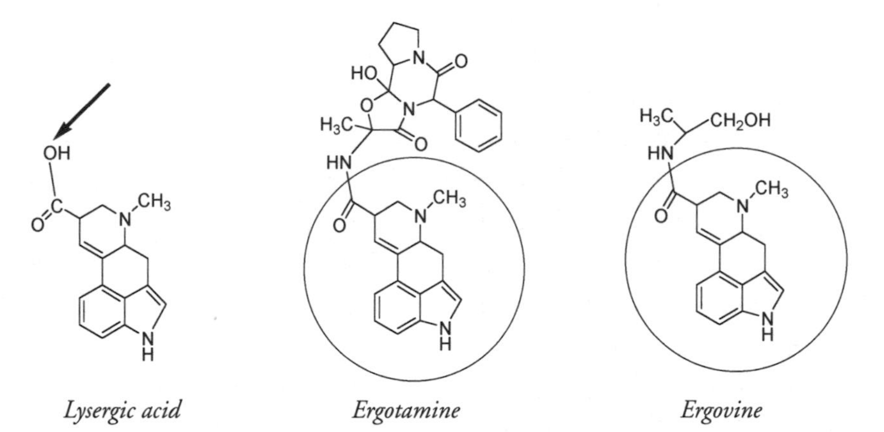

Lysergic acid *Ergotamine* *Ergovine*

In 1938, having already prepared a number of synthetic derivatives of lysergic acid of which some had proved useful, Albert Hofmann, a chemist working in the research laboratories of the Swiss pharmaceutical company Sandoz, in Basel, prepared another derivative. It was the twenty-fifth derivative he had made, so he called lysergic acid diethylamide LSD-25, now known, of course, as just LSD. Nothing exceptional was noted about the properties of LSD.

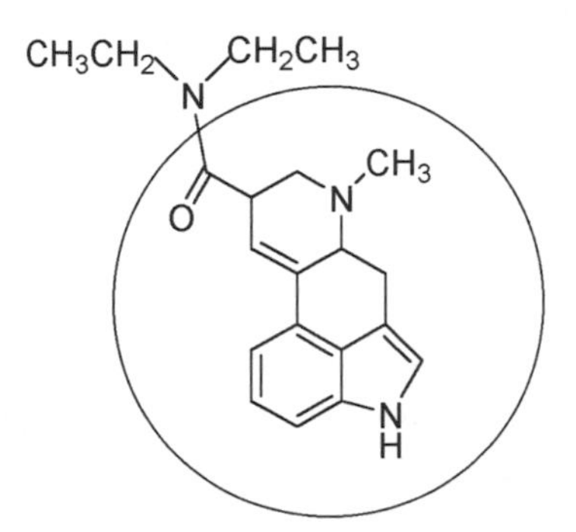

Lysergic acid diethylamide (LSD-25), or LSD as it was to become known. The lysergic acid part is circled.

It was not until 1943, when Hofmann again made this derivative, that he inadvertently experienced the first of what was to become known in the 1960s as an acid trip. LSD is not absorbed through the skin, so Hofmann probably transferred LSD from his fingers to his mouth. Even a slight trace would have produced what he described as an experience of "an uninterrupted stream of fantastic pictures, extraordinary shapes with intense kaleidoscope play of colors."

Hofmann decided to deliberately take LSD to test his assumption that this was the compound producing the hallucinations. The medical dosage for lysergic acid derivatives such as ergotamine was at least a few milligrams. Thinking, no doubt, that he was being cautious, he swallowed only a quarter of a milligram, an amount at least five times that needed to produce the now well-known hallucinogenic effects. LSD is ten thousand times more potent as a hallucinogen than naturally occurring mescaline, found in the peyote cactus of Texas and northern Mexico and used for centuries by native Americans in their religious ceremonies.

Rapidly growing dizzy, Hofmann asked his assistant to accompany him as he rode his bicycle home through the streets of Basel. For the next few hours he went through the full range of experiences that later users came to know as a bad trip. As well as having visual hallucinations, he became paranoid, alternated between feelings of intense restlessness and paralysis, babbled incoherently, feared choking, felt he had left his body, and perceived sounds visually. At some point Hofmann even considered the possibility that he might have suffered permanent brain damage. His symptoms gradually subsided, though his visual disturbances persisted for some time. Hofmann awoke the morning after this experience feeling totally normal, with a complete memory of what had happened but seemingly with no side effects.

In 1947 the Sandoz company began to market LSD as a tool in psychotherapy and in particular for the treatment of alcoholic schizophrenia. In the 1960s LSD became a popular drug for young people around the world. It was promoted by Timothy Leary, a psychologist and onetime member of the Harvard University Center for Research in Personality, as the religion of the twenty-first century and the way to spiritual and creative fulfillment. Thousands followed his advice to "turn on, tune in, drop out." Was this alkaloid-induced escape from everyday life in the twentieth century so different from that experienced by women accused of witchcraft a few hundred years before? Though centuries apart, the psychedelic experiences were not always positive. For the flower children of the 1960s, taking the alkaloid-derivative LSD could lead to flashbacks, permanent psychoses, and in extreme cases suicide; for the witches of Europe, absorbing the alkaloids atropine and scopolamine from their flying salves could lead to the stake.

The atropine and ergot alkaloids did not cause witchcraft. Their effects, however, were interpreted as evidence against large numbers of innocent women, usually the poorest and most vulnerable in society. Accusers would make a chemical case against the witch: "She must be a witch, she says she can fly" or "she must be guilty, the whole village is bewitched." The attitudes that had allowed four centuries of persecution of women as witches did not change immediately once the burnings were stopped. Did these alkaloid molecules contribute to a perceived heritage of prejudice against women—a view that may still linger in our society?

In medieval Europe the very same women who were persecuted kept alive the important knowledge of medicinal plants, as did native people in other parts of the world. Without these herbal traditions we might never have produced our present-day range of pharmaceuticals. But today, while we no longer execute those who value potent remedies from the plant world, we are eliminating the plants instead. The continuing loss of the world's tropical rain forests, now estimated at almost two million hectares each year, may deprive us of the discovery of other alkaloids that would be even more effective in treating a variety of conditions and diseases.

We may never know that there are molecules with antitumor properties, that are active against HIV, or that could be wonder drugs for schizophrenia, for Alzheimer's or Parkinson's disease in the tropical plants that are daily becoming closer to extinction. From a molecular point of view, the folklore of the past may be a key to our survival in the future.

13. MORPHINE, NICOTINE, AND CAFFEINE

GIVEN THE HUMAN tendency to desire those things that make us feel good, it is not surprising that three different alkaloid molecules—morphine from the opium poppy, nicotine in tobacco, and caffeine in tea, coffee, and cocoa—have been sought out and prized for millennia. But for every benefit these molecules have brought to mankind, they have also offered danger. Despite, or maybe because of, their addictive nature, they have affected many different societies in many ways. And all three came together unexpectedly at one intersection in history.

THE OPIUM WARS

Although it is nowadays mainly associated with the Golden Triangle—the border region of the countries of Burma, Laos, and Thailand—the opium poppy, *Papaver somniferum,* is native to the eastern Mediterranean region. The products of the opium poppy may have been gathered and appreciated since prehistoric times. Evidence suggests that more than five thousand years ago the properties of opium were known in the Euphrates River delta, generally credited as the site of the first recognizable human civilization. Archaeological indications of the use of opium at least three thousand years ago have been unearthed in Cyprus. Opium was included in the herbal lists and healing remedies of the Greeks, Phoenicians, Minoans, Egyptians, Babylonians, and other civilizations of antiquity. Supposedly around 330 B.C., Alexander the Great took opium to Persia and India, from where cultivation slowly spread eastward and reached China in about the seventh century.

For hundreds of years opium remained a medical herb, either drunk as a bitter infusion or swallowed as a rolled pellet. By the eighteenth and particularly the nineteenth centuries artists, writers, and poets in Europe and the United States used opium to achieve a dreamlike state of mind that was thought to enhance creativity. Being less expensive than alcohol, opium also found a use by the poor as a cheap intoxicant. During these years its habit-forming qualities, if recognized, were seldom a concern. So pervasive was its use that even small babies and teething infants were dosed with opium preparations that were advertised as soothing syrups and cordials and that contained as much as 10 percent morphine. Laudanum, a solution of opium in alcohol often recommended for women, was widely consumed and available at any pharmacy without a prescription. It was a socially acceptable form of opium until it was prohibited in the early twentieth century.

In China opium had been a respected medicinal herb for hundreds of years. But the introduction of a new alkaloid-bearing plant, tobacco, changed the role of opium in Chinese society. Smoking was unknown in Europe until Christopher Columbus, at the end of his second voyage in 1496, brought tobacco back from the New World, where he had seen it in use. Tobacco use spread rapidly, despite severe penalties for its possession or importation in many Asian and Middle Eastern countries. In China in the middle of the seventeenth century the last emperor of the Ming dynasty prohibited the smoking of tobacco. Possibly the Chinese started to smoke opium as a substitute for the banned tobacco, as some reports suggest. Other historians credit the Portuguese from small trading posts on Formosa (now Taiwan) and Amoy in the East China Sea with introducing Chinese merchants to the idea of mixing opium with tobacco.

The effect of alkaloids such as morphine and nicotine, absorbed directly into the bloodstream through smoke inhaled into the lungs, is extraordinarily rapid and intense. When taken in this manner, opium quickly becomes addictive. By the beginning of the eighteenth century the smoking of opium was widespread throughout China. In 1729 an imperial edict banned the importation and sale of opium in China, but it was probably too late. An opium-smoking culture and a vast opium-related network of distribution and marketing already existed.

This is where our third alkaloid, caffeine, enters the story. Traders from Europe had previously found little satisfaction in trading with China. There were few commodities that China was willing to buy from the West, least of all the manufactured goods that the Dutch, British, French, and other European trading nations wanted to sell. But Chinese exports were in demand in Europe, particularly tea. Probably caffeine, the mildly addictive alkaloid molecule in tea, fueled the insatiable appetite of the West for the dried leaves of a shrub that had been grown since antiquity in China.

The Chinese were quite prepared to sell their tea, but they wanted to be paid in silver coin or bullion. For the British, buying tea with valuable silver was not their definition of trade. It soon became apparent that there was one commodity, though illegal, that the Chinese wanted and did not have. Thus Britain entered the opium business. Opium, cultivated in Bengal and other parts of British India by agents of the British East India Company, was sold to independent traders. It was then resold to Chinese importers, often under the protection of bribed Chinese officials. In 1839 the Chinese government attempted to halt this outlawed but flourishing trade. It confiscated and destroyed a year's supply of opium located in warehouses in Canton (present-day Guangzhou) and in British ships awaiting unloading in Canton's harbor. Only days later a group of drunken British sailors was accused of killing a local farmer, giving the British an excuse to declare war on China. British victory in what is now called the First Opium War (1839–1842) changed the balance of trade between the nations. China was required to pay a very large amount in reparations, to open five Chinese ports to British trade, and to cede Hong Kong as a British crown colony.

Nearly twenty years later another Chinese defeat in the Second Opium War, involving the French as well as the British, wrung further concessions from China. More ports were opened to foreign trade, Europeans were allowed the right of residence and travel, freedom of movement was given to Christian missionaries, and ultimately the opium trade was legalized. Opium, tobacco, and tea became responsible for breaking down centuries of Chinese isolation. China entered a period of upheaval and change that culminated in the Revolution of 1911.

In the Arms of Morpheus

Opium contains twenty-four different alkaloids. The most abundant one, morphine, makes up about 10 percent of crude opium extract, a sticky, dried secretion from the poppy flowerpod. Pure morphine was first isolated from this poppy latex in 1803 by a German apothecary, Friedrich Serturner. He named the compound that he obtained morphine, after Morpheus, the Roman god of dreams. Morphine is a narcotic, a molecule that numbs the senses (thus removing pain) and induces sleep.

Intense chemical investigation followed Serturner's discovery, but the chemical structure of morphine was not finally determined until 1925. This 122-year delay should not be seen as unproductive. On the contrary, organic chemists often view the actual deciphering of the structure of morphine as equally beneficial to mankind as the well-known pain-relieving effects of this molecule. Classical methods of structure determination, new laboratory procedures, an understanding of the three-dimensional nature of carbon compounds, and new synthetic techniques were just some of the results of the unraveling of this marathon chemical puzzle. Structures of other important compounds have been deduced because of the work done on the composition of morphine.

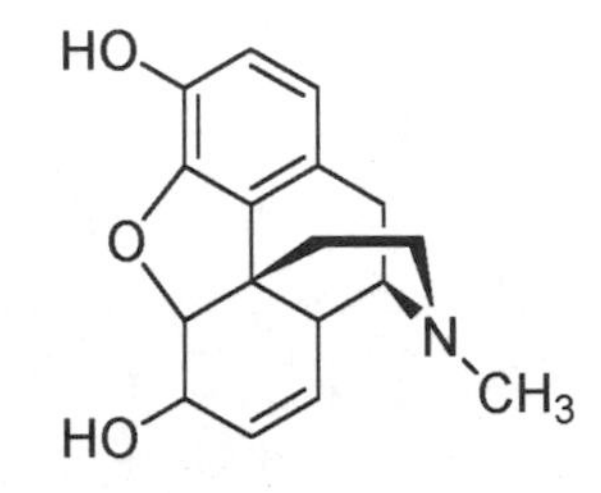

The structure of morphine. The darker lines of the wedge-shaped bonds point out of (above) the plane of the paper.

Today morphine and its related compounds are still the most effective painkillers known. Unfortunately, the painkilling or analgesic effect seems to be correlated with addiction. Codeine, a similar compound found in much smaller quantities (about 0.3 to 2 percent) in opium, is less addictive but is also a less powerful analgesic. The difference in structure is very small; codeine has a CH_3O that replaces the HO at the position shown by the arrow on the structure below.

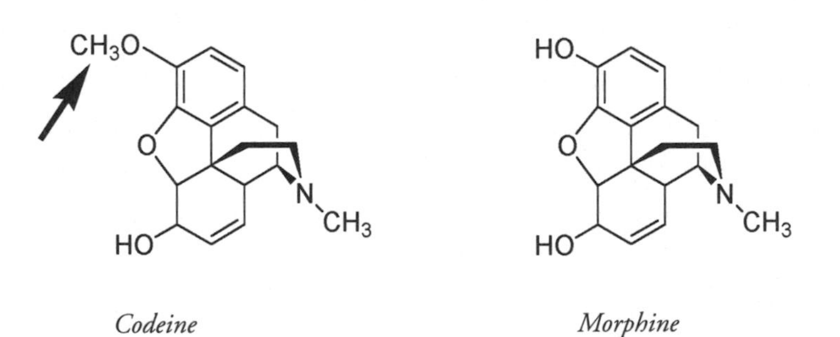

The structure of codeine. The arrow points to the only difference between codeine and morphine.

Well before the complete structure of morphine was known, attempts were made to modify it chemically in the hope of producing a compound that was a better pain reliever without addictive properties. In 1898, at the laboratory of Bayer and Company, the German dye manufacturer where, five years before, Felix Hofmann had treated his father with acetyl salicylic acid, chemists subjected morphine to the same acylation reaction that had converted salicylic acid to aspirin. Their reasoning was logical. Aspirin had turned out to be an excellent analgesic and a lot less toxic than salicylic acid.

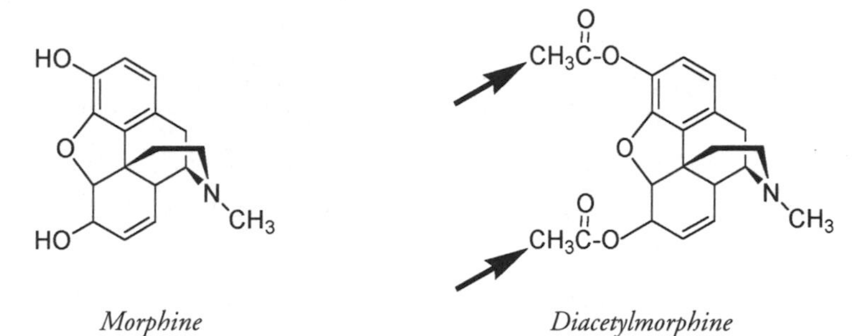

The diacetyl derivative of morphine. The arrows indicate where CH_3CO has replaced the Hs in the two HOs of morphine, producing heroin.

The product of replacing the Hs of the two OH groups of morphine with CH_3CO groups was, however, a different matter. At first the results seemed promising. Diacetylmorphine was an even more powerful narcotic than morphine, so effective that extremely low doses could be given. But its effectiveness masked a major problem, obvious when the commonly accepted name for diacetylmorphine is known. Originally marketed as Heroin—the name refers to a "hero" drug—it is one of the most powerfully addictive substances known. The physiological effects of morphine and heroin are the same; inside the brain the diacetyl groups of heroin are converted back to the original OH groups of morphine. But the heroin molecule is more easily transported across the blood-brain barrier than is morphine, producing the rapid and intense euphoria craved by those who become addicted.

Bayer's Heroin, initially thought to be free of the common side effects of morphine like nausea and constipation and therefore assumed also to be free of addiction, was marketed as a cough suppressant and a remedy for headaches, asthma, emphysema, and even tuberculosis. But as the side effects of their "super aspirin" became obvious, Bayer and Company quietly stopped advertising it. When the original patents for acetyl salicylic acid expired in 1917 and other companies began producing aspirin, Bayer sued for breach of copyright over the name. Not surprisingly, Bayer has never sued for copyright violation of the Heroin trade name for diacetylmorphine.

Most countries now ban the importation, manufacture, or possession of heroin. But this has done little to stop the illegal trade in this molecule. Laboratories that are set up to manufacture heroin from morphine often have a major problem disposing of acetic acid, one of the side products of the acylation reaction. Acetic acid has a very distinctive smell, that of vinegar, which is a 4 percent solution of this acid. This smell often alerts authorities to the existence of an illicit heroin manufacturer. Specially trained police dogs can detect faint traces of vinegar odor below the level of human sensitivity.

Investigation into why morphine and similar alkaloids are such effective pain relievers suggests that morphine does not interfere with nerve signals to the brain. Instead it selectively changes how the brain receives these messages—that is, how the brain perceives the pain being signaled. The morphine molecule seems able to occupy and block a pain receptor in the brain, a theory that correlates with the idea that a certain shape of chemical structure is needed to fit into a pain receptor.

Morphine mimics the action of endorphins, compounds found in very low concentrations in the brain that serve as natural pain relievers and that increase in concentration in times of stress. Endorphins are polypeptides, compounds made from joining amino acids together, end to end. This is the same peptide formation that is responsible for the structure of proteins such as silk (see Chapter 6). But whereas a silk molecule has hundreds or even thousands of amino acids, endorphins consist of only a few. Two endorphins that have been isolated are pentapeptides, meaning that they contain five amino acids. Both of these endorphin pentapeptides and morphine have a structural feature in common: they all contain a β-phenylethylamine unit, the same chemical construction that is thought responsible for affecting the brain in LSD, in mescaline, and in some other hallucinogenic molecules.

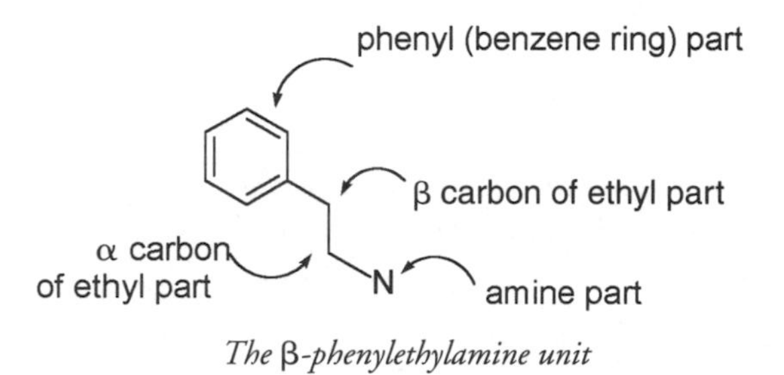

The β-phenylethylamine unit

Though the pentapeptide endorphin molecules are otherwise quite unlike the morphine molecule, this structural similarity is thought to account for the common binding site in the brain.

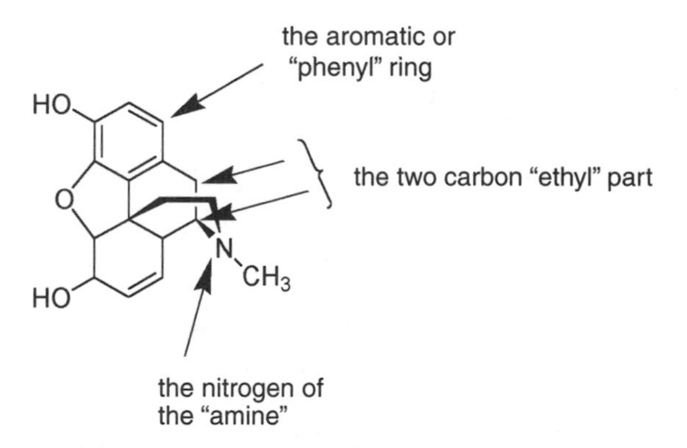

Structure of the morphine molecule, showing the β-phenylethylamine unit

But morphine and its analogs differ in biological activity from other hallucinogens in that they also have narcotic effects—pain-relief, sleep-inducing, and addictive components. These are thought to be due to another combination found in the chemical structure of, taken in order: (1) a phenyl or aromatic ring, (2) a quaternary carbon atom; that is, a carbon atom directly attached to four other carbon atoms, (3) a

CH_2-CH_2 group attached to (4) a tertiary N atom (a nitrogen atom directly attached to three other carbon atoms).

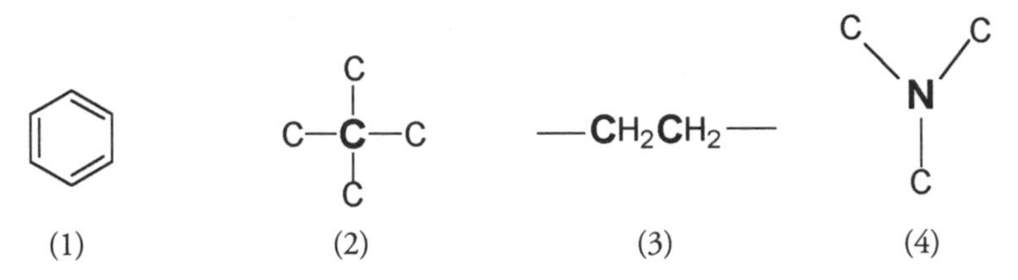

(1) The benzene ring, (2) quaternary carbon atom (bolded), (3) the two CH_2 groups with the carbons bolded, and (4) the tertiary nitrogen atom (bolded)

Combined, this set of requirements—known as the morphine rule—looks like:

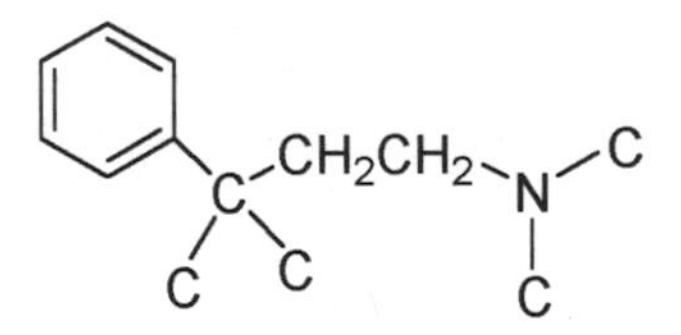

Essential components for the morphine rule

You can see in the diagrams of morphine that all four requirements are present, as they also are in codeine and heroin.

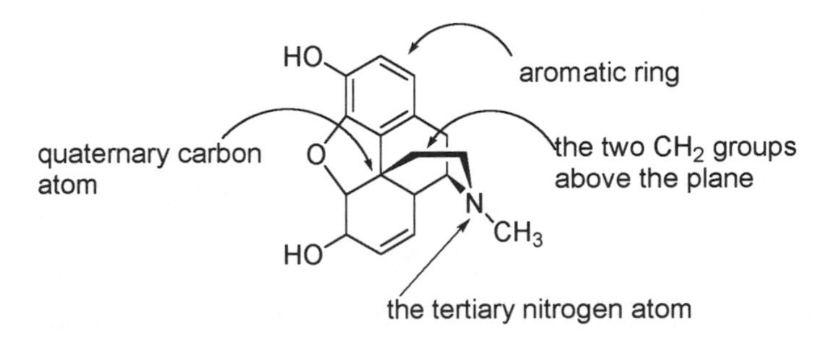

Morphine structure, showing how it fits the morphine rule for biological activit.

The discovery that this part of the molecule might account for narcotic activity is another example of serendipity in chemistry. Investigators injecting a man-made compound, meperidine, into rats noted that it caused them to hold their tails in a certain way, an effect previously seen with morphine.

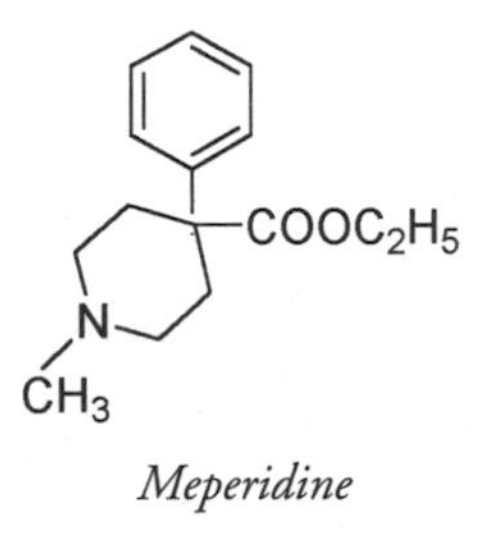

Meperidine

The meperidine molecule was not particularly similar to the morphine molecule. What meperidine and morphine did have in common were (1) an aromatic or phenyl ring attached to (2) a quaternary carbon, followed by (3) the CH_2-CH_2 group and then a tertiary nitrogen; in other words, the same arrangement that was to become known as the morphine rule.

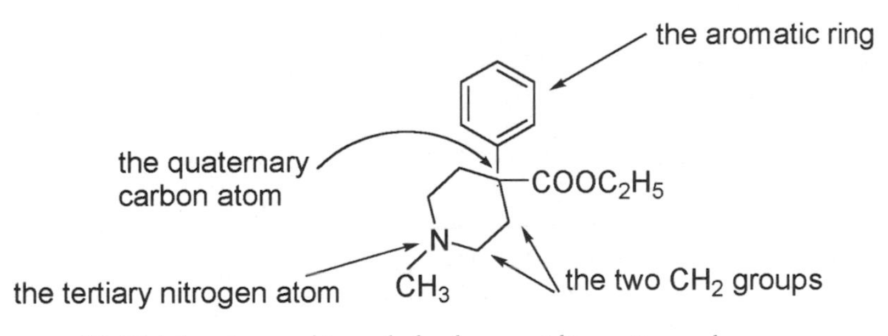

Highlighting the morphine rule for the meperidine or Demerol structure

Testing of meperidine revealed that it had analgesic properties. Usually known by the trade name Demerol, it is often used instead of morphine as, although it is less effective, it is unlikely to cause nausea. But it is still addictive. Another synthetic and very potent analgesic, methadone, like heroin and morphine, depresses the nervous system but does not produce the drowsiness or the euphoria associated with opiates. The structure of methadone is not a complete match to the requirements for the morphine rule. There is a CH_3 group attached to the second carbon atom of the $-CH_2-CH_2-$. This very small change in structure is presumed responsible for the difference in biological activity.

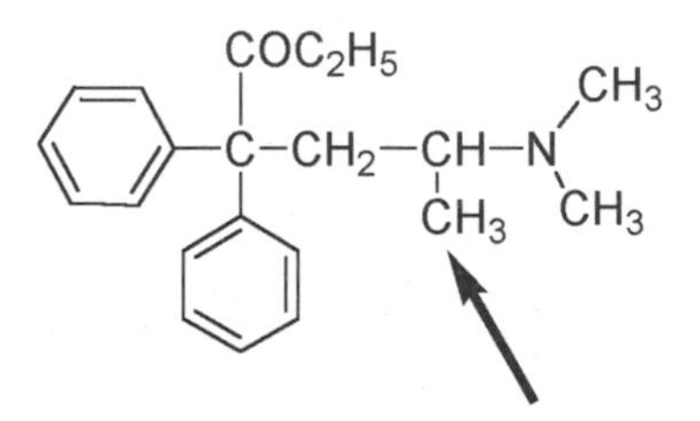

The structure of methadone. The arrow shows the position of the CH_3 group, the only deviation from the morphine rule but enough to change the physiological effect.

Methadone is, however, still addictive. Dependence on heroin can be transferred to dependence on methadone, but whether this is a reasonable method of dealing with the problems associated with heroin addiction is still a subject of debate.

DRINKING SMOKE

Nicotine, the second alkaloid associated with the Opium Wars, was unknown in Europe when Christopher Columbus landed in the New World. There he saw both men and women who "drank" or inhaled the smoke of rolls of burning leaves inserted into their nostrils. Smoking, snuffing (inhaling a powdered form into the nose), and chewing leaves of tobacco plants, species of the genus *Nicotiana,* was widespread among the Indians of South America, Mexico, and the Caribbean. The use of tobacco was mainly ceremonial; tobacco smoke sucked from pipes or from rolled leaves, or inhaled directly from foliage strewn over glowing embers, was said to cause trances or hallucinations in participants. This would have meant that their tobacco had significantly higher concentrations of active ingredients than are found in the species *Nicotiana tabacum* that was introduced to Europe and the rest of the world. The tobacco noted by Columbus was most probably *Nicotiana rustica,* the tobacco of the Mayan civilization, which is known to be a more potent variety.

The use of tobacco spread quickly throughout Europe, and tobacco cultivation soon followed. Jean Nicot, the French ambassador to Portugal whose name is commemorated in the botanical name of the plant and the name of the alkaloid, was a tobacco enthusiast, as were other notable sixteenth-century figures: Sir Walter Raleigh in England and

An engraving from Brazil (around 1593) is the first copperplate showing smoking in South America. A plant is smoked through a long tube at this Tupi Indian feast. (Courtesy of John G. Lord Collection)

Catherine de Médicis, Queen of France. Smoking did not, however, meet with universal approval. Papal edicts banned the use of tobacco in church, and King James I of England is said to have authored a 1604 pamphlet decrying the "custome loathesome to the eye, hatefull to the nose, harmefull to the brain and daungerous to the lungs."

In 1634 smoking was outlawed in Russia. Punishment for breaking this law was extremely harsh—slitting of the lips, flogging, castration, or exile. Around fifty years later the ban was removed as Tsar Peter the Great, a smoker, promoted the use of tobacco. Just as Spanish and Portuguese sailors took chili peppers containing the capsaicin alkaloid around the world, so they introduced tobacco and the nicotine alkaloid to every port they visited. By the seventeenth century tobacco smoking had become widespread throughout the East, and draconian penalties, including torture, did little to stop its popularity. Although various countries, including Turkey, India, and Persia, at times prescribed the ultimate cure for addiction to tobacco—the death penalty—smoking is just as widespread in these places today as anywhere else.

From the beginning the supply of tobacco cultivated in Europe could not meet the demand. Spanish and English colonies in the New World soon started growing tobacco for export. Tobacco cultivation was highly labor intensive; weeds had to be kept under control, tobacco plants trimmed to the right height, suckers pruned, pests removed, and leaves manually harvested and prepared for drying. This work, done on plantations mainly by slaves, means that nicotine joins glucose, cellulose, and indigo as another molecule involved in slavery in the New World.

There are at least ten alkaloids in tobacco, the major one being nicotine. The content of nicotine in tobacco leaves varies from 2 to 8 percent, depending on the method of culture, climate, soil, and the process used to cure the leaves. In very small doses nicotine is a stimulant of the central nervous system and the heart, but eventually, or with larger doses, it acts as a depressant. This apparent paradox is explained by nicotine's ability to imitate the role of a neurotransmitter.

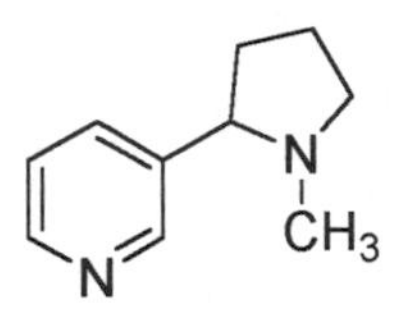

The structure of nicotine

The nicotine molecule forms a bridge at the junction between nerve cells, which initially heightens the transmission of a neurological impulse. But this connection is not readily cleared between impulses, so eventually the transmission site becomes obstructed. The stimulating effect of nicotine is lost, and muscle activity, particularly the heart, is slowed. Thus the blood circulation slows, and oxygen is delivered to the body and the brain at a lower rate, resulting in an overall sedative effect. This accounts for nicotine users who speak of needing a cigarette to calm their nerves, but nicotine is actually counterproductive in situations where an alert mind is required. As well, longtime tobacco users are more susceptible to infections such as gangrene that thrive in the low oxygen conditions from poor circulation.

In larger doses nicotine is a lethal poison. Absorbing a dose as small as fifty milligrams can kill an adult in just a few minutes. But its toxicity depends not only on amount but also on how the nicotine enters the body. Nicotine is about a thousand times more potent when absorbed through the skin as when taken orally. Stomach acids presumably break down the nicotine molecule to some extent. In smoking much of the alkaloid content in tobacco is oxidized to less toxic products by the high temperature of burning. This does not mean tobacco smoking is harmless, just that if this oxidation of most of the nicotine and other tobacco alkaloids did not occur, smoking would invariably be fatal with only a few cigarettes. As it is, the nicotine that remains in tobacco smoke is particularly hazardous, being absorbed directly from the lungs into the bloodstream.

Nicotine is a potent natural insecticide. Many millions of pounds of nicotine were produced for use as an insecticide in the 1940s and 1950s before synthetic pesticides were developed. Yet nicotinic acid and pyridoxine, with similar structures to nicotine, are not poisons. They are, in fact, beneficial—they are both B vitamins, essential nutrients for our health and survival. Once again a small change in chemical structure makes an enormous difference in properties.

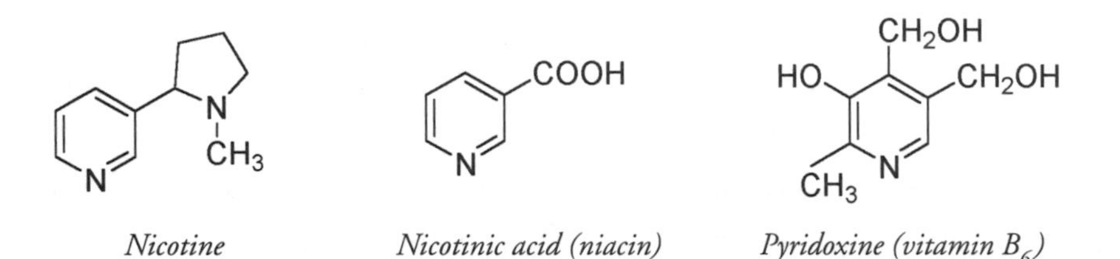

Nicotine *Nicotinic acid (niacin)* *Pyridoxine (vitamin B_6)*

In humans a dietary deficiency of nicotinic acid (also known as niacin) results in the disease pellagra, which is characterized by a set of three symptoms: dermatitis, diarrhea, and dementia. It is prevalent where the diet is almost entirely composed of corn and originally was thought to be an infectious disease, possibly a form of leprosy. Until pellagra was identified as caused by a lack of niacin, many of its victims were institutionalized in insane asylums. Pellagra was common in the southern United States in the early part of the twentieth century, but efforts by Joseph Goldberger, a doctor with the U.S. Public Health Service, convinced the medical community that it was indeed a deficiency disease. The name *nicotinic acid* was changed to *niacin* when commercial bakers did not want their vitamin-enriched white bread to bear a name that sounded too similar to nicotine.

THE STIMULATING STRUCTURE OF CAFFEINE

Caffeine, the third alkaloid connected to the Opium Wars, is also a psychoactive drug, but it is freely available almost everywhere in the world and is unregulated to such an extent that drinks laden with extra caffeine are manufactured and advertised as such. The structures of caffeine and the very closely related alkaloids theophylline and theobromine are shown below.

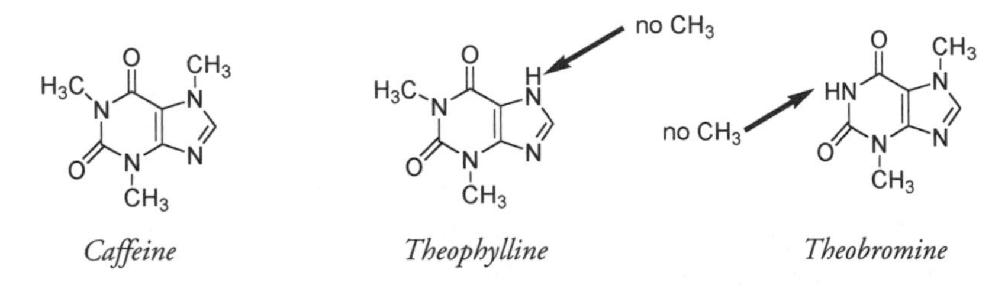

Caffeine *Theophylline* *Theobromine*

Theophylline, found in tea, and theobromine, in cocoa, differ from caffeine only in the number of CH_3 groups attached to the rings of the structure; caffeine having three and theophylline and theobromine each having two but in slightly different positions. This very small change of molecular structure accounts for the different physiological effect of these molecules. Caffeine is found naturally in coffee beans, tea leaves, and to a lesser extent cacao pods, cola nuts, and other plant sources mainly from South America, such as maté leaves, guarana seeds, and yoco bark.

Caffeine is a powerful central nervous stimulant and one of the most studied drugs in the world. The latest of numerous theories that have been suggested over the years to explain its effects on human physiology is that caffeine blocks the effect of adenosine in the brain and in other parts of the body. Adenosine is a neuromodulator, a molecule that decreases the rate of spontaneous nerve firing and thus slows the release of other neurotransmitters; therefore it can induce sleep. Caffeine cannot be said to wake us up, although it may feel like it does; its effect is really to hinder the normal role of adenosine in making us sleepy. When caffeine occupies adenosine receptors in other parts of the body, we experience a caffeine buzz: heartbeat rate increases, some blood vessels constrict while others open, and certain muscles more easily contract.

Caffeine is used medicinally to relieve and prevent asthma, to treat migraines, to increase blood pressure, as a diuretic, and for a host of other conditions. It is often found in both over-the-counter and prescription medications. Numerous studies have looked for possible negative side effects of caffeine, including its relationship to various forms of cancer, heart disease, osteoporosis, ulcers, liver disease, premenstrual syndrome, kidney disease, sperm motility, fertility, fetal development, hyperactivity, athletic performance, and mental dysfunction. So far there is no clear evidence that any of these can be linked to moderate amounts of caffeine consumption.

But caffeine is toxic; a fatal dose is estimated at about ten grams taken orally by an average-sized adult. Since the caffeine content for a cup of coffee varies between 80 to 180 milligrams, depending on the method of preparation, you would have to drink something like 55 to 125 cups, all at one time, in order to receive a lethal dose. Obviously, caffeine poisoning by this method is most unlikely, if not absolutely impossible. By dry weight, tea leaves have twice as much caffeine as coffee beans, but because less tea is used per cup and less caffeine is extracted by the normal method of making tea, a cup of tea ends up with about half the caffeine of a cup of coffee.

Tea also contains small amounts of theophylline, a molecule that has a similar effect to caffeine. Theophylline is widely used today in the treatment of asthma. It is a better bronchodilator, or bronchial tissue relaxant, than caffeine, while having less of an effect on the central nervous system. The cacao pod, the source of cocoa and chocolate, contains 1 to 2 percent of theobromine. This alkaloid molecule stimulates the central nervous system even less than theophylline, but as the amount of theobromine in cacao products is seven or eight times higher than the caffeine concentration, the effect is still apparent. Like morphine and nicotine, caffeine (and theophylline and theobromine) are addictive compounds; withdrawal symptoms include headaches, fatigue, drowsiness, and even—when caffeine intake has been excessive—nausea and vomiting. The good news is that caffeine clears the body relatively quickly, a week at most—though few of us have any intention of giving up on the world's favorite addiction.

Caffeine-containing plants were probably known to prehistoric man. They were almost certainly used in ancient times, but it is not possible to know whether tea, cacao, or coffee was the first. Legend has it that Shen Nung, the mythical first emperor of China, introduced the practice of boiling the drinking water for his court as a precaution against illness. One day he noticed that leaves from a nearby bush had fallen into the boiling water being prepared by his servants. The resulting infusion was supposedly the first of what must now be trillions of cups of tea that have been enjoyed in the five thousand years since. Although legends refer to drinking of tea in earlier times, Chinese literature does not mention tea, or its ability to "make one think better," until the second century B.C. Other traditional Chinese stories indicate that tea may have been introduced from northern India or from Southeast Asia. Wherever the origin, tea has been a part of Chinese life for many centuries. In many Asian countries, particularly Japan, tea also became an important part of the national culture.

The Portuguese, with a trading post in Macao, were the first Europeans to establish a limited trade with China and become tea drinkers. But it was the Dutch who brought the first bale of tea to Europe at the beginning of the seventeenth century. Tea was initially very expensive, affordable only by the wealthy. As the volume of imported tea increased and import duties were gradually lowered, the price slowly fell. By the early eighteenth century tea was beginning to replace ale as the national beverage of England, and the stage was set for the role that tea (with its caffeine) would play in the Opium Wars and the opening of trade with China.

Tea is often seen as a major contributor to the American Revolution, although its role was more symbolic than real. By 1763 the British had successfully expelled the French from North America and were negotiating treaties with the natives, controlling expansion of settlements, and regulating trade. The colonists' unhappiness with control by the

British Parliament of what they saw as local matters threatened to turn from irritation to rebellion. Particularly galling was the high level of taxation on both internal and external trade. Though the Stamp Act of 1764–1765, which raised money by requiring revenue stamps for almost every type of document, was withdrawn, and though the duties on sugar, paper, paint, and glass were eliminated, tea was still subject to a heavy customs duty. On December 16, 1773, a cargo of tea was dumped into Boston Harbor by a group of irate citizens. The protest was really about "taxation without representation" rather than about tea, but the Boston Tea Party, as it was called, is sometimes considered the start of the American Revolution.

Archaeological discoveries indicate that the cacao bean was the first source of caffeine in the New World. It was used in Mexico as early as 1500 B.C.; the later Mayan and Toltec civilizations also cultivated this Mesoamerican source of the alkaloid. Columbus, upon returning from his fourth voyage to the New World in 1502, presented cacao pods to King Ferdinand of Spain. But it was not until 1528, when Hernán Cortés drank the bitter drink of the Aztecs in the court of Montezuma II, that Europeans recognized the stimulating effect of its alkaloids. Cortés referred to cacao by the Aztec description "drink of the gods," from which was to come the name of the predominant alkaloid, theobromine, found in the seeds (or beans) of the foot-long pods of the tropical tree *Theobroma cacao.* The names are from the Greek *theos,* meaning "god," and *broma,* meaning "food."

For the rest of the sixteenth century the drinking of chocolate, as it came to be called, remained the preserve of the wealthy and aristocratic in Spain, spreading eventually to Italy, France, and Holland and then to the rest of Europe. Thus the caffeine in cacao, though present in smaller concentrations, predates European caffeine use from either tea or coffee.

Chocolate contains another interesting compound, anandamide, which has been shown to bind to the same receptor in the brain as the phenolic compound tetrahydrocannabinol (THC), the active ingredient in marijuana, even though the structure of anandamine is quite differ-

Cacao pods from the Theobroma cacao *tree.* (Photo by Peter Le Couteur)

ent from the structure of THC. If anandamide is responsible for the feel-good appeal that many people claim for chocolate, then we could ask a provocative question: What is it that we want to outlaw, the THC molecule or its mood-altering effect? If it is the mood-altering effect, should we be considering making chocolate illegal?

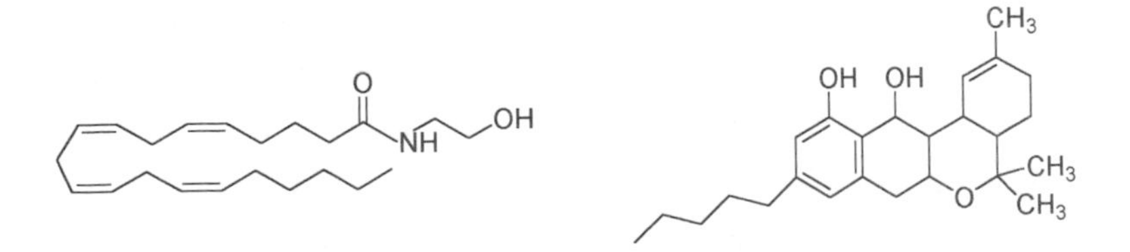

Anandamide (left) from chocolate and THC (right) from marijuana are structurally different.

Caffeine was introduced to Europe through chocolate. It was at least a century later that a more concentrated brew of the alkaloid arrived in the form of coffee, but by then coffee had been used in the Middle East for hundreds of years. The earliest surviving record of coffee drinking is from Rhazes, an Arabian physician of the tenth century. Doubtless coffee was known well before this time as the Ethiopian myth of Kaldi, the goatherd, suggests. Kaldi's goats, nibbling on the leaves and berries of a tree he had never noticed before, became frisky and started to dance on their hind legs. Kaldi decided to taste the bright red berries himself and found the effects as exhilarating as had his goats. He took a sample to a local Islamic holy man who, disapproving of their use, threw the berries on a fire. A wonderful aroma arose from the flames. The roasted beans were retrieved from the embers and used to make the first cup of coffee. Although it is a nice story, there is little evidence that Kaldi's goats were the true discoverers of caffeine from the *Coffea arabica* tree, but coffee may have originated somewhere in the highlands of Ethiopia and spread across northeastern Africa and into Arabia. Caffeine in the form of coffee was not always accepted, and sometimes it was even forbidden; nevertheless by the end of the fifteenth century Muslim pilgrims had carried it to all parts of the Islamic world.

A similar pattern followed the introduction of coffee to Europe during the seventeenth century. The lure of caffeine eventually won over the initial apprehension of officials of the Church and government as well as physicians. Sold on the streets of Italy, in cafés in Venice and Vienna, in Paris and Amsterdam, in Germany and Scandinavia, coffee has been credited with bringing greater sobriety to the population of Europe. To a certain extent it took the place of wine in southern Europe and beer in the north. No longer did workingmen consume ale for breakfast. By 1700 there were over two thousand coffeehouses in London; patronized exclusively by men, many came to be associated with a specific religion or trade or profession. Sailors and merchants gathered at Edward Lloyd's coffeehouse to peruse shipping lists, an activity that eventually led to the underwriting of trade voyages and the establishment of the famous insurance company Lloyd's of London. Various banks, newspapers, and magazines as well as the stock exchange are all supposed to have started life in London's coffeehouses.

The cultivation of coffee played an enormous role in the development of regions of the New World, especially Brazil and countries in Central America. Coffee trees were first grown in Haiti in 1734. Fifty years later half of the world's coffee derived from this source. The political and economic circumstances of Haitian society today are often attributed to the long and bloody slave uprising that started in 1791 as a revolt against the appalling conditions imposed on slaves toiling to produce coffee and sugar. When the coffee trade declined in the West Indies, plantations in other countries—Brazil, Colombia, the Central American states, India, Ceylon, Java, and Sumatra—rushed their product to the rapidly growing world market.

In Brazil in particular coffee cultivation came to dominate agriculture and commerce. Huge areas of land that had already been established as sugar plantations switched to growing coffee trees in the expectation of huge profits to be reaped from the bean. In Brazil the abolition of slavery was delayed through the political power of coffee growers, who needed cheap labor. Not until 1850 was the importation of new slaves into Brazil banned. From 1871 all children born to slaves were considered legally free, ensuring the country's eventual, although gradual, abolition of slavery. In 1888, years after other Western nations, slavery in Brazil was finally completely outlawed.

Coffee cultivation fueled economic growth for Brazil as railways were built from coffee-growing regions to major ports. When slave labor vanished, thousands of new immigrants, mainly poor Italians, arrived to work on the coffee plantations, thus changing the ethnic and cultural face of the country.

Continued coffee growing has radically changed Brazil's environment. Huge swaths of land have been cleared, natural forest cut down or burned, and native animals destroyed for the vast coffee plantations that cover the countryside. Grown as a monoculture, the coffee tree quickly exhausts soil fertility, requiring new land to be developed as the old becomes less and less productive. Tropical rain forests can take centuries to regenerate; without suitable plant covering erosion can remove what little soil is present, effectively destroying any hope of forest renewal. Overreliance on one crop generally means local populations forgo planting more traditional necessities, making them even more vulnerable to the vagaries of world markets. Monoculture is also highly susceptible to devastating pest infestations, like coffee leaf rust, that can wipe out a plantation in a matter of days.

A similar pattern of exploitation of people and the environment occurred in most of the coffee-growing countries of Central America. Starting in the last decades of the nineteenth century, the indigenous Mayan people in Guatemala, El Salvador, Nicaragua, and Mexico were systematically forced from their lands as coffee monoculture spread up the hillsides, which offered perfect conditions for cultivation of the coffee shrub. Labor was provided through coercion of the displaced population; men, women, and children worked long hours for a pittance and, as forced laborers, had few rights. The elite—the coffee plantation owners—controlled the wealth of the state and directed government policies in the pursuit of profit, fomenting decades of bitterness over social inequality. The history of political unrest and violent revolution in these countries is partly a legacy of people's desire for coffee.

From its beginning as a valuable medicinal herb of the eastern Mediterranean, the opium poppy spread throughout Europe and Asia. Today profit from illegal trafficking in opium continues to finance organized crime and international terrorism. The health and happiness of millions have been destroyed, directly or indirectly, by alkaloids from the opium poppy, yet at the same time many millions more have benefited from the judicious medical application of their amazing pain-relieving properties.

Just as opium has been alternately sanctioned and prohibited, so has nicotine been both encouraged and forbidden. Tobacco was once considered to have advantageous health effects and was employed as a cure for numerous maladies, but at other times and places the use of tobacco was outlawed as a dangerous and depraved habit. For the first half of the twentieth century tobacco use was more than tolerated—it was promoted in many societies. Smoking was upheld as a symbol of the emancipated woman and the sophisticated man. At the beginning of the twenty-first century the pendulum has swung the other way, and in many places nicotine is being treated more like the alkaloids from opium: controlled, taxed, proscribed, and banned.

In contrast, caffeine—although once subject to edicts and religious injunctions—is now readily available. There are no laws or regulations to keep children or teenagers from consuming this alkaloid. In fact, parents in many cultures routinely provide their children with caffeinated drinks. Governments now restrict opium alkaloid use to regulated medical purposes, but they reap large tax benefits from the sale of caffeine and nicotine, making it unlikely that they will give up such a lucrative and reliable source of income and ban either of these two alkaloids.

It was the human desire for three molecules—morphine, nicotine, and caffeine—that initiated the events leading to the Opium Wars of the mid-1800s. The results of these conflicts are now seen as the beginning of the transformation of a social system that had been the basis of Chinese life for centuries. But the role these compounds have played in history has been even greater. Grown in lands far from their origins, opium, tobacco, tea, and coffee have had a dramatic effect on local populations and on those people who have cultivated these plants. In many cases the ecology of these regions changed dramatically as native flora were destroyed to make way for acres of poppies, fields of tobacco, and verdant hillsides covered with tea bushes or coffee trees. The alkaloid molecules in these plants have spurred trade, generated fortunes, fueled wars, propped up governments, funded coups, and enslaved millions—all because of our eternal craving for a quick chemical fix.

14. OLEIC ACID

A CHEMICAL EXPLANATION of the prime condition for the trading of goods is "highly desired molecules unevenly distributed throughout the world." Many of the compounds we have considered—those in spices, tea, coffee, opium, tobacco, rubber, and dyes—fit this definition, as does oleic acid, a molecule found in abundance in the oil pressed from the small green fruit of the olive tree. Olive oil, a valued trade item for thousands of years, has been called the lifeblood of the societies that developed around the Mediterranean Sea. Even as civilizations rose and fell in the region, the olive tree and its golden oil were always at the base of their prosperity and at the heart of their culture.

The Lore of the Olive

Myths and legends about the olive tree and its origin abound. Isis, goddess of the ancient Egyptians, allegedly introduced the olive and its bountiful harvest to humanity. Roman mythology credits Hercules with bringing the olive tree from North Africa; the Roman goddess Minerva supposedly taught the art of cultivation of the olive tree and extraction of its oil. Another legend claims the olive goes back to the first man; the first olive tree is said to have grown out of the ground on Adam's tomb.

The ancient Greeks told of a contest between Poseidon, the god of the sea, and Athena, goddess of peace and wisdom. The victor would be the one who produced the most useful gift for the people of the newly built city in the region known as Attica. Poseidon struck a rock with his trident and a spring appeared. Water began to flow, and from the spring the horse appeared—a symbol of strength and power and an invaluable aid in war. When Athena's turn came, she threw her spear into the ground, and it turned into the olive tree—a symbol of peace and a provider of food and fuel. Athena's gift was considered the greater, and the new city, Athens, was named in her honor. The olive is still considered a divine gift. An olive tree still grows atop the Acropolis in Athens.

The geographic origin of the olive tree is debatable. Fossil evidence of what is believed to be an ancestor of the modern olive tree has been found in both Italy and Greece. The first cultivation of olive trees is usually ascribed to lands around the eastern Mediterranean—to various regions in the present-day countries of Turkey, Greece, Syria, Iran, and Iraq. The olive tree, *Olea europaea,* the only species of the *Olea* family grown for its fruit, has been cultivated for at least five thousand and probably as long as seven thousand years.

From the eastern shores of the Mediterranean, olive cultivation spread to Palestine and on to Egypt. Some authorities believe this cultivation started in Crete, where by 2000 B.C. a flourishing industry exported the oil to Greece, to North Africa, and to Asia Minor. As their colonies grew, the Greeks took the olive tree to Italy, France, Spain and Tunisia. When the Roman Empire expanded, the olive culture also expanded throughout the whole Mediterranean basin. For

Olive tree atop the Acropolis in Athens. (Photo by Peter Le Couteur)

centuries, olive oil was the most important trading commodity of the region.

As well as its obvious role supplying valuable calories as a food, olive oil was used in many other aspects of everyday life by the people living around the Mediterranean. Lamps filled with olive oil lit the dark evenings. The oil was used for cosmetic purposes; both the Greeks and Romans rubbed it into their skin after bathing. Athletes considered olive oil massages essential to keep muscles supple. Wrestlers added a coating of sand or dust to their layer of oil to allow their opponent a grip. Rituals after athletic events involved bathing and more olive oil, massaged into the skin to soothe and heal abrasions. Women used olive oil to keep their skin looking young and their hair shiny. It was thought to help prevent baldness and promote strength. Compounds responsible for fragrance and flavor in herbs are very often soluble in oils, so bay, sesame, rose, fennel, mint, juniper, sage, and other leaves and flowers were used to infuse olive oil, producing exotic and highly prized scented mixtures. Physicians in Greece prescribed olive oil or some of these blends for numerous ailments, including nausea, cholera, ulcers, and insomnia. Numerous references to olive oil, taken internally or applied externally, appear in early Egyptian medical texts. Even the leaves of the olive tree were used to reduce fevers and provide relief from malaria. These leaves, we now know, contain salicylic acid, the same molecule as in the willow tree and meadowsweet plant from which Felix Hofmann developed aspirin in 1893.

The importance of olive oil to the people of the Mediterranean is reflected in their writings and even their laws. The Greek poet Homer called it "liquid gold." The Greek philosopher Democritus believed a diet of honey and olive oil could allow a man to live to be a hundred, an extremely old age in a time where life expectancy hovered around forty years. In the sixth century B.C. the Athenian legislator Solon—whose other deeds included establishing a humane code of laws, popular courts, the right of assembly, and a senate—introduced laws protecting olive trees. In a grove only two trees could be removed each year. Breaking this law incurred severe penalties, including execution.

There are more than a hundred references in the Bible to olives and olive oil. For example: the dove brings an olive sprig back to Noah after the flood, Moses is instructed to prepare an anointing mixture of spices and olive oil, the Good Samaritan pours wine and olive oil into the wounds of the robbers' victim, and the wise virgins keep their lamps filled with olive oil. We have the Mount of Olives at Jerusalem. The Hebrew king David appointed guards to protect his olive groves and warehouses. The Roman historian Pliny, in the first century A.D., referred to Italy having the best olive oil in the Mediterranean. Virgil praised the olive—"Thus you shall cultivate the rich olive, beloved of Peace."

With this integration of lore of olives into religion, mythology, and poetry as well as everyday life, it is not surprising that the olive tree came to be a symbol for many cultures. In ancient Greece, presumably because a plentiful supply of olive oil for food and lamps implied prosperity that was absent during times of war, the olive became synonymous with times of peace. We still talk about extending the olive branch when we mean an attempt to make peace. The olive was also considered a symbol of victory, and winners at the original Olympic Games were awarded a wreath made of olive leaves as well as a supply of its oil. Olive groves were often targeted during war, as destruction of an enemy's olive groves not only eliminated a major food source but also inflicted a devastating psychological blow. The olive tree represented wisdom and renewal as well; olive trees that appeared to have been destroyed by fire or felling often sprouted new shoots and eventually bore fruit again.

Finally, the olive represented strength (an olive trunk was the staff of Hercules) and sacrifice (the cross to which Christ was nailed was supposedly made of olive wood). At various times and in various cultures the olive has symbolized power and wealth, virginity, and fertility. Olive oil has been used for centuries to anoint kings, emperors, and bishops at their coronations or ordinations. Saul, the first king of Israel, had olive oil rubbed onto his forehead at his crowning. Hundreds of years later, on the other side of the Mediterranean, the first king of the Franks, Clovis, was anointed with olive oil at his coronation, becoming Louis the First. Thirty-four more French monarchs were anointed with oil from the same pear-shaped vial, until it was destroyed during the French Revolution.

The olive tree is remarkably hardy. It needs a climate with a short cold winter to set the fruit and no blossom-killing spring frosts. A long, hot summer and a mellow fall allow the fruit to ripen. The Mediterranean Sea cools its African coast and warms its northern shores, making the region ideally suited for the cultivation of the olive. Inland, away from the moderating effect of a large body of water, the olive does not grow. Olive trees can survive where there is very little rainfall. Their long taproot penetrates deeply to find water, and the leaves are narrow and leathery with a slightly fuzzy, silvery underside—adaptations that prevent loss of water through evaporation. The olive can survive periods of drought and can grow in rocky soil and on stony terraces. Extreme frost and ice storms may snap branches and crack trunks, but the tenacious olive, even if it appears to be destroyed by the cold, will regenerate and send out fresh green suckers the following spring. Little wonder that the people who were dependent on the olive tree for thousands of years came to venerate it.

THE CHEMISTRY OF OLIVE OIL

Oils have been extracted from many plants: walnuts and almonds, corn, sesame seeds, flax seeds, sunflower seeds, coconuts, soybeans, and peanuts, to name a few. Oils—and fats, their chemically very close cousins usually from animal sources—have long been valued for food, for lighting, and for cosmetic and medicinal purposes. But no other oil or fat has ever become so much part of the culture and economy, so intertwined in the hearts and minds of the people, or so important to the growth of Western civilization as has the oil from the fruit of the olive tree.

The chemical difference between olive oil and any other oil or fat is very small. But once again a very small difference accounts for a large part of the course of human history. We do not think it too speculative to assert that without oleic acid—named after the olive and the molecule that differentiates olive oil from other oils or fats—the development of Western civilization and democracy might have followed a very different path.

Fats and oils are known as triglycerides. They are all compounds formed from a glycerol (also called glycerin) molecule and three molecules of fatty acid.

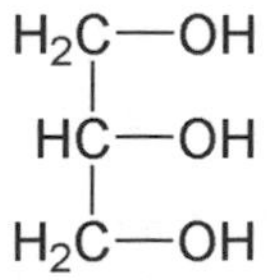

$$
\begin{array}{l}
H_2C-OH \\
\;\;\;\;| \\
\;\;HC-OH \\
\;\;\;\;| \\
H_2C-OH
\end{array}
$$

The glycerol molecule

Fatty acids are long chains of carbon atoms with an acid group, COOH (or HOOC), at one end:

$HO-\underset{\underset{O}{\|}}{C}-CH_2-CH_2-CH_2-CH_2-CH_2-CH_2-CH_2-CH_2-CH_2-CH_2-CH_3$

A twelve-carbon-atom fatty-acid molecule. The acid group, on the left, is circled.

Although they are simple molecules, fatty acids have a number of carbon atoms, so it is often clearer to represent them in zigzag format, where every intersection and each end of a line represents a carbon atom and most of the hydrogen atoms are not shown at all.

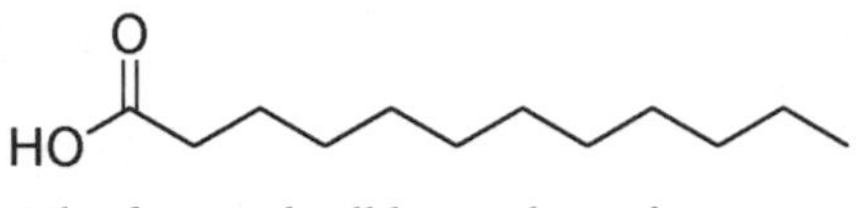

This fatty acid still has twelve carbon atoms.

When three molecules of water (H_2O) are eliminated between the H from each of the three OH groups on glycerol and OH from the HOOC of three different fatty-acid molecules, a triglyceride molecule is formed. This condensation process—the joining of molecules through the loss of H_2O—is similar to the formation of polysaccharides (discussed in Chapter 4).

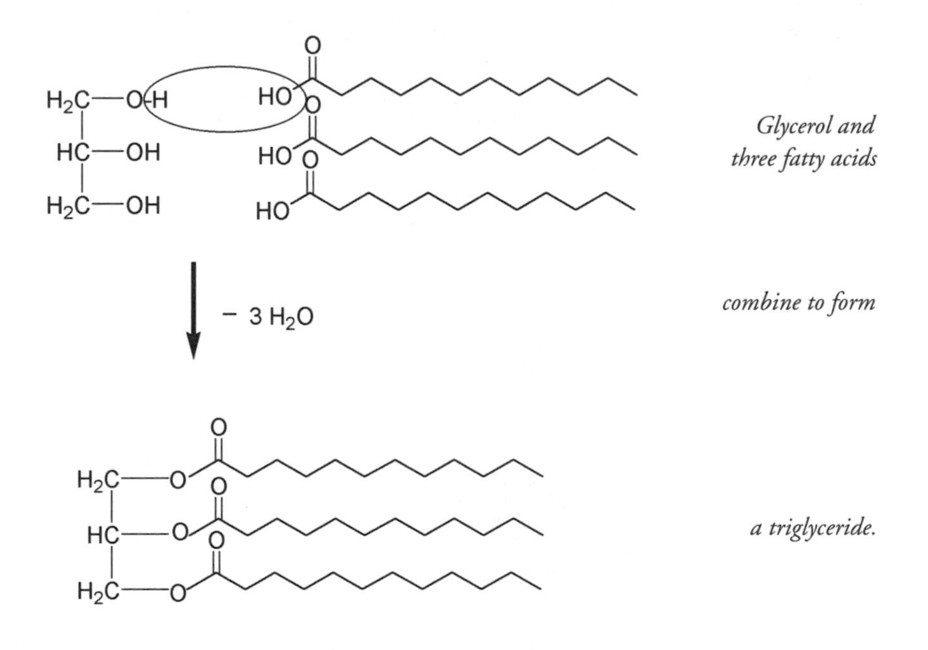

Glycerol and three fatty acids

combine to form

a triglyceride.

The triglyceride molecule shown above has all three fatty-acid molecules the same. But it is also possible for only two of the fatty-acid molecules to be the same. Or they can all be different. Fats and oils have the same glycerol portion; it is the fatty acids that vary. In the previous example we used what is known as a saturated fatty acid. *Saturated* in this case means saturated with hydrogen; no more hydrogen can be added to the fatty-acid portion of the molecule, as there are no carbon-to-carbon double bonds that can be broken to allow the attachment of more hydrogen atoms. If such bonds are present in the fatty acid, it is termed unsaturated. Some common saturated fatty acids are:

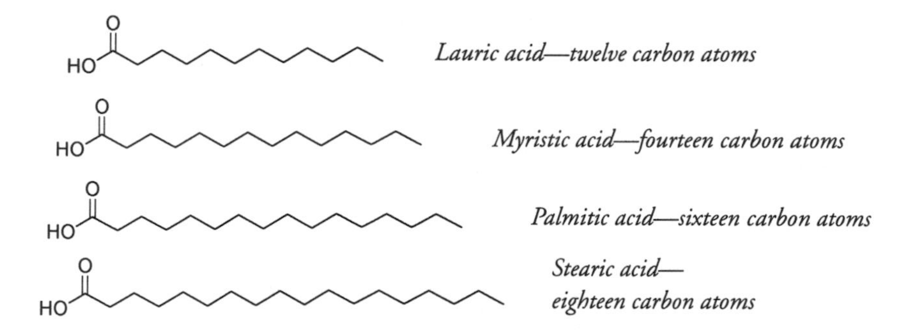

Lauric acid—twelve carbon atoms

Myristic acid—fourteen carbon atoms

Palmitic acid—sixteen carbon atoms

Stearic acid—eighteen carbon atoms

From their names it is not hard to guess that the main source of stearic acid is beef tallow and that palmitic acid is a component of palm oil.

Almost all fatty acids have an even number of carbon atoms. The examples above are the most common fatty acids, although others do exist. Butter contains butyric acid (named from butter), with only four carbon atoms, and caproic acid, also present in butter and in fat from goat's milk—*caper* is Latin for goat—has six carbon atoms.

Unsaturated fatty acids contain at least one carbon-to-carbon double bond. If there is only one of these double bonds, the acid is referred to as *monounsaturated;* with more than one double bond it is *polyunsaturated.* The triglyceride shown below is formed from two monounsaturated fatty acids and one saturated fatty acid. The double bonds are cis in arrangement, as the carbon atoms of the long chain are on the same side of the double bond.

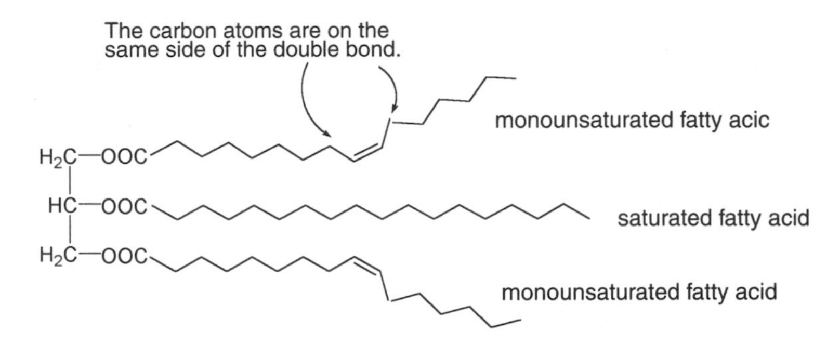

Triglyceride from two monounsaturated and one saturated fatty acid

This puts a kink in the chain, so such triglycerides cannot pack together as closely as triglycerides composed of saturated fatty acids (below).

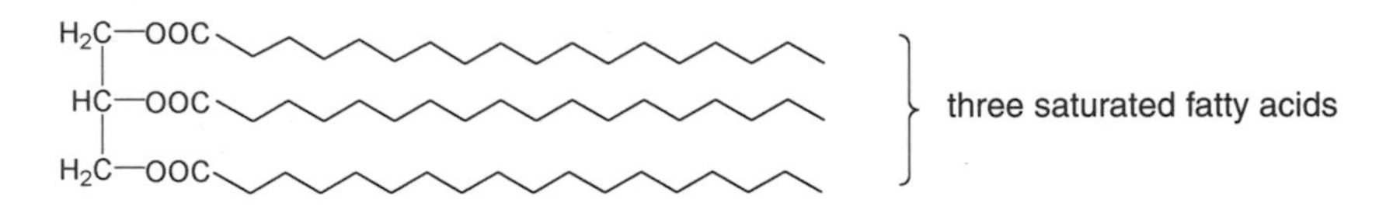

Triglyceride from three saturated fatty acids

The more double bonds in a fatty acid, the more bent it is and the less efficient its packing. Less efficient packing requires less energy to overcome the attractions holding the molecules together, and they can therefore be separated at lower temperatures. Triglycerides with a higher proportion of unsaturated fatty acids tend to be liquids at room temperature rather than solids. We call them oils; they are most often of plant origin. Saturated fatty acids that can pack closely together require more energy to separate individual molecules and so melt at higher temperatures. Triglycerides from animal sources, with a higher proportion of saturated fatty acids than oils, are solid at room temperature. We call them fats.

Some common unsaturated fatty acids are:

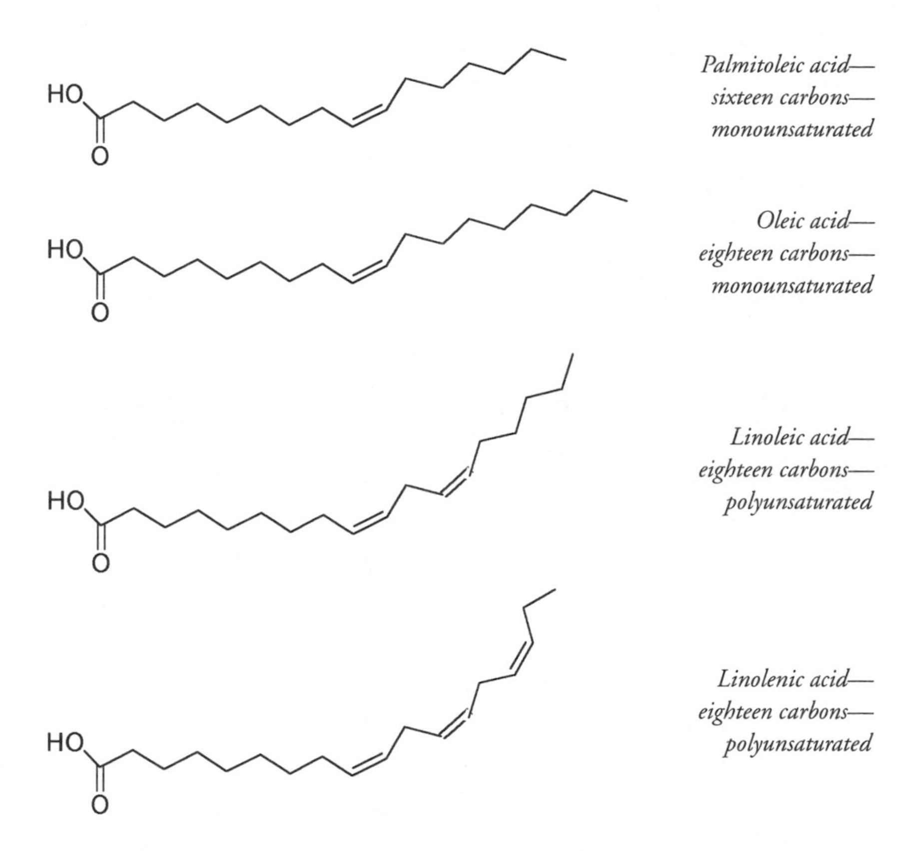

Palmitoleic acid—sixteen carbons—monounsaturated

Oleic acid—eighteen carbons—monounsaturated

Linoleic acid—eighteen carbons—polyunsaturated

Linolenic acid—eighteen carbons—polyunsaturated

The monounsaturated, eighteen-carbon oleic acid is the major fatty acid of olive oil. Although oleic acid is found in other oils and also in many fats, olive oil is the most important source of this fatty acid. Olive oil contains a larger proportion of monounsaturated fatty acid than any other oil. The percentage of oleic acid in olive oil varies from about 55 to 85 percent, depending on the variety and the growing conditions, cooler areas producing a higher oleic acid content than warmer areas. There is now convincing evidence that a diet with a high proportion of saturated fat can contribute to the development of heart disease. The incidence of heart disease is lower in the Mediterranean region, where a lot of olive oil—and oleic acid—is consumed. Saturated fats are known to increase serum cholesterol concentrations, whereas polyunsaturated fats and oils lower these levels. Monounsaturated fatty acids, like oleic acid, have a neutral effect on the serum cholesterol level (the level of cholesterol in the blood).

The relationship between heart disease and fatty acids also involves another factor: the ratio of high-density lipoprotein (known as HDL) to low-density lipoprotein (known as LDL). A lipoprotein is a water-insoluble accumulation of cholesterol, protein, and triglycerides. High-density lipoproteins—often called the "good" lipoproteins—transport cholesterol from cells that have accumulated too much of this compound back to the liver for disposal. This prevents excess cholesterol from depositing on artery walls. The "bad" lipoproteins, LDLs, transport cholesterol from the liver or small intestine to newly formed or growing cells. While this is a necessary function, too much cholesterol in the bloodstream ultimately ends up as deposits of plaque on the artery walls, leading to narrowing of the arteries. If the coronary arteries leading to the heart muscles become clogged, the resultant decreased blood flow can cause chest pain and heart attacks.

It is the ratio of HDL to LDL, as well as the total cholesterol level, that is important in determining the risk of heart disease. Although polyunsaturated triglycerides have the positive effect of reducing serum cholesterol levels, they also lower the HDL:LDL ratio, a negative effect. Monounsaturated triglycerides like olive oil, while not reducing serum cholesterol levels, increase the HDL:LDL ratio, that is, the proportion of good lipoprotein to bad lipoprotein. Among the saturated fatty acids, palmitic (C_{16}) and lauric (C_{12}) acids raise LDL levels appreciably. The so-called tropical oils—coconut, palm, and palm kernel—which have high proportions of these fatty acids, are particularly suspect in heart disease because they increase both serum cholesterol and LDL levels.

Although the healthy properties of olive oil were prized by ancient Mediterranean societies and were considered to account for longevity, there was no knowledge of the chemistry behind these beliefs. In fact, in times when the main dietary problem would have been simply to obtain enough calories, serum cholesterol levels and HDL:LDL ratios would have been irrelevant. For centuries, for the vast majority of the population of northern Europe, where the main source of triglycerides in the diet was animal fat and life expectancy was less than forty years, hardening of the arteries was not a problem. Only with increased life expectancy and the higher intake of saturated fatty acids accompanying economic prosperity has coronary heart disease become a major cause of death.

Another aspect of the chemistry of olive oil also accounts for its importance in the ancient world. As the number of carbon-to-carbon double bonds in a fatty acid increases, so does the tendency for the oil to oxidize—become rancid. The proportion of polyunsaturated fatty acids in olive oil is much lower than in other oils, usually less than 10 percent, giving olive oil a longer shelf life than almost any other oil. As well, olive oil contains small amounts of polyphenols and of vitamins E and K, antioxidant molecules that play a critical role as natural preservatives. The traditional cold-press method of extracting oil from olives helps retain these antioxidant molecules, which can be destroyed by high temperatures.

Today one method of improving the stability and increasing the shelf life of oils is the elimination of some of the double bonds by hydrogenation, a process of adding hydrogen atoms to the double bonds of unsaturated fatty acids. The result is also a more solid triglyceride; this is the method used to convert oils into butter substitutes like margarine. But the process of hydrogenation also changes the remaining double bonds from the cis configuration to the trans configuration, where the carbon atoms of the chain are on opposite sides of the double bond.

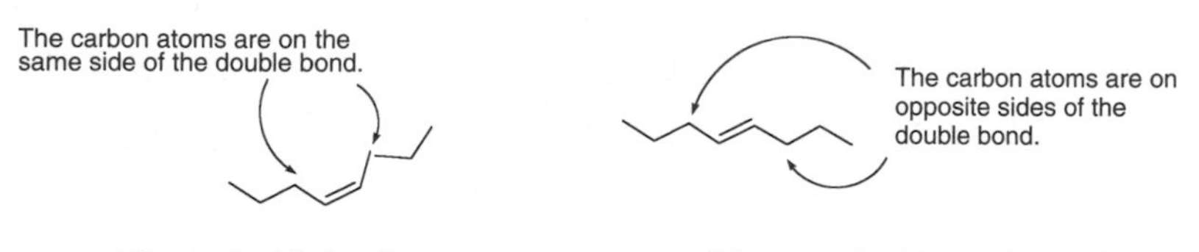

The cis double bond *The trans double bond*

Trans-fatty acids are known to elevate LDL levels but not as much as saturated fatty acids.

Trade in Olive Oil

The natural preservatives present as antioxidants in olives would have been of paramount importance to the oil traders of ancient Greece. This civilization was a loose association of city-states, with a common language, a common culture, and a common agricultural economic base: wheat, barley, grapes, figs, and olives. For centuries the land around the shores of the Mediterranean Sea was more wooded than it is now; the soil was more fertile, and more water was available from springs. As the population grew, cultivation of crops spread from the original small valleys up the sides of coastal mountains. With its ability to grow on steep and stony ground and withstand drought, the olive tree became increasingly more important. Its oil was even more valuable as an export commodity, for in the sixth century B.C., along with the strict laws against uncontrolled felling of olive trees, Solon of Athens also mandated olive oil as the only agricultural product that could be exported. As a result, coastal forests were cut down and more olive trees were planted. Where grain crops once grew, olive groves thrived.

The economic value of olive oil was readily apparent. City-states became centers of commerce. Large ships, powered by sail or oars and built to carry hundreds of amphorae of oil, traded throughout the whole Mediterranean Sea, returning with metals, spices, fabrics, and other goods available from far-flung ports. Colonization followed trade, and by the end of the sixth century B.C. the Hellenic world had expanded well beyond the Aegean: to Italy, Sicily, France, and the Balearic Islands in the west, around the Black Sea to the east, and even to coastal Libya on the southern shores of the Mediterranean.

But Solon's method of enhancing the production of olive oil had environmental consequences that are still apparent in Greece today. The woodlands that were destroyed and the grains that were no longer planted had fibrous root systems that had drawn water from near the surface and had served to hold the surrounding earth together. The long taproot of the olive tree drew water from layers deep below the surface and had no binding effect on the topsoil. Gradually springs dried up, soil washed away, and the land eroded. Fields that once grew cereals and slopes that bore vines could no longer support these crops. Livestock became scarce. Greece was awash in olive oil, but more and more other foodstuffs had to be imported—a significant factor in governing a large empire. Many reasons have been given for the decline of classical Greece: internal strife among warring city-states, decades of war, lack of effective leadership, the collapse of religious traditions, attacks from outside. Maybe we can add another: the loss of valuable agricultural land to the demands of the olive oil trade.

SOAP FROM OLIVE OIL

Olive oil may have been a factor in the collapse of classical Greece, but around the eighth century A.D. the introduction of a product from olive oil, soap, may have had even more important consequences for European society. Today soap is such a common item that we don't recognize what a significant role it has played in human civilization. Try to imagine, for a moment, life without soap—or detergents, shampoos, laundry powders, and similar products. We take for granted the cleaning ability of soap, yet without it the megacities of the present day would hardly be possible. Dirt and disease would make living hazardous under such conditions and maybe not even viable. The filth and squalor of medieval towns, which had far fewer inhabitants than today's big cities, cannot be blamed entirely on lack of soap, but without this essential compound maintaining cleanliness would have been extremely difficult.

For centuries humankind has made use of the cleansing power of some plants. Such plants contain saponins, glycosidic (sugar-containing) compounds such as those from which Russell Marker extracted the sapogenins that became the basis for birth control pills, and the cardiac glycosides like digoxin and other molecules used by herbalists and supposed witches.

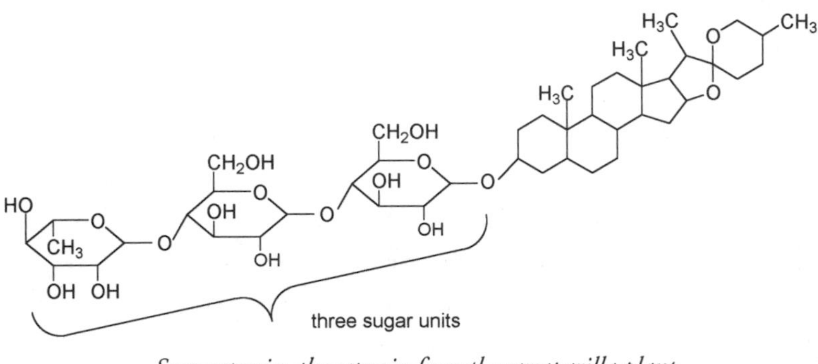

Sarsasaponin, the saponin from the sarsaparilla plant

Plant names like soapwort, soapberry, soap lily, soap bark, soapweed, and soaproot give a clue to the properties of the diverse range of saponin-bearing plants. These include members of the lily family, bracken, campions, yuccas, rues, wattles, and the genus *Sapindus.* Saponin extracts from some of these plants are still used today to wash delicate fabrics or as hair shampoos; they create a very fine lather and have a very gentle cleansing effect.

The process of making soap was most probably an accidental discovery. Those cooking over wood fires might have noticed that fats and oils that dripped from the food into the ashes produced a substance that formed a foamy lather in water. It would not have taken long to realize that this substance was a useful cleaning agent and that it could be deliberately manufactured using fats or oils and wood ash. Such discoveries no doubt occurred in many parts of the world, as there is evidence of soap production from many civilizations. Clay cylinders containing a type of soap and instructions for its manufacture have been found in excavations from Babylonian times, nearly five thousand years ago. Egyptian records dating from 1500 B.C. show that soaps were made from fats and wood ash, and through the centuries there are references to the use of soap in the textile and dyeing industries. The Gauls are known to have used a soap made from goat fat and potash, to brighten or redden their hair. Another use of this soap was as a type of pomade to stiffen hair—an early hair gel. The Celts have also been credited with the discovery of soap making and for using it to bathe and to wash clothes.

Roman legend attributes the discovery of soap making to women washing clothes in the Tiber River downstream from the temple on Mount Sapo. Fats from animals sacrificed at the temple combined with ashes from sacrificial fires. When it was raining, these wastes would run down the hill and enter the Tiber as a soapy steam, which could be used by the washerwomen of Rome. The chemical term for the reaction that occurs when triglycerides of fats and oils react with alkalis—from ashes—is *saponification,* the word derived from the name of Mount Sapo, as is the word for soap in a number of languages.

Although soap was manufactured in Roman times, it was mainly used for washing clothes. As with the ancient Greeks, personal hygiene for most Romans usually involved rubbing the body with a mixture of olive oil and sand, which was then removed with a scraper made especially for this purpose and known as a strigil. Grease, dirt, and dead skin were removed by this method. Soap gradually came to be used for bathing during the later centuries of Roman times. Soap and soap making would have been associated with the public baths, a common feature of Roman cities that spread throughout the Roman Empire. With the decline of Rome, soap making and soap using appears to have also declined in western Europe, although it was still made and used in the Byzantine Empire and the Arab world.

In Spain and France during the eighth century there was a revival of the art of soap making, using olive oil. The resulting soap, known as "castile" after a region of Spain, was of very high quality, pure, white, and shiny. Castile soap was exported to other parts of Europe, and by the thirteenth century Spain and southern France had become famous for this luxury item. The soaps of northern Europe were based on animal fat or fish oils; the soaps they produced were of poor quality and were used mainly for washing fabric.

The chemical reaction for making soap—saponification—breaks a triglyceride into its component fatty acids and glycerol through the use of an alkali, or base, such as potassium hydroxide (KOH) or sodium hydroxide (NaOH).

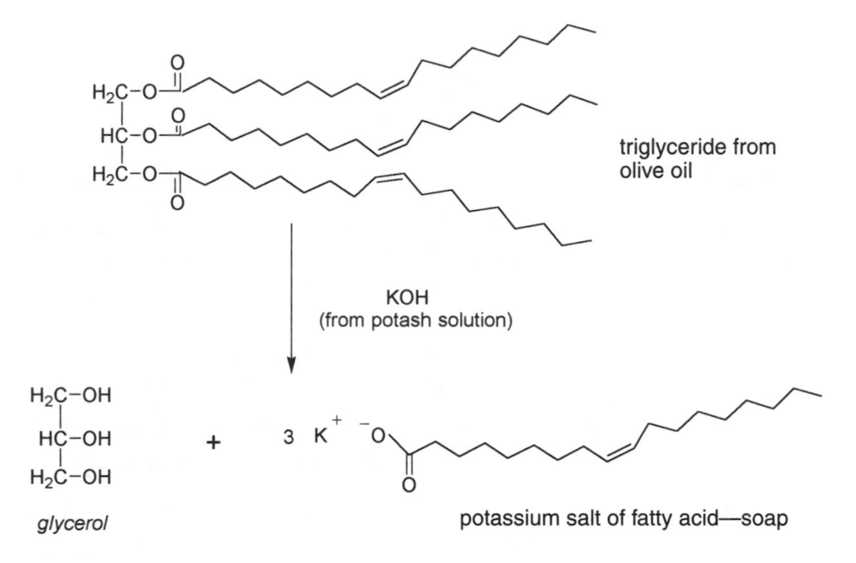

The saponification reaction of a triglyceride molecule of oleic acid, forming glycerol and three molecules of soap

Potassium soaps are soft; those made with sodium are hard. Originally most soaps would have been potassium soaps, as wood ash from burning timber and peat was the most readily available source of alkali. Potash (literally, the ashes from a fire pot) is potassium carbonate (K_2CO_3), and in water forms a mildly alkaline solution. Where soda ash (sodium carbonate, Na_2CO_3) was available, hard soaps were produced. A major source of income in some coastal regions—Scotland and Ireland in particular—was collecting kelp and other seaweeds, which were burned to make soda ash. Soda ash dissolved in water also produces an alkaline solution.

In Europe the practice of bathing declined along with the Roman Empire, although public baths still existed and were used in many towns until late in the Middle Ages. During the plague years, starting in the fourteenth century, city authorities began closing public baths, fearing that they contributed to the spread of the Black Death. By the sixteenth century bathing had become not only unfashionable but was even considered dangerous or sinful. Those who could afford it covered body odors with liberal applications of scents and perfumes. Few homes had baths. A once-a-year bath was the norm; the stench of unwashed bodies must have been dreadful. Soap, however, was still in demand during these centuries. The rich had their clothes and linens laundered. Soap was used to clean pots and pans, dishes and cutlery, floors and counters. Hands and possibly faces were washed with soap. It was washing the whole body that was frowned upon, particularly naked bathing.

Commercial soap making began in England in the fourteenth century. As in most northern European countries, soap was made mainly from cattle fat or tallow, whose fatty acid content is approximately 48 percent oleic acid. Human fat has about 46 percent oleic acid; these two fats contain some of the highest percentages of oleic acid in the animal world. By comparison, the fatty acids in butter are about 27 percent oleic acid and in whale blubber about 35 percent. In 1628, when Charles I ascended to the throne of England, soap making was an important industry. Desperate for a source of revenue—Parliament refused to approve his proposals for increased taxation—Charles sold monopoly rights to the production of soap. Other soap makers, incensed at the loss of their livelihood, threw their support behind Parliament. Thus it has been said that soap was one of the causes of the English Civil War of 1642–1652, the execution of Charles I, and the establishment of the only republic in English history. This claim seems somewhat far-fetched, as the support of soap makers can hardly have been a crucial factor; disagreements on policies of taxation, religion, and foreign policy, the major issues between the king and Parliament, are more likely causes. In any event, the overthrow of the king was of little advantage to soap makers, since the Puritan regime that followed considered toiletries frivolous, and the Puritan leader, Oliver Cromwell, Lord Protector of England, imposed heavy taxes on soap.

Soap can, however, be considered responsible for the reduction in infant mortality in England that became evident in the later part of the nineteenth century. From the start of the Industrial Revolution in the late eighteenth century, people flocked to towns seeking work in factories. Slum housing conditions followed this rapid growth of the urban population. In rural communities, soap making was mainly a domestic craft; scraps of tallow and other fats saved from the butchering of farm animals cooked up with last night's ashes would produce a coarse but affordable soap. City dwellers had no comparable source of fat. Beef tallow had to be purchased and was too valuable a food to be used to make soap. Wood ashes were also less obtainable. Coal was the fuel of the urban poor, and the small amounts of coal ash available were not a good source of the alkali needed to saponify fat. Even if the ingredients were on hand, the living quarters of many factory workers had, at best, only rudimentary kitchen facilities and little space or equipment for soap making. Thus soap was no longer made at home. It had to be purchased and was generally beyond the means of factory workers. Standards of hygiene, not high to start with, fell even lower, and filthy living conditions contributed to a high infant death rate.

At the end of the eighteenth century, though, a French chemist, Nicolas Leblanc, discovered an efficient method of making soda ash from common salt. The reduced cost of this alkali, an increased availability of fat, and finally in 1853 the removal of all taxes on soap lowered the price so that widespread use was possible. The decline in infant mortality dating from about this time has been attributed to the simple but effective cleansing power of soap and water.

Soap molecules clean because one end of the molecule has a charge and dissolves in water, whereas the other end is not soluble in water but does dissolve in substances such as grease, oil, and fat. The structure of the soap molecule is

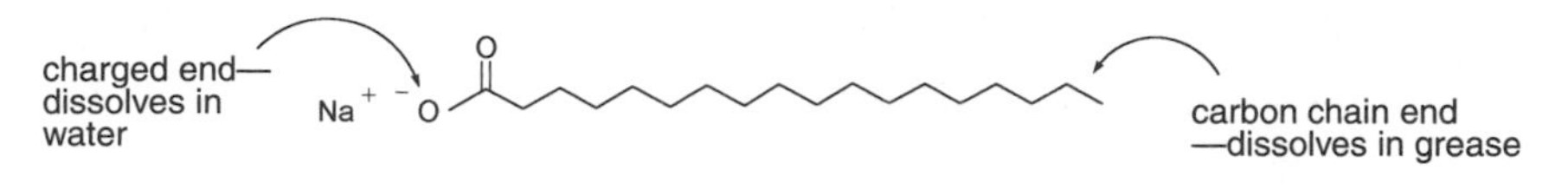

A molecule of sodium stearate—a soap molecule from beef tallow

and can also be represented as

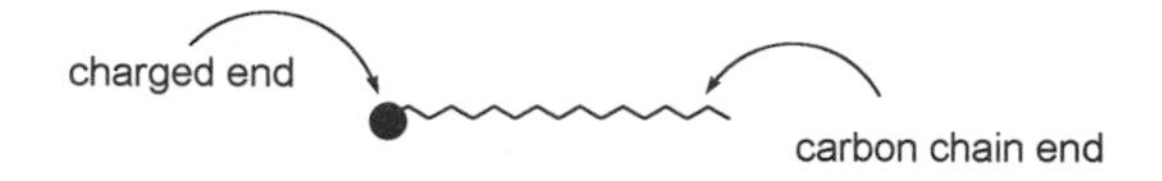

The next diagram shows the carbon chain end of many of these molecules penetrating a grease particle and forming a cluster known as a *micelle.* Soap micelles, with the negatively charged ends of the soap molecules on the outside, repel each other and are washed away by water, taking the grease particle with them.

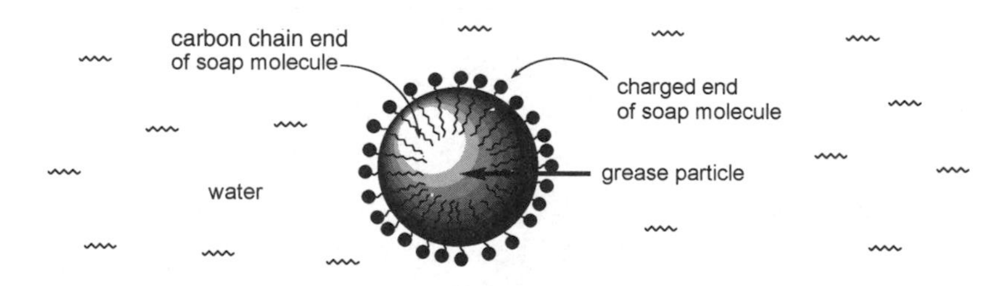

A soap micelle in water. The charged ends of the soap molecules remain in the water; the carbon chain ends are embedded in the grease.

Although soap had been made for thousands of years and commercially manufactured for hundreds of years, the chemical principles of its formation were not understood for a very long time. Soap could be made from what seemed like a wide variety of different substances—olive oil, tallow, palm oil, whale oil, pork fat—and as the chemical structures of these products were not known until the early nineteenth century, the essential similarity of their triglyceride structures was not realized. It was well into the nineteenth century before soap chemistry was appreciated. By then social changes in attitudes toward bathing, the gradually increasing prosperity of the working classes, and an understanding of the relationship between disease and cleanliness meant that soap had become an essential item of everyday living. Fine toilet soaps made from different fats and oils challenged the long-established supremacy of castile soap made from olive oil, but it was castile soap—and hence olive oil—that had been mainly responsible for maintaining some degree of personal hygiene for almost a millennium.

Today olive oil is generally recognized for its positive effects on cardiac health and for the delicious flavor it imparts to food. Its role in keeping alive the tradition of soap making and thus combating dirt and disease during medieval times is less well known. But the wealth that olive oil brought to ancient Greece ultimately allowed the development of many of the ideals of that culture that we still value today. The roots of present-day Western civilization are found in ideas fostered in the political culture of classical Greece: concepts of democracy and self-government, philosophy, logic and the beginning of rational inquiry, scientific and mathematical investigations, education, and the arts.

The affluence of Greek society permitted the participation of thousands of citizens in the process of inquiry, in rigorous debate, and in political choices. More than any other ancient society, men (women and slaves were not citizens) were involved in decisions that affected their lives. Trade in olive oil provided much of the society's prosperity; education and civic involvement followed. The glories that were Greece—now considered to be the foundation of today's democratic societies—would not have been possible without the triglyceride of oleic acid.

15. SALT

THE HISTORY OF common salt—sodium chloride, with a chemical formula of NaCl—parallels the history of human civilization. So valued is salt, so needed and so important, that it has been a major player not only in global trade but in economic sanctions and monopolies, wars, the growth of cities, systems of social and political control, industrial advances, and the migration of populations. Today salt is something of an enigma. It is absolutely essential to life—we die without it—but we are told to watch our salt intake as salt can kill. Salt is cheap; we produce and use enormous quantities of it. Yet for almost all of recorded history and probably for centuries before any history was recorded, salt was a precious commodity and often very expensive. The average person at the beginning of the nineteenth century would have had great difficulty in believing that we now routinely throw mounds of salt on roads to eliminate ice.

The price of many other molecules has dropped through the efforts of chemists, either because we can now synthesize the compound in laboratories and factories (ascorbic acid, rubber, indigo, penicillin) or because we can make artificial substitutes, compounds whose properties are so similar that the natural product is less important (textiles, plastics, aniline dyes). Today we rely on newer chemicals (refrigerants) for the preservation of food, so spice molecules no longer command the price they once did. Other chemicals—pesticides and fertilizers—have increased crop yields and hence the supply of such molecules as glucose, cellulose, nicotine, caffeine, and oleic acid. But of all compounds, salt has probably had the largest increase in production coupled with the most precipitous drop in price.

Getting Salt

Throughout history humans have collected or produced salt. Three main methods of salt production—evaporating seawater, boiling down salt solutions from brine springs, and mining rock salt—were all used in ancient times and are still in use today. Solar evaporation of seawater was (and still is) the most common method of salt production in tropical coastal regions. The process is slow but cheap. Originally seawater was thrown onto burning coals, and the salts were scraped off when the fire was extinguished. Larger quantities could be harvested from the sides of coastal rock pools. It would not have taken much imagination to realize that artificial shallow lakes or "pans," constructed in areas where tidal flow could be used to fill the pans as needed, could provide much greater quantities of salt.

Raw sea salt is of much lower quality than either brine salt or rock salt. Although seawater is about 3.5 percent dissolved salts, only about two-thirds of this is sodium chloride; the rest is a mixture of magnesium chloride ($MgCl_2$) and calcium chloride ($CaCl_2$). As these latter two chlorides are both more soluble and less abundant than sodium chloride, NaCl crystallizes out of solution first, so it is possible to remove most of the $MgCl_2$ and $CaCl_2$ by draining them away in the residual brine. But enough remains to give sea salt a sharper taste, which is attributable to these impurities. Both magnesium and calcium chloride are deliquescent, meaning they absorb water from the air, and when this happens, salt containing these additional chlorides clumps and is difficult to pour.

The evaporation of seawater was most effective in hot, dry climates, but brine springs, underground sources of highly concentrated solutions of salt—sometimes ten times more concentrated than seawater—were also an excellent source of salt in any climate, if there was wood for the fires necessary to boil off the water in the brine solutions. Wood demand for salt production helped deforest parts of Europe. Brine salt, uncontaminated by magnesium and calcium chloride, which lessened the effectiveness of food preservation, was more desirable than sea salt but also more expensive.

Deposits of rock salt or halite—the mineral name of the NaCl found in the ground—are found in many parts of the world. Halite is the dried remains of old oceans or seas and has been mined for centuries, particularly where such deposits occur near the earth's surface. But salt was so valuable that as early as the Iron Age people in Europe turned to underground mining, creating deep shafts, miles of tunnels, and large caverns hollowed out by the removal of salt. Settlements grew up around these mines, and the continued extraction of salt led to the establishment of towns and cities, which grew wealthy from the salt economy.

Salt making or mining was important in many places in Europe throughout the Middle Ages; so valued was salt that it was known as "white gold." Venice, center of the spice trade for centuries, started as a community that obtained a living by extracting salt from the brines of the marshy lagoons in the area. Names of rivers, towns and cities in Europe—Salzburg, Halle, Hallstatt, Hallein, La Salle, Moselle—commemorate their links with salt mining or salt production, as the Greek word for salt is *hals* and the Latin is *sal. Tuz,* the Turkish name for salt, shows up in Tuzla, a town in a salt-producing region of Bosnia-Herzegovina, as well as in coastal communities in Turkey with the same or similar names.

Today, through tourism, salt is still the source of wealth for some of these old salt towns. In Salzburg, Austria, salt mines are a major tourist attraction, as they are at Wieliczka, near Cracow in Poland, where, in the great caverns hollowed out by salt removal, a dance hall, a chapel with an altar, religious statues carved from salt, and an underground lake now enchant thousands of visitors. The largest *salar,* or saltpan, in the world is the Salar de Uyuni in Bolivia, where tourists can stay at a nearby hotel made entirely from salt.

The salt hotel near Salar de Uyuni in Bolivia. (Photo by Peter Le Couteur)

TRADING SALT

That salt has been a trade commodity from earliest times is shown in records from ancient civilizations. The ancient Egyptians traded for salt, an essential ingredient in the mummification process. The Greek historian Herodotus reported visiting a salt mine in the Libyan desert in 425 B.C. Salt from the great salt plain at Danakil in Ethiopia was traded to the Romans and Arabs and exported as far as India. The Romans established a large coastal saltworks at Ostia, which was then at the mouth of the River Tiber, and around 600 B.C. built a road, the Via Salaria, to transport salt from the coast to Rome. One of the main thoroughfares in present-day Rome is still known as Via Salaria—the salt road. Forests were felled to provide fuel for the saltworks at Ostia, and subsequent soil erosion washed increasing amounts of sediment into the Tiber. Extra sediment hastened the expansion of the delta at the river mouth. Centuries later Ostia was no longer on the coast, and the saltworks had to be moved out to the shoreline again. This has been cited as one of the first examples of the impact of human industrial activity on the environment.

Salt was the basis for one of the world's great trade triangles and coincidently for the spread of Islam to the west coast of Africa. The extremely arid and inhospitable Sahara Desert was for centuries a barrier between the northern African countries bordering the Mediterranean and the rest of the continent to the south. Though there were enormous deposits of salt in the desert, south of the Sahara salt was in great demand. In the eighth century Berber merchants from North Africa began to trade grains and dried fruit, textiles and utensils, for slabs of halite mined from the great salt deposits of the Sahara (in present-day Mali and Mauritania). So abundant was salt at these sites that entire cities such as Teghaza (city of salt), built from blocks of salt, grew up around the mines. The Berber caravans, often comprising thousands of camels at a time, now laden with slabs of salt, would continue across the desert to Timbuktu, originally a small camp on the southern edge of the Sahara on a tributary of the Niger River.

By the fourteenth century, Timbuktu had become a major trading post, exchanging gold from West Africa for salt from the Sahara. It also became a center for the expansion of Islam, which was brought to the region by the Berber traders. At the height of its power—most of the sixteenth century—Timbuktu boasted an influential Koranic university, great mosques and towers, and impressive royal palaces. Caravans leaving Timbuktu carried gold, and sometimes slaves and ivory, back to the Mediterranean coast of Morocco and thence to Europe. Over the centuries many tons of gold were shipped to Europe through the Saharan gold/salt trade route.

Saharan salt was also shipped to Europe as the demand there for salt increased. Freshly caught fish must be preserved quickly, and while smoking and drying were rarely possible at sea, salting was. The Baltic and North Seas teemed with herring, cod, and haddock, and from the fourteenth century onward millions of tons of these fish, salted at sea or in nearby ports, were sold throughout Europe. In the fourteenth and fifteenth centuries the Hanseatic League, an organization of north German towns, controlled the trade in salt fish (and almost everything else) in the countries bordering the Baltic Sea.

The North Sea trade was centered in Holland and the east coast of England. But with salt available to preserve the catch, it became possible to fish even farther afield. By the end of the fifteenth century fishing boats from England, France, Holland, the Basque region of Spain, Portugal, and other European countries were regularly sailing to fish the Grand Banks off Newfoundland. For four centuries fishing fleets plundered the vast schools of cod in this region of the North Atlantic, cleaning and salting the fish as they were caught and returning to port with millions of tons of what seemed an inexhaustible supply. Sadly this was not the case; Grand Banks cod were brought to the brink of extinction in the 1990s. Today a moratorium on cod fishing, introduced by Canada in 1992, is being observed by many, but not all, of the traditional fishing nations.

With salt in such demand, it is hardly surprising that it was often considered a prize of war rather than a commodity of trade. In ancient times settlements around the Dead Sea were conquered specifically for their precious supplies of salt. In the Middle Ages the Venetians waged war against neighboring coastal communities who threatened their all-important salt monopoly. Capturing an enemy's supply of salt was long considered a sound wartime tactic. During the American Revolution salt shortages resulted from a British embargo of imports from Europe and the West Indies into the former colony. The British destroyed salt works along the New Jersey coast to maintain the hardship affecting the colonists as a result of the high prices for imported salt. The 1864 capture of Saltville, Virginia, by Union forces during the American Civil War was seen as a vital step in reducing civilian morale and defeating the Confederate army.

It has even been suggested that a lack of dietary salt might have prevented wartime wounds from healing and was thus responsible for the death of thousands of Napoleon's soldiers during the 1812 retreat from Moscow. Lack of ascorbic acid (and the subsequent onset of scurvy) seems as likely a culprit as lack of salt under these circumstances, so both these compounds could join tin and lysergic acid derivatives as chemicals that thwarted Napoleon's dreams.

THE STRUCTURE OF SALT

Halite, with a solubility of about 36 grams in every 100 grams of cold water, is far more soluble in water than are other minerals. As life is thought to have developed in the oceans and as salt is essential for life, without this solubility of salt life as we know it would not exist.

The Swedish chemist Svante August Arrhenius first proposed the idea of oppositely charged ions as an explanation for the structure and properties of salts and their solutions in 1887. For over a century scientists had been mystified by a particular property of salt solutions—their ability to conduct electrical currents. Rainwater shows no electrical conductivity, yet saline solutions and solutions of other salts are excellent conductors. Arrhenius's theory accounted for this conductivity; his experiments showed that the more salt dissolves into solution, the greater the concentration of the charged species—the ions—needed to carry the electrical current.

The concept of ions, as proposed by Arrhenius, also explained why acids, despite seemingly different structures, have similar properties. In water all acids produce hydrogen ions (H^+) which are responsible for the sour taste and chemical reactivity of acid solutions. Although Arrhenius's ideas were not accepted by many conservative chemists of the time, he displayed a commendable degree of perseverance and diplomacy in campaigning determinedly for the soundness of the ionic model. His critics were eventually convinced, and Arrhenius received the 1903 Nobel Prize in chemistry for his electrolytic dissociation theory.

By this time there was both a theory and practical evidence for how ions form. British physicist Joseph John Thomson in 1897 had demonstrated that all atoms contain *electrons,* the negatively charged fundamental particle of electricity that had been first proposed in 1833 by Michael Faraday. Thus if one atom lost an electron or electrons, it became a positively charged ion; if another atom gained an electron or electrons, a negatively charged ion was formed.

Solid sodium chloride is composed of a regular array of two different ions—positively charged sodium ions and negatively charged chloride ions—held together by strong attractive forces between the negative and positive charges.

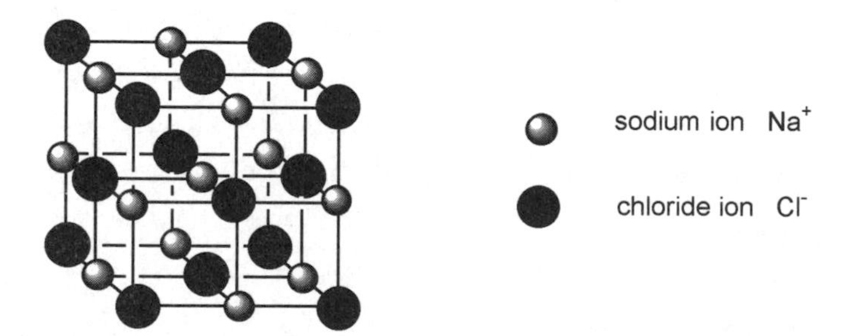

The three-dimensional structure of solid sodium chloride. The lines joining the ions are nonexistent—they are included here to show the cubic arrangement of the ions.

Water molecules, although not consisting of ions, are partially charged. One side of a water molecule (the hydrogen side) is slightly positive, and the other side (the oxygen side) is slightly negative. This is what allows sodium chloride to dissolve in water. Although the attraction between a positive sodium ion and the negative end of water molecules (and the attraction between negative chloride ions and the positive end of water molecules) are similar to the attractive force between Na^+ ions and Cl^- ions, what ultimately accounts for the solubility of salt is the tendency for these ions to disperse randomly. If ionic salts do not dissolve to any extent in water, it is because the attractive forces between the ions are greater than the water-to-ion attractions.

Representing the water molecule as:

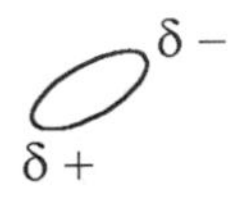

with δ– indicating the partial negative end of the molecule and δ+ the partial positive end of the molecule, we can show the negative chloride ions in aqueous solution as surrounded by the slightly positive end of water molecules:

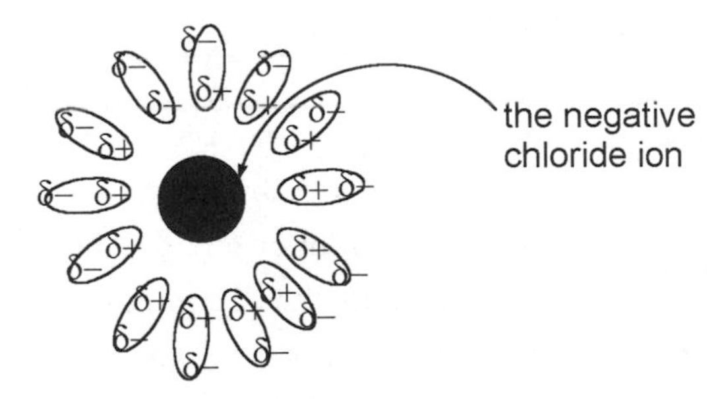

and the positive sodium ion in aqueous solution as surrounded by the slightly negative end of water molecules:

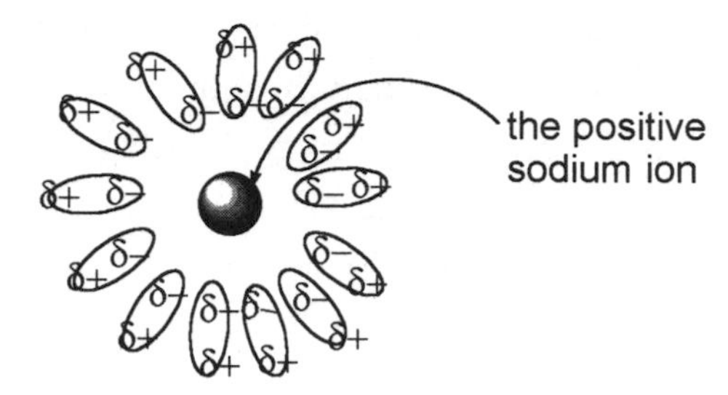

It is this solubility of sodium chloride that makes salt—by attracting water molecules—such a good preservative. Salt preserves meat and fish by removing water from the tissues; in conditions of much-reduced water levels and a high salt content, the bacteria that cause decay are unable to survive. A lot more salt was used in this manner to keep food from decaying than was deliberately added to enhance flavors. In regions where dietary salt came mainly from meat, additional salt for food preservation was an essential factor in maintaining life. The other traditional methods of food preservation, smoking and drying, very often required the use of salt as part of the process. Food would be soaked in a brine solution prior to the actual smoking or drying. Communities without a local source of salt were dependent on supplies obtained by trade.

THE BODY'S NEED FOR SALT

From earliest times, even if it was not needed for food preservation, humans recognized the necessity to obtain salt for their diet. Ions from salt play an essential role in the human body, maintaining the electrolyte balance between cells and the fluid surrounding the cells. Part of the process that generates the electrical impulses transmitted along neurons in the nervous system involves what is called the sodium-potassium pump. More Na^+ (sodium) ions are forced out of a cell than K^+ (potassium) ions are pumped into it, resulting in a net negative charge of the cytoplasm inside the cell compared with the outside of the cell membrane. Thus a difference in charge—known as a membrane potential—is generated, which powers electrical impulses. Salt is therefore vital for the functioning of nerves and ultimately muscle movement.

Cardiac glycoside molecules, such as the digoxin and digitoxin found in foxglove, inhibit the sodium-potassium pump, giving a higher level of Na^+ ions inside the cell. This ultimately increases the contractive force of the heart muscles and accounts for the activity of these molecules as heart stimulants. The chloride ion from salt is also needed in the body to produce hydrochloric acid, an essential component of the digestive juices in the stomach.

Salt concentration in a healthy person varies within a very narrow range. Lost salt must be replaced; excess salt must be excreted. Salt deprivation causes loss of weight and appetite, cramps, nausea, and inertia and can, in extreme cases of depletion of body salt—such as in marathon runners—lead to vascular collapse and death. Excess sodium ion intake, however, is known to contribute to high blood pressure, a significant factor for cardiovascular disease, and to kidney and liver disorders.

The average human body contains about four ounces of salt; we are continuously losing salt, mainly through perspiration and excretion in urine, and so we have to replace it on a daily basis. Prehistoric man filled his dietary need for salt from the meat of the largely herbivorous animals he hunted, as raw meat is an excellent source of salt. As agriculture developed and grains and vegetables became a larger part of the diet, supplementary salt was needed. While carnivorous animals do not seek out salt licks, herbivorous animals need to do so. Humans in parts of the world where little meat is eaten and vegetarians require additional salt. Supplemental salt, a necessity as soon as humans adopted a settled agrarian way of life, must be obtained locally or through trade.

Taxing Salt

The human need for salt, together with its specific methods of production, have historically made this mineral peculiarly fitted for political control, monopoly, and taxation. For a government, a tax on salt would produce a reliable income. There was no substitute for salt, and everyone needed it, so everyone would have to pay. Salt sources were known; the production of salt is difficult to hide, salt itself is bulky and hard to conceal, and its transportation can be easily regulated and taxed. From 2000 B.C. in China, where the Emperor Hsia Yu ordered that the imperial court would be supplied by salt from Shantung Province, down through the ages, salt has been profitable for governments through taxes, tolls, and tariffs. In biblical times salt, considered a spice and taxed as such, was subject to customs duty at the many stopping places along caravan routes. After the death of Alexander the Great in 323 B.C., officials in Syria and Egypt continued to collect a salt tax that had originally been imposed by the Greek administration.

Throughout all these centuries the process of gathering taxes required tax collectors, many of whom became wealthy by increasing tax rates, adding extra duties, and selling exemptions. Rome was no exception. Originally the Ostia saltworks on the Tiber delta were taken over by the Roman state, so that salt could be supplied at reasonable rates to everyone. Such largesse did not last. The revenues from taxing salt offered too great a temptation, and a salt tariff was imposed. As the Roman Empire expanded, so did salt monopolies and salt taxes. Tax gatherers, independent agents supervised by the governor of each Roman province, levied taxes wherever they could. For those who lived far from salt-producing areas, the high cost of salt not only reflected transportation costs but tariffs, taxes, and duties at every step of the way.

Throughout the Middle Ages in Europe the taxation of salt continued, often in the form of tolls imposed on barges or carts carrying salt from salt mines or coastal production plants. It reached its height in

France with the infamous, oppressive, and much-hated salt tax known as the *gabelle.* Reports on the origin of the gabelle vary. Some accounts say Charles of Anjou in Provence imposed it in 1259, others that it started as a general tax applied to commodities like wheat, wine, and salt in the late thirteenth century to help maintain a permanent army. Whatever its true origin, by the fifteenth century the gabelle had become one of France's main national levies, and the name referred only to the tax on salt.

But the gabelle was not just a tax on salt. It also carried the requirement that every man, woman, and child over the age of eight purchase a weekly amount of salt, at a price set by the king. Not only could the salt tax itself be raised, but the obligatory ration could also be increased at the monarch's whim. What was once intended as a uniform tax on the population soon extracted higher penalties on some regions of France than others. In general, those provinces that obtained their salt from Atlantic salt works were subject to the *grande gabelle,* more than twice what was paid in regions—known as Les Provinces des Petites Gabelles—where salt was supplied from Mediterranean saltworks. Through political influence or treaty arrangements some areas were exempt from the gabelle or paid only a fraction; at times there was no gabelle in Brittany and a special low rate in Normandy. At its height the gabelle increased the price of salt more than twenty times its real cost for those citizens in Les Provinces des Grandes Gabelles.

Salt tax collectors—referred to as gabelle farmers, as they harvested the taxes from the people—would monitor the per capita use of salt to ensure consumption obligations were met. Attempts to smuggle salt were rife despite severe penalties for being discovered with contraband salt; a common punishment in such cases was a sentence to the galleys. Peasant farmers and poor city dwellers were the most severely affected by the harsh and unfairly applied gabelle. Appeals to the king for relief from this onerous tax fell on deaf ears, and it has been suggested that the gabelle was one of the main grievances responsible for the French Revolution. It was abolished at the height of the revolution in 1790, and more than thirty gabelle collectors were executed. But the abolition did not last. In 1805 Napoleon reintroduced the gabelle, a measure he claimed was necessary to pay for his war against Italy. It was not finally eliminated until after World War II.

France was not the only country where such taxes on a necessity of life were burdensome. In coastal Scotland, particularly around the Firth of Forth, salt had been produced for centuries before it was ever taxed. Solar evaporation was not feasible in the cool, damp climate; seawater was boiled in large vessels, which were originally wood-fired and later coal-fired. By the 1700s there were more than 150 such saltworks in Scotland, plus numerous others that were peat-fired. The salt industry was so important to the Scots that the Eighth Article of the 1707 Treaty of Union between Scotland and England guaranteed Scotland a seven-year exemption from English salt taxes and after that a reduced rate forever. England's salt industry was based on the extraction of salt from brine as well as the mining of rock salt. Both methods were a great deal more efficient and profitable than the coal-fired seawater method of Scottish production. The industry in Scotland needed relief from English salt taxes in order to survive.

In 1825 the United Kingdom became the first country to abolish its salt tax, not so much because of the resentment this tax had generated among the working class throughout the centuries as because of the recognition of the changing role of salt. The Industrial Revolution is usually thought of as a mechanical revolution—the development of the flying shuttle, the spinning jenny, the water frame, the steam engine, the power loom—but it was also a chemical revolution. Large-scale production of chemicals was required for the textile industry, for bleaching, soap making, glassmaking, potteries, the steel industry, tanneries, paper manufacturing, and the brewing and distilling industries. Manufacturers and factory owners pushed for the repeal of the salt tax because salt was becoming vastly more important as a starting material in manufacturing processes than as a preservative and food culinary supplement. Removal of the salt tax, sought by generations of the poor, became a reality only when salt was recognized as a key raw material for industrial prosperity in Britain.

Britain's enlightened stance on salt taxes did not extend to her colonies. In India a British-imposed salt tax became a symbol of colonial oppression seized upon by Mahatma Gandhi as he led the struggle for Indian independence. The salt tax in India was more than a tax. As many conquerors had found over the centuries, control of salt supplies meant political and economic control. Government regulations in British India made the nongovernmental sale or production of salt a criminal offense. Even collecting salt formed through natural evaporation around rock pools at the seacoast was outlawed. Salt, sometimes imported from England, had to be purchased from government agents at prices established by the British. In India, where the diet is mainly vegetarian and the often intensely hot climate promotes salt loss through sweating, it was especially important that salt be added to food. Under colonial rule the population was forced to pay for a mineral that millions had traditionally been able to gather or produce themselves at little or no cost.

In 1923, almost a century after Britain had eliminated the salt tax on its own citizens, the salt tax in India was doubled. In March 1930, Gandhi and a handful of supporters started a 240-mile march to the small village of Dandi, on the northwest coast of India. Thousands joined his pilgrimage, and once they reached the shoreline, they began to collect salt incrustations from the beach, to boil seawater, and to sell the salt they produced. Thousands more joined in breaking the salt laws; illegal salt was sold in villages and cities all over India and was frequently confiscated by the police. Gandhi's supporters were often brutalized by the police, and thousands were imprisoned. Thousands more took their places making salt. Strikes, boycotts, and demonstrations followed. By the following March the draconian salt laws of India had been modified: local people were permitted to collect salt or make it from local sources and sell it to others in their villages. Although a commercial tax still applied, the British government's salt monopoly was broken. Gandhi's ideals of nonviolent civil disobedience had proven effective, and the days of the British Raj were numbered.

SALT AS A STARTING MATERIAL

The removal of the salt tax in Britain was important not only to those industries that used salt as part of their manufacturing processes but also to companies that made inorganic chemicals, where salt was a major starting material. It was particularly significant for another sodium compound, sodium carbonate (Na_2CO_3), known as soda ash or washing soda. Soda ash, used in soap making and needed in large quantities as the demand for soap increased, came mainly from naturally occurring deposits, often incrustations around drying alkaline lakes or from residues from burning kelp and other seaweeds. Soda ash from these sources was impure and supplies were limited, so the possibility of producing sodium carbonate from the plentiful supplies of sodium chloride attracted attention. In the 1790s Archibald Cochrane, the ninth earl of Dundonald—now considered one of the leaders of Britain's chemical revolution and a founder of the chemical alkali industry—whose modest family estate on Scotland's Firth of Forth bordered on numerous coal-fired saltpans, took out a patent for converting salt to "artificial alkali," but his process was never a commercial success. In France in 1791, Nicolas Leblanc developed a method of making sodium carbonate from salt, sulfuric acid, coal, and limestone. The onset of the French Revolution delayed establishment of Leblanc's process, and it was in England where the profitable manufacture of soda ash began.

In Belgium in the early 1860s, the brothers Ernest and Alfred Solvay developed an improved method of converting sodium chloride to sodium carbonate using limestone ($CaCO_3$) and ammonia gas (NH_3). The key steps were the formation of a precipitate of sodium bicarbonate ($NaHCO_3$) from a concentrated solution of brine, infused with ammonia gas and carbon dioxide (from limestone):

$$\underset{\text{sodium chloride}}{NaCl_{(aq)}} + \underset{\text{ammonia}}{NH_{3(g)}} + \underset{\text{carbon dioxide}}{CO_{2(g)}} + \underset{\text{water}}{H_2O_{(l)}} \rightarrow \underset{\text{sodium bicarbonate}}{NaHCO_{3(s)}} + \underset{\text{ammonium chloride}}{NH_4Cl_{(aq)}}$$

and then production of sodium carbonate by heating the sodium bicarbonate:

$$\underset{\text{sodium bicarbonate}}{2\ NaHCO_{3(s)}} \rightarrow \underset{\text{sodium carbonate}}{Na_2CO_{3(s)}} + \underset{\text{carbon dioxide}}{CO_{2(g)}}$$

Today the Solvay process remains the main method of preparing synthetic soda ash, but discoveries of massive deposits of natural soda ash—the Green River basin of Wyoming, for example, has soda ash resources estimated at over ten billion tons—have decreased the demand for its preparation from salt.

Another sodium compound, caustic soda (NaOH), has also long been in demand. Industrially, caustic soda or sodium hydroxide is made by passing an electric current through a solution of sodium chloride—a process known as *electrolysis.* Caustic soda, one of the ten most produced chemicals in the United States, is essential in extracting aluminum metal from its ore and in the manufacture of rayon, cellophane, soaps, detergents, petroleum products, paper, and pulp. Chlorine gas, also produced in the electrolysis of brine, was originally considered a by-product of the process, but it was soon discovered that chlorine was an excellent bleaching agent and a potent disinfectant. Today the production of chlorine is as much a reason for commercial electrolysis of NaCl solutions as the production of NaOH. Chlorine is now used in the manufacture of many organic products, such as pesticides, polymers, and pharmaceuticals.

From fairy tales to biblical parables, from Swedish folk myths to North American Indian legends, different societies around the world tell stories of salt. Salt is used in ceremonies and rites, it symbolizes hospitality and good fortune, and it protects against evil spirits and ill luck. The important role of salt in shaping human culture is also seen in language. We earn a salary—the derivation of the word comes from the fact that Roman soldiers were often paid in salt. Our words for salad (originally dressed only with salt), sauce and salsa, sausage and salami all come from the same Latin root. As in other languages, our everday speech is "salted" with metaphors: "salt of the earth," "old salt," "worth his salt," "below the salt," "with a grain of salt," "back to the salt mine."

The ultimate irony in the story of salt is that despite all the wars fought over it, despite the battles and protests over taxation and tolls imposed on it, despite migrations in search of it and the despair of hundreds of thousands imprisoned for smuggling it, by the time the discovery of new underground salt deposits and modern technology had vastly decreased its price, the need for salt in food preservation was already greatly diminished—refrigeration had become the standard method of preventing decomposition of food. This compound that throughout history has been honored and revered, desired and fought over, and sometimes been valued more highly than gold, is nowadays not only cheap and readily available but is considered commonplace.

16. CHLOROCARBON COMPOUNDS

IN 1877 THE ship *Frigorifique* sailed from Buenos Aires to the French port of Rouen with a cargo of Argentinian beef in its hold. While today this passage would be seen as routine, it was in fact a historic voyage. The ship carried a cooled cargo that signaled the beginning of the era of refrigeration and the end of food preservation by molecules of spice and salt.

Keeping Cool

Since at least 2000 B.C., people have used ice to keep things cool, relying on the principle that solid ice absorbs heat from its surroundings as it melts. The liquid water produced drains away, and a new supply of ice is added. Refrigeration, on the other hand, involves not solid and liquid phases but liquid and vapor phases. As liquid evaporates, it absorbs heat from its surroundings. The vapor produced by evaporation is then returned to the liquid state by compression. This compression stage is what puts the *re* in *refrigeration*—vapor is returned to a liquid, then re-evaporates causing cooling, and the whole cycle repeats. A key component of the cycle is an energy source to drive the mechanical compressor. The old fashioned icebox, where ice had to be added continually, was technically not a refrigerator. Today we often use the word *refrigerate* to mean "to make or keep cool" without considering how it is done.

A true refrigerator needs a refrigerant—a compound that undergoes the evaporation-compression cycle. As early as 1748 ether was used to demonstrate the cooling effect of a refrigerant, but it was more than a hundred years before a compressed ether machine was employed as a refrigerator. Around 1851, James Harrison, a Scotsman who had immigrated to Australia in 1837, built an ether-based vapor-compression refrigerator for an Australian brewery. He and an American, Alexander Twining, who had made a similar vapor-compression refrigeration system at about the same time, are considered to be among the first developers of commercial refrigeration.

Ammonia was used as a refrigerant in 1859, by Ferdinand Carré of France—another claimant to the title of first commercial developer of refrigeration. Methyl chloride and sulfur dioxide were also used in these early days; sulfur dioxide was the cooling agent for the world's first artificial skating rink. These small molecules effectively ended the reliance on salt and spices for food preservation.

$C_2H_5—O—C_2H_5$	NH_3	CH_3Cl	SO_2
ether (diethyl ether)	ammonia	methyl chloride	sulfur dioxide

In 1873, after successfully establishing land-based refrigeration for the Australian meat-packing industry as well as the breweries, James Harrison decided to transport meat on a refrigerated ship from Australia to Britain. But his ether-based evaporation-compression mechanical system failed at sea. Then in early December 1879 the S.S. *Strathleven,* equipped by Harrison, left Melbourne and arrived in London two months later with forty tons of still-frozen beef and mutton. Harrison's refrigeration process was proven. In 1882 a similar system was installed on the S.S. *Dunedin,* and the first cargo of New Zealand lamb was shipped to Britain. Though the *Frigorifique* is often referred to as the world's first refrigerated ship, technically the claim better fits Harrison's 1873 attempt. It was not, however, the first *successful* voyage of a refrigerated ship. This latter title more rightly belongs to the S.S. *Paraguay,* which arrived in Le Havre, France, in 1877 with a cargo of frozen beef from Argentina. The *Paraguay*'s refrigeration system was designed by Ferdinand Carré and used ammonia as a refrigerant.

On the *Frigorifique* the "refrigeration" was maintained by water that was cooled by ice (stored in a well-insulated room) and then pumped around the ship in pipes. The ship's pump broke down on the journey from Buenos Aires, and the meat was spoiled before it arrived in France. So although the *Frigorifique* predated the S.S. *Paraguay* by a number of months, it was not a true refrigerated ship; it was only an insulated ship, keeping food chilled or frozen with stored ice. What the *Frigorifique* can claim to be is a pioneer in transporting chilled meat across the ocean, even if it was not a successful pioneer.

Irrespective of whose claim to the first refrigerated ship is most valid, by the 1880s the mechanical compression-evaporation process was set to solve the problem of transporting meat from the producing areas of the world to the larger markets of Europe and the eastern United States. Ships from Argentina and the even more distant cattle and sheep pastures of Australia and New Zealand faced a two- or three-month journey through the warm temperatures of the tropics. The simple ice system of the *Frigorifique* would not have been effective. Mechanical refrigeration began to increase in reliability, giving ranchers and farmers a new means of getting their products to world markets. Refrigeration thus played a major role in the economic development of Australia, New Zealand, Argentina, South Africa, and other countries, where great distances from markets reduced their natural advantages of abundant agricultural production.

Fabulous Freons

The ideal refrigerant molecule has special practical requirements. It must vaporize within the right temperature range; it must liquefy by compression—again within the required temperature range; and it must absorb relatively large amounts of heat as it vaporizes. Ammonia, ether, methyl chloride, sulfur dioxide, and similar molecules satisfied these technical requirements as good refrigerants. But they either decomposed, were fire hazards, were poisonous, or smelled terrible—sometimes all of these.

Despite the problems with refrigerants, the demand for refrigeration, both commercial and domestic, grew. Commercial refrigeration, developed to meet the demand of trade, preceded home refrigeration by fifty or more years. The first refrigerators for in-home use became available in 1913 and by the 1920s had begun to replace the more traditional icebox, supplied with ice from industrial ice plants. In some early home refrigerators the noisy compressor unit was installed in the basement, separate from the food box.

Looking for an answer to concerns about toxic and explosive refrigerants, mechanical engineer Thomas Midgley, Jr.—already successful as the developer of tetraethyl lead, a substance added to gasoline to reduce engine knock—and chemist Albert Henne, working at the Frigidaire Division of General Motors, considered compounds that were likely to have boiling points within the defined range of a refrigeration cycle. Most of the known compounds that fitted this criterion were already in use or had been eliminated as impractical, but one possibility, compounds of fluorine, had not been considered. The element fluorine is a highly toxic and corrosive gas, and few organic compounds containing fluorine had ever been prepared.

Midgley and Henne decided to prepare a number of different molecules containing one or two carbon atoms and a varying number of fluorine and chlorine atoms instead of hydrogen atoms. The resulting compounds, chlorofluorocarbons (or CFCs, as they are now known), admirably fulfilled all the technical requirements of a refrigerant and were also very stable, nonflammable, nontoxic, inexpensive to manufacture, and nearly odorless.

In a very dramatic manner Midgley demonstrated the safety of his new refrigerants at a 1930 meeting of the American Chemical Society in Atlanta, Georgia. He poured some liquid CFC into an open container, and as the refrigerant boiled, he put his face in the vapor, opened his mouth, and took a deep breath. Turning to a previously lit candle he slowly exhaled the CFC, extinguishing the candle flame—a remarkable and unusual demonstration of the nonexplosive and nonpoisonous properties of this chlorofluorocarbon.

A number of different CFC molecules were then put into use as refrigerants: dichlorodifluoromethane, which was more usually known by its Du Pont Corporation trade name of Freon 12; trichlorofluoromethane, or Freon 11; and 1,2-dichloro-1,1,2,2,-tetrafluoroethane, or Freon 114.

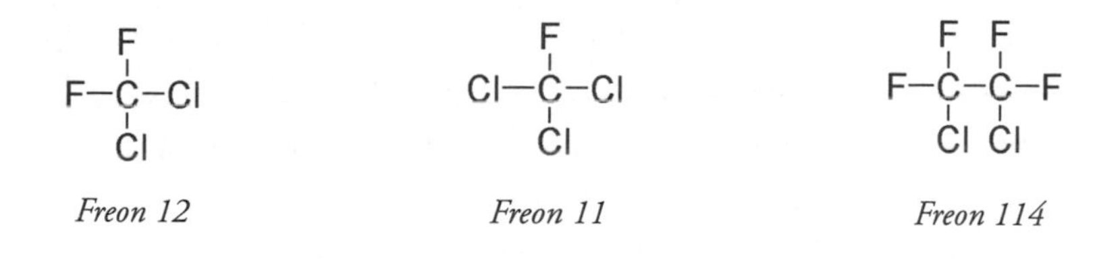

Freon 12 *Freon 11* *Freon 114*

The numbers in the Freon names were a code developed by Midgley and Henne. The first digit is the number of carbon atoms minus one. If this is zero it does not get written; thus Freon 12 is really Freon 012. The next number is the number of hydrogen atoms (if any) plus one. The last number is the number of fluorine atoms. Any remaining atoms are chlorine.

CFCs were the perfect refrigerants. They revolutionized the refrigeration business and became the basis for a huge increase in home refrigeration especially as more and more homes were connected to electricity. By the 1950s a refrigerator was considered a standard home appliance in the developed world. Shopping for fresh foods on a daily basis was no longer necessary. Perishable items could be stored safely and meals readied ahead of time. The frozen food industry blossomed; new products were developed; ready-to-eat meals—TV dinners—were introduced. CFCs changed how we bought food, how we prepared food, and even what food we ate. Refrigeration allowed heat-sensitive antibiotics, vaccines, and other medications to be stored and shipped around the world.

A plentiful supply of safe refrigerant molecules also gave people the means of cooling something other than food—their surroundings. For centuries capturing natural breezes, moving air by means of fans, and using the cooling effect of evaporating water had been the main ways of coping with the temperature of hot climates. Once CFCs arrived on the scene, the fledgling air-conditioning industry expanded rapidly. In tropical regions and other places where summers were extremely hot, air-conditioning made homes, hospitals, offices, factories, malls, cars—anywhere people lived and worked—more comfortable.

Other uses for CFCs were also found. As they reacted with virtually nothing, they made ideal propellants for virtually everything that could be applied through a spray can. Hair sprays, shaving foams, colognes, suntan lotions, whipped cream toppings, cheese spreads, furniture polish, carpet cleaners, bathtub mildew removers, and insecticides are just a few of the huge variety of products that were forced through the tiny holes of aerosol cans by expanding CFC vapor.

Some CFCs were perfect for foaming agents in the manufacture of the very light and porous polymers used as packing materials, as insulating foam in buildings, as fast food containers, and as "Styrofoam" coffee cups. The solvent properties of other CFCs, such as Freon 113, made them ideal cleaners for circuit boards and other electronic parts. Substitution of a bromine atom for a chlorine or a fluorine in the CFC molecule produced heavier compounds of higher boiling point, such as Freon 13B1 (the code is adjusted to indicate bromine), just right for use in fire extinguishers.

Freon 113

Freon 13B1

By the early 1970s almost a million tons of CFCs and related compounds were being produced annually. It seemed that these molecules were indeed ideal, perfectly suited to their roles in the modern world, without a drawback or a downside. They seemed to make the world a better place.

FREONS REVEAL THEIR DARK SIDE

The glow around CFCs lasted until 1974, when disturbing findings were announced by researchers Sherwood Rowland and Mario Molina at another meeting of the American Chemical Society in Atlanta. They had found that the very stability of CFCs presented a totally unexpected and extremely disturbing problem.

Unlike less stable compounds, CFCs don't break down by ordinary chemical reactions, a property that had originally made them so appealing. CFCs released into the lower atmosphere drift around for years or even decades, then eventually rise to the stratosphere, where they are ruptured by solar radiation. Within the stratosphere there is a stratum stretching from about fifteen to thirty kilometers above the surface of the earth known as the *ozone layer.* This may sound like a fairly thick cover, but if this same ozone layer were to exist at sea level pressures, it would measure only millimeters. In the rarefied region of the stratosphere, the air pressure is so low that the ozone layer is vastly expanded.

Ozone is an elemental form of oxygen. The only difference between these forms is the number of atoms of oxygen in each molecule—oxygen is O_2 and ozone is O_3—but the two molecules have very different properties. High above the ozone layer intense radiation from the sun breaks the bond in an oxygen molecule, producing two oxygen atoms:

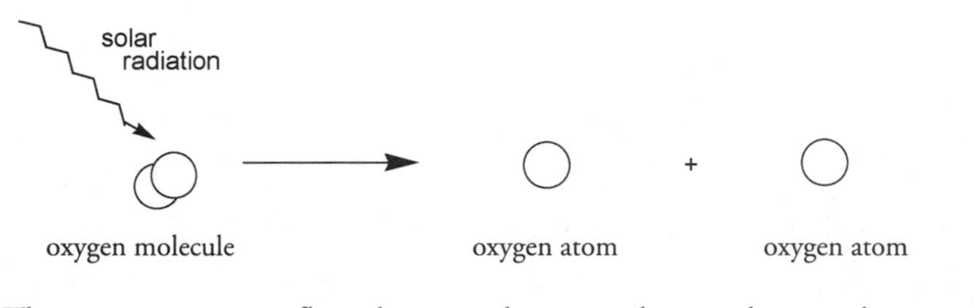

These oxygen atoms float down to the ozone layer, where each reacts with another oxygen molecule to form ozone:

+ 

oxygen atom oxygen molecule ozone molecule

Within the ozone layer ozone molecules are broken up by high-energy ultraviolet radiation to form an oxygen molecule and an oxygen atom.

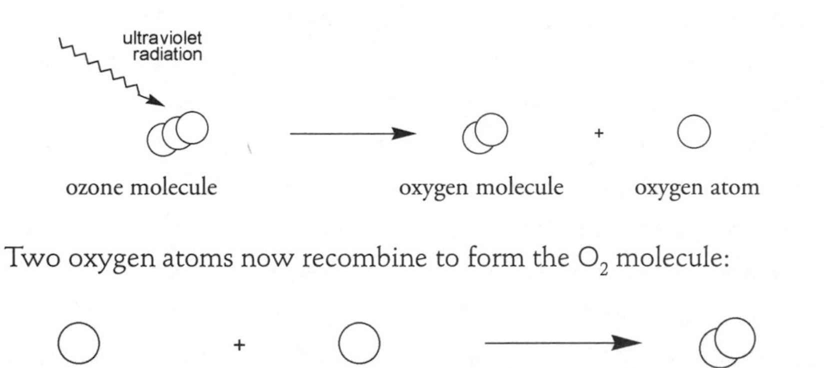

Two oxygen atoms now recombine to form the O_2 molecule:

+

oxygen atom oxygen atom oxygen molecule

Thus in the ozone layer ozone is constantly being made and constantly being broken down. Over millennia these two processes have achieved a balance, so that the concentration of ozone in the Earth's atmosphere remains relatively constant. This arrangement has important consequences for life on earth; ozone in the ozone layer absorbs the portion of the ultraviolet spectrum from the sun that is most harmful to living things. It has been said that we live under an umbrella of ozone that protects us from the sun's deadly radiation.

But Rowland and Molina's research findings showed that chlorine atoms increase the rate of breakdown of ozone molecules. As a first step, a chlorine atom collides with ozone to form a chlorine monoxide molecule (ClO), and leaves behind an oxygen molecule:

In the next step ClO reacts with an oxygen atom to form an oxygen molecule and regenerates the chlorine atom:

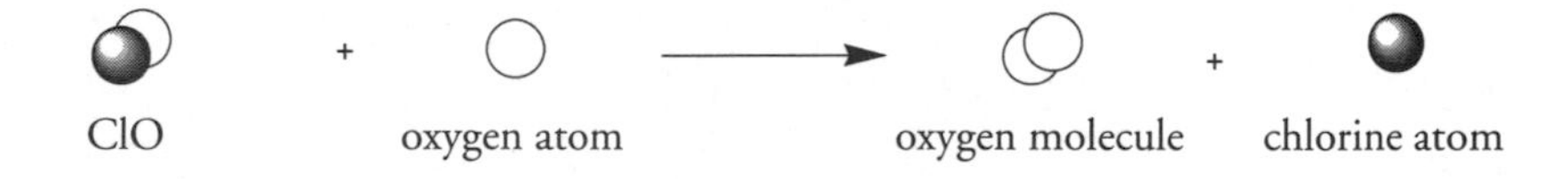

Rowland and Molina suggested that this overall reaction could upset the balance between ozone and oxygen molecules, as chlorine atoms hasten the breakup of ozone but have no effect on the making of ozone. A chlorine atom, used up in the first step of the ozone breakdown and produced anew in the second step, acts as a catalyst; that is, it increases the rate of reaction but is not itself consumed. This is the most alarming aspect of the effect of chlorine atoms on the ozone layer—not just that ozone molecules are being destroyed by chlorine but that the same chlorine atom can catalyze this breakdown again and again. One estimate is that, on average, every chlorine atom that finds its way to the upper atmosphere via a CFC molecule will destroy a hundred thousand ozone molecules before it is deactivated. For every 1 percent of ozone layer depletion, an additional 2 percent of damaging ultraviolet radiation might penetrate the Earth's atmosphere.

Based on their experimental results, Rowland and Molina predicted that chlorine atoms from CFCs and related compounds would, on reaching the stratosphere, start the decomposition of the ozone layer. At the time of their research billions of CFC molecules were being released into the atmosphere every day. The news that CFCs posed a real and immediate threat of depletion of the ozone layer and to the health and safety of all living things prompted some concerned reaction, but it was a number of years—and further studies, reports, task forces, voluntary phaseouts, and partial bans—before CFCs were completely banned.

Data from an entirely unexpected source provided the political will to ban CFCs. In 1985 studies from the Antarctic showed a growing depletion in the ozone layer above the South Pole. That the largest so-called "hole" in the ozone layer could appear in winter above a virtually uninhabited continent—there was little call for refrigerants or aerosol hair sprays in Antarctica—was baffling. It obviously meant that the release of CFCs into the environment was a global concern and not just a local problem. In 1987 a high-altitude research plane flying above the south polar region found chlorine monoxide (ClO) molecules in the low-ozone areas—experimental verification of the predictions of Rowland and Molina (who eight years later shared in the 1995 Nobel Prize in chemistry for their recognition of the long-term effects of CFCs on the stratosphere and the environment).

In 1987 an agreement called the Montreal Protocol required all the nations who signed it to commit to a phaseout of the use of CFCs and ultimately a complete ban. Today hydrofluorocarbon and hydrochlorofluorocarbon compounds are used as refrigerants instead of chlorofluorocarbons. These substances either do not contain chlorine or are more easily oxidized in the atmosphere; few reach the high stratospheric levels that the less reactive CFCs did. But the newer replacements for CFCs are not as effective refrigerants, and they require up to 3 percent more energy for the refrigeration cycle.

There are still billions of CFC molecules in the atmosphere. Not every country has signed the Montreal Protocol, and even in those countries that have there are still millions of CFC-containing refrigerators in use and probably hundreds of thousands of old abandoned appliances leaking CFCs into the atmosphere, where they will join the rest of the CFCs on the slow but inevitable journey upward to wreak havoc on the ozone layer. The effect of these once-lauded molecules may be felt for hundreds of years to come. If the intensity of high-energy ultraviolet radiation reaching the Earth's surface increases, the potential for damage to cells and their DNA molecules—leading to higher levels of cancer and greater rates of mutation—also increases.

The Dark Side of Chlorine

Chlorofluorocarbons are not the only chemical group that were considered wonder molecules when first discovered but later revealed an unexpected toxicity or potential for environmental or social damage. What is perhaps surprising, however, is that organic compounds containing chlorine have shown this "dark side" more than any other group of organic compounds. Even elemental chlorine displays the dichotomy. Millions of people around the world depend on chlorination of their water supplies, and while other chemicals may be as effective as chlorine in purifying water, they are a lot more expensive.

One of the major public health advances of the past century has been the effort to bring clean drinking water to all parts of the world—something we have still to achieve. Without chlorine we would be a lot further from this goal; yet chlorine is poisonous, a fact well understood by Fritz Haber, the German chemist whose work on synthesizing ammonia from nitrogen in the air, and on gas warfare, was described in Chapter 5. The first poisonous compound used in World War I was the yellowish-green chlorine gas, whose initial effects include choking and difficulty breathing. Chlorine is a powerful irritant to cells and can cause fatal swelling of tissues in the lungs and airways. Mustard gas and phosgene, compounds used in later poisonous gas releases, are also chlorine-containing organic compounds, with effects as horrifying as those of chlorine gas. Although the mortality rate for exposure to mustard gas is not high, it does cause permanent eye damage and severe, lasting respiratory impairment.

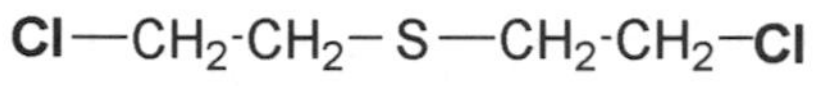

$$\mathbf{Cl}-CH_2\text{-}CH_2-S-CH_2\text{-}CH_2-\mathbf{Cl}$$

Mustard gas

$$(\mathbf{Cl})_2C=O$$

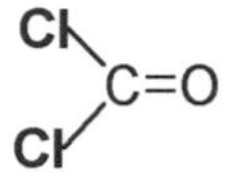

Phosgene

Poisonous gas molecules used in World War I. The chlorine atoms are bolded.

Phosgene gas is colorless and highly toxic. It is the most insidious of these poisons; it is not immediately irritating, so fatal concentrations may be inhaled before its presence is detected. Death usually results from severe swelling of tissues in the lungs and airways, which leads to suffocation.

PCBs—Further Trouble from Chlorinated Compounds

Still more chlorocarbon compounds that were initially greeted as wonder molecules have, like CFCs, turned out to pose a serious health hazard. Industrial production of polychlorinated biphenyls, or PCBs as they are most commonly known, began in the late 1920s. These compounds were considered ideal for use as electrical insulators and coolants in transformers, reactors, capacitors, and circuit breakers, where their extreme stability, even at high temperatures, and their lack of flammability were highly prized. They were employed as plasticizers—flexibility-enhancing agents—in the manufacture of various polymers, including those used for wrapping in the food industry, for liners in baby bottles, and for polystyrene coffee cups. PCBs also found a use in the manufacture of various inks in the printing business, carbonless copying paper, paints, waxes, adhesives, lubricants, and vacuum pump oils.

Polychlorinated biphenyls are compounds where chlorine atoms have been substituted for hydrogen atoms on the parent biphenyl molecule.

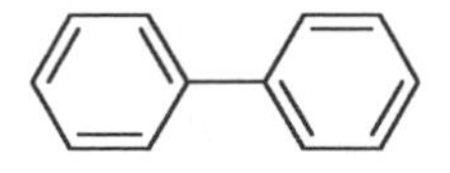

The biphenyl molecule

This structure has many possible arrangements, depending on how many chlorine atoms are present and where they are placed on the biphenyl rings. The following examples show two different trichlorinated biphenyls, each of which has three chlorines, and one pentachlorinated biphenyl with five chlorines. More than two hundred different combinations are possible.

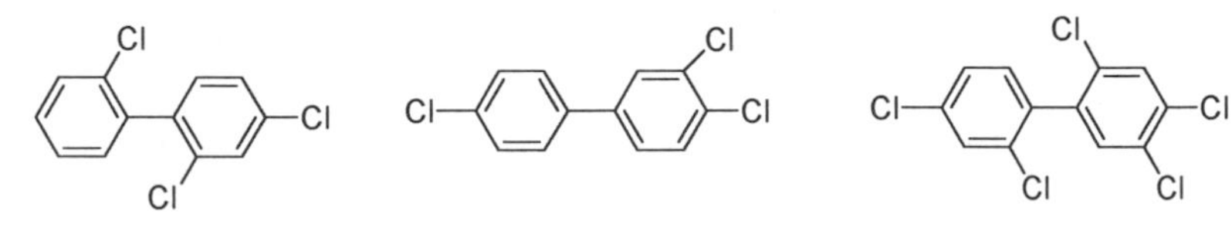

Trichlorinated biphenyl *Trichlorinated biphenyl* *Pentachlorinated biphenyl*

Not long after the manufacture of PCBs began, reports of health problems among workers at PCB plants emerged. Many reported a skin condition now known as chloracne, where blackheads and suppurating pustules appear on the face and body. We now know that chloracne is one of the first symptoms of systemic PCB poisoning and can be followed by damage to the immune, nervous, endocrine, and reproductive systems, and by liver failure and cancer. PCBs are anything but a wonder molecule and, in fact, are among the most dangerous compounds ever synthesized. Their menace lies not only in their direct toxicity to humans and other animals but, like CFCs, in the very stability that made them so useful in the first place. PCBs persist in the environment; they are subject to the process of bioaccumulation (or biomagnification), where their concentration increases along the food chain. Animals at the top of the food chain, such as polar bears, lions, whales, eagles, and humans, can build up high concentrations of PCBs in the fat cells of their bodies.

In 1968 a devastating episode of human PCB poisoning epitomized the direct effects of ingestion of these molecules. Thirteen hundred residents of Kyushu, Japan, became ill—initially with chloracne and respiratory and vision problems—after eating rice-bran oil that had accidentally become contaminated with PCBs. The long-term consequences included birth defects and liver cancer fifteen times the normal rate. In 1977 the United States banned the discharge of PCB-containing materials into waterways. Their manufacture was finally outlawed in 1979, well after numerous studies had reported the toxic effects of these compounds on human health and the health of our planet. Despite regulations controlling PCBs, there are still millions of pounds of these molecules in use or awaiting safe disposal. They still leak into the environment.

CHLORINE IN PESTICIDES— FROM BOON TO BANE TO BANNED

Other chlorine-containing molecules have not just leaked into the environment; they have been deliberately put there in the form of pesticides, sometimes in huge amounts, over decades, and in many countries. Some of the most effective pesticides ever invented contain chlorine. Very stable pesticide molecules—those that persist in the environment—were originally thought to be desirable. The effects of one application could perhaps last for years. This indeed has turned out to be true, but unfortunately the consequences were not always as foreseen. The use of chlorine-containing pesticides has been of great value to humanity but has also caused, in some cases, totally unsuspected and very harmful side effects.

More than any other chlorine-containing pesticide, the DDT molecule illustrates the conflict between potential benefit and hazard. DDT is a derivative of 1,1-diphenylethane; *DDT* is an abbreviation from the name *dichloro-diphenyl-trichloroethane.*

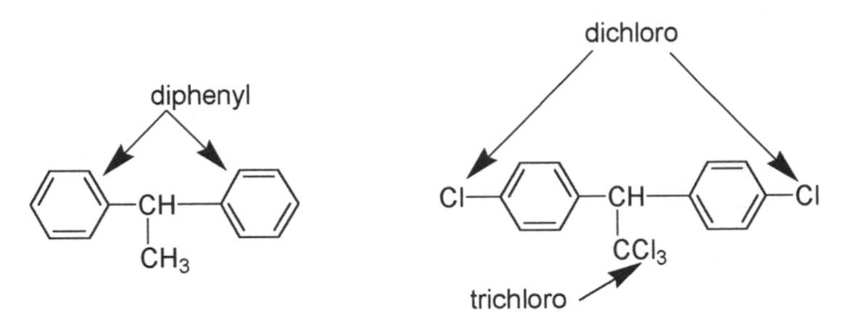

1,1-diphenylethane *Dichloro-diphenyl-trichloroethane, or DDT*

DDT was first prepared in 1874. That it was a potent insecticide was not realized until 1942, just in time for its use in World War II as a delousing powder to stop the spread of typhus and to kill the larvae of disease-carrying mosquitoes. "Bug bombs," made from aerosol cans filled with DDT, were used extensively by the U.S. military in the South Pacific. These delivered a double blow to the environment, releasing large amounts of CFCs along with clouds of DDT.

Even before 1970, by which time three million tons of DDT had been manufactured and used, concerns about its effect on the environment and the development of insect resistance to it had surfaced. The effect of DDT on wildlife, particularly birds of prey such as eagles, falcons, and hawks that are at the top of food chains, are attributed not directly to DDT but instead to its main breakdown product. Both DDT and the breakdown product are fat-soluble compounds that accumulate in animal tissues. In birds, however, this breakdown product inhibits the enzyme that supplies calcium to their eggshells. Thus birds exposed to DDT will lay eggs with very fragile shells that often break before hatching. Starting in the late 1940s, a steep decline in the population of eagles, hawks, and falcons was noted. Major disturbances to the balance between useful and harmful insects, outlined by Rachel Carson in her 1962 book *Silent Spring*, were traced to increasingly heavy use of DDT.

During the Vietnam War, from 1962 to 1970, millions of gallons of Agent Orange—a mixture of chlorine-containing herbicides 2,4-D and 2,4,5-T—were sprayed over areas of Southeast Asia to destroy guerrilla-concealing foliage.

2,4-D *2,4,5-T*

Although these two compounds are not particularly toxic, 2,4,5-T contains traces of a side product that has been implicated in the wave of birth defects, cancers, skin diseases, immune system deficiencies, and other serious health problems that affect Vietnam to this day. The compound responsible has the chemical name 2,3,7,8-tetrachlorodibenzodioxin—now commonly known as dioxin, though the word actually refers to a class of organic compounds that do not necessarily share the harmful properties of 2,3,7,8-tetrachlorodibenzodioxin.

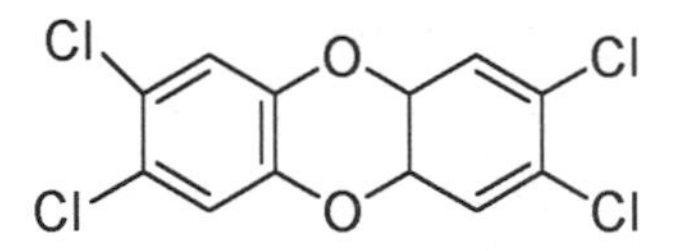

2,3,7,8-tetrachlorodibenzodioxin, or dioxin

Dioxin is considered the most lethal compound made by man, although it is still a million times less deadly than nature's most toxic compound, botulinum toxin A. In 1976 an industrial explosion in Seveso, Italy, allowed the release of a quantity of dioxin, with devastating results—chloracne, birth defects, cancer—for local people and animals. Afterward widespread media reporting of the event firmly established all compounds referred to as dioxins as villains in the mind of the public.

Just as unexpected human health problems accompanied the use of a defoliant herbicide, so too did unexpected human health problems appear with another chlorinated molecule, hexachlorophene, an extremely effective germicide product used extensively in the 1950s and 1960s in soaps, shampoos, aftershave lotions, deodorants, mouthwashes, and similar products.

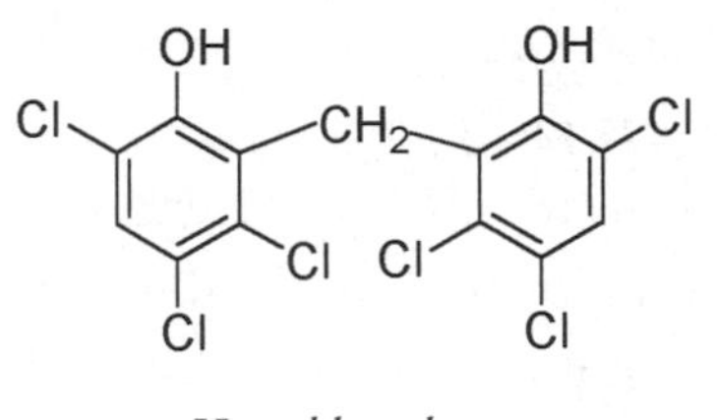

Hexachlorophene

Hexachlorophene was also routinely used on infants and added to diapers, talcum powders, and other baby toiletries. But in 1972 tests showed that its use led to brain and nervous system damage in laboratory animals. Hexachlorophene was subsequently banned from over-the-counter preparations and baby products, but because it is so effective against certain bacteria, it still has a limited use, despite its toxicity, in prescription acne medications and for surgical scrub preparations.

MOLECULES THAT PUT YOU TO SLEEP

Not all chlorocarbon molecules have been disastrous for human health. Beyond the antiseptic properties of hexachlorophene, one small chlorine-containing molecule proved to be a boon for medicine. Until the mid-1800s, surgery was performed without anesthesia—but sometimes with the administration of copious amounts of alcohol, in the belief that this would numb the agony. Some surgeons supposedly also imbibed in order to fortify themselves before inflicting such pain. Then in October 1846 a Boston dentist, William Morton, successfully demonstrated the use of ether as a way to induce narcosis—a temporary unconsciousness—for surgical procedures. Word of ether's ability to allow painless surgery spread rapidly, and other compounds were soon being investigated for anesthetic properties.

James Young Simpson, a Scottish physician and professor of medicine and midwifery at the University of Edinburgh Medical School, developed a unique way of testing compounds as possible anesthetics. He would allegedly ask his dinner guests to join him in inhaling various substances. Chloroform ($CHCl_3$), first synthesized in 1831, evidently passed the test. Simpson came to on the dining-room floor after the experiment with this compound, surrounded by his still comatose visitors. Simpson lost no time in employing chloroform on his patients.

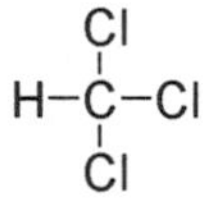

$$\begin{array}{c} Cl \\ | \\ H-C-Cl \\ | \\ Cl \end{array}$$

Chloroform

$$H_3C-CH_2-O-CH_2-CH_3$$

Ether (diethyl ether)

The use of this chlorocarbon compound as an anesthetic had a number of advantages over ether: chloroform worked faster and smelled better, and less of it was required. As well, recovery from a procedure in which chloroform had been administered was faster and less unpleasant than from surgery using ether. The extreme flammability of ether was also a problem. It formed an explosive mixture with oxygen, and the smallest spark during a surgical procedure, even from metal instruments clanking together, could cause ignition.

Chloroform anesthesia was readily accepted for surgical cases. Even though some patients died, the associated risks were considered small. As surgery was often a last resort and as patients sometimes died from shock during surgery without anesthetics anyway, the death rate was deemed acceptable. Because surgical procedures were performed rapidly—a practice that had been essential without anesthesia—patients were not exposed to chloroform for any great length of time. It has been estimated that during the American Civil War almost seven thousand battlefield surgeries were performed under chloroform, with fewer than forty deaths due to the use of the anesthetic.

Surgical anesthesia was universally recognized as a great advance, but its use in childbirth was controversial. The reservations were partly medical; some physicians rightly expressed concerns about the effect of chloroform or ether on the health of an unborn child, citing observations of reduced uterine contractions and lowered rates of infant respiration with a delivery under anesthesia. But the issue was about more than just infant safety and maternal well-being. Moral and religious views upheld the belief that the pain of labor was necessary and righteous. In the Book of Genesis women, as Eve's descendants, are condemned to suffer during childbirth as punishment for her disobedience in Eden: "In sorrow thou shalt bring forth children." According to strict interpretation of this biblical passage, any attempt to alleviate labor pain was contrary to the will of God. A more extreme view equated the travails of childbirth with atonement for sin—presumably the sin of sexual intercourse, the only means of conceiving a child in the mid–nineteenth century.

But in 1853 in Britain, Queen Victoria delivered her eighth child, Prince Leopold, with the assistance of chloroform. Her decision to use this anesthetic again at her ninth and final confinement—that of Princess Beatrice in 1857—hastened the acceptance of the practice, despite criticism leveled against her physicians in *The Lancet,* the respected British medical journal. Chloroform became the anesthetic of choice for childbirth in Britain and much of Europe; ether remained more popular in North America.

In the early part of the twentieth century a different method of pain control in childbirth gained rapid acceptance in Germany and quickly spread to other parts of Europe. Twilight Sleep, as it was known, consisted of administration of scopolamine and morphine, compounds that were discussed in Chapters 12 and 13. A very small amount of morphine was administered at the beginning of labor. It reduced pain, although not completely, especially if the labor was long or difficult. Scopolamine induced sleep and, more important for the doctors endorsing this combination of drugs, ensured that a woman had no memory of her delivery. Twilight Sleep was seen as the ideal solution for the pain of childbirth, so much so that a public campaign promoting its use began in the United States in 1914. The National Twilight Sleep Association published booklets and arranged lectures extolling the virtues of this new approach.

Serious misgivings expressed by members of the medical community were labeled as excuses for callous and unfeeling doctors to retain control over their patients. Twilight Sleep became a political issue, part of the larger movement that eventually gained women the right to vote. What seems so bizarre about this campaign now is that women believed the claims that Twilight Sleep removed the agony of childbirth, allowing the mother to awaken refreshed and ready to welcome her new baby. In reality women suffered the same pain, behaving as if no drugs had been administered, but the scopolamine-induced amnesia blocked any memory of the ordeal. Twilight Sleep provided a false picture of a tranquil and trouble-free maternity.

Like the other chlorocarbons in this chapter, chloroform—for all its blessings to surgical patients and the medical profession—also turned out to have a dark side. It is now known to cause liver and kidney damage, and high levels of exposure increase the risk of cancer. It can damage the cornea of the eye, cause skin to crack, and result in fatigue, nausea, and irregular heartbeat, along with its anesthetic and narcotic actions. When exposed to high temperatures, air, or light, chloroform forms chlorine, carbon monoxide, phosgene, and/or hydrogen chloride, all of which are toxic or corrosive. Nowadays working with chloroform requires protective clothing and equipment, a far cry from the splash-happy days of the original anesthetic providers. But even if its negative properties had been recognized more than a century ago, chloroform would still have been considered a godsend rather than a villain by the hundreds of thousands who thankfully inhaled its sweet-smelling vapors before surgery.

There is no doubt that many chlorocarbons deserve the role of villain, although perhaps that label would be better applied to those who have knowingly disposed of PCBs in rivers, argued against the banning of CFCs even after their effects on the ozone layer were demonstrated, indiscriminately applied pesticides (both legal and illegal) to land and water, and put profit ahead of safety in factories and laboratories around the world.

We now make hundreds of chlorine-containing organic compounds that are not poisonous, do not destroy the ozone layer, are not harmful to the environment, are not carcinogenic, and have never been used in gas warfare. These find a use in our homes and industries, our schools and hospitals, and our cars and boats and planes. They garner no publicity and do no harm, but they cannot be described as chemicals that changed the world.

The irony of chlorocarbons is that those that have done the most harm or have the potential to do the most harm seem also to have been the very ones responsible for some of the most beneficial advances in our society. Anesthetics were essential to the development of surgery as a highly skilled branch of medicine. The development of refrigerant molecules for use in ships, trains, and trucks opened new trade opportunities; growth and prosperity followed in undeveloped parts of the world. Food storage is now safe and convenient with home refrigeration. We take the comfort of air-conditioning for granted, and we assume our drinking water is safe and that our electrical transformers will not burst into flames. Insect-borne diseases have been eliminated or greatly reduced in many countries. The positive impact of these compounds cannot be discounted.

17. MOLECULES VERSUS MALARIA

THE WORD *MALARIA* means "bad air." It comes from the Italian words *mal aria,* because for many centuries this illness was thought to result from poisonous mists and evil vapors drifting off low-lying marshes. The disease, caused by a microscopic parasite, may be the greatest killer of humanity for all time. Even now there are by conservative estimates 300 million to 500 million cases a year worldwide, with two to three million deaths annually, mainly of children in Africa. By comparison, the 1995 Ebola virus outbreak in Zaire claimed 250 lives in six months; more than twenty times that number of Africans die of malaria each day. Malaria is transmitted far more rapidly than AIDS. Calculations estimate that HIV-positive patients infect between two and ten others; each infected malaria patient can transmit the disease to hundreds.

There are four different species of the malaria parasite (genus *Plasmodium*) that infect humans: *P. vivax, P. falciparum, P. malariae,* and *P. ovale.* All four cause the typical symptoms of malaria—intense fever, chills, terrible headache, muscle pains—that can recur even years later. The most lethal of these four is falciparum malaria. The other forms are sometimes referred to as "benign" malarias, although the toll they take on the overall health and productivity of a society is anything but benign. Malaria fever is usually periodic, spiking every two or three days. With the deadly falciparum form this episodic fever is rare, and as the disease progresses, the infected patient becomes jaundiced, lethargic, and confused before lapsing into a coma and dying.

Malaria is transmitted from one human to another through the bite of the anopheles mosquito. A female mosquito requires a meal of blood before laying her eggs. If the blood she obtains comes from a human infected with malaria, the parasite is able to continue its life cycle in the mosquito gut and be passed on when another human supplies the next meal. It then develops in the liver of the new victim; a week or so later it invades the bloodstream and enters the red blood corpuscles, now available to another bloodsucking anopheles.

We now consider malaria to be a tropical or semitropical disease, but until very recently it was also widespread in temperate regions. References to a fever—most probably malaria—occur in the earliest written records of China, India, and Egypt from thousands of years ago. The English name for the disease was "the ague." It was very common in the low-lying coastal regions of England and the Netherlands—areas with extensive marshlands and the slow-moving or stagnant waters ideal for the mosquito to breed. The disease also occurred in even more northern communities: in Scandinavia, the northern United States, and Canada. Malaria was known as far north as the areas of Sweden and Finland near the Gulf of Bothnia, very close to the Arctic Circle. It was endemic in many countries bordering the Mediterranean Sea and the Black Sea.

Wherever the anopheles mosquito thrived, so did malaria. In Rome, notorious for its deadly "swamp fever," each time a papal conclave was held, a number of the attending cardinals would die from the disease. In Crete and the Peloponnesus peninsula of mainland Greece, and other parts of the world with marked wet and dry seasons, people would move their animals to the high hill country during the summer months. This may have been as much to escape malaria from the coastal marshes as to find summer pastures.

Malaria struck the rich and famous as well as the poor. Alexander the Great supposedly died of malaria, as did the African explorer David Livingstone. Armies were particularly vulnerable to malaria epidemics; sleeping in tents, makeshift shelters, or out in the open gave night-feeding mosquitoes ample opportunity to bite. Over half the troops in the American Civil War suffered from annual bouts of malaria. Can we possibly add malaria to the woes suffered by Napoleon's troops—at least in the late summer and fall of 1812, as they began their great push to Moscow?

Malaria remained a worldwide problem well into the twentieth century. In the United States in 1914 there were more than half a million cases of malaria. In 1945 nearly two billion people in the world were living in malarial areas, and in some countries 10 percent of the population was infected. In these places malaria-related absenteeism in the workforce could be as high as 35 percent and up to 50 percent for schoolchildren.

QUININE—NATURE'S ANTIDOTE

With statistics like these it is little wonder that for centuries a number of different methods have been used to try to control the disease. They have involved three quite different molecules, all of which have interesting and even surprising connections to many of the molecules mentioned in earlier chapters. The first of these molecules is quinine.

High in the Andes, between three thousand and nine thousand feet above sea level, there grows a tree whose bark contains an alkaloid molecule, without which the world would be a very different place today. There are about forty species of this tree, all of which are members of the *Cinchona* genus. They are indigenous to the eastern slopes of the Andes, from Colombia south to Bolivia. The special properties of the bark were long known to the local inhabitants, who surely passed on the knowledge that a tea brewed from this part of the tree was an effective fever cure.

Many stories tell how early European explorers in the area found out about the antimalarial effect of cinchona bark. In one a Spanish soldier suffering a malarial episode drank water from a pond surrounded by cinchona trees, and his fever miraculously disappeared. Another account involves the countess of Chinchón, Doña Francisca Henriques de Rivera, whose husband, the count of Chinchón, was the Spanish viceroy to Peru from 1629 to 1639. In the early 1630s Doña Francisca became very ill from malaria. Traditional European remedies were ineffectual, and her physician turned to a local cure, the cinchona tree. The species was named (although misspelled) after the countess, who survived thanks to the quinine present in its bark.

These stories have been used as evidence that malaria was present in the New World before the arrival of Europeans. But the fact that the Indians knew that the *kina* tree—a Peruvian word, which in Spanish became *quina*—cured a fever does not prove that malaria was indigenous to the Americas. Columbus arrived on the shores of the New World well over a century before Doña Francisca took the quinine cure, more than enough time for malarial infection to find its way from early explorers into local anopheles mosquitoes and spread to other inhabitants of the Americas. There is no evidence that the fevers treated by *quina* bark in the centuries before the conquistadors arrived were malarial. It is now generally accepted among medical historians and anthropologists that the disease was brought from Africa and Europe to the New World. Both Europeans and African slaves would have been a source of infection. By the mid–sixteenth century the slave trade to the Americas from West Africa, where malaria was rife, was already well established. In the 1630s, when the countess of Chinchón contracted malaria in Peru, generations of West Africans and Europeans harboring malarial parasites had already established an enormous reservoir of infection awaiting distribution throughout the New World.

Word that the bark of the *quina* tree could cure malaria quickly spread to Europe. In 1633 Father Antonio de la Calaucha recorded the amazing properties of the bark of the "fever tree," and other members of the Jesuit order in Peru began using *quina* bark to both cure and prevent malaria. In the 1640s Father Bartolomé Tafur took some of the bark to Rome, and word of its miraculous properties spread through the clergy. The papal conclave of 1655 was the first without a death from malaria among the attending cardinals. The Jesuits were soon importing large amounts of the bark and selling it throughout Europe. Despite its excellent reputation in other countries, "Jesuit's powder"—as it became known—was not at all popular in Protestant England. Oliver Cromwell, refusing to be treated by a papists' remedy, succumbed to malaria in 1658.

Another remedy for malaria gained prominence in 1670, when Robert Talbor, a London apothecary and physician, warned the public to beware of the dangers associated with Jesuit's powder and started promoting his own secret formulation. Talbor's cure was taken to the royal courts of both England and France; his own king, Charles II, and the son of Louis XIV, the French king, both survived severe bouts of malaria thanks to Talbor's amazing medication. It was not until after Talbor's death that the miraculous ingredient in his formulation was revealed; it was the same cinchona bark as in Jesuit's powder. Talbor's deceit, while no doubt making him wealthy—presumably his main motive—did save the lives of Protestants who refused to be associated with a Catholic cure. That quinine cured the disease known as ague is taken as evidence that this fever, which had plagued much of Europe for centuries, was indeed malaria.

Through the next three centuries malaria—as well as indigestion, fever, hair loss, cancer, and many other conditions—was commonly treated with bark from the cinchona tree. It was not generally known what plant the bark came from until 1735, when a French botanist,

Joseph de Jussieu, while exploring the higher elevations of the South American rain forests, discovered that the source of the bitter bark was various species of a broad-leafed tree that grew as high as sixty-five feet. It was a member of the *Rubiaceae,* the same family as the coffee tree. There was always great demand for the bark, and its harvesting became a major industry. Although it was possible to gather some of the bark without killing the tree, greater profits could be made if the tree was felled and all the bark stripped. By the end of the eighteenth century an estimated 25,000 *quina* trees were being cut down each year.

With the cost of cinchona bark high and the source tree possibly becoming endangered, isolating, identifying, and manufacturing the antimalarial molecule became an important objective. Quinine is thought to have been first isolated, although probably in an impure form, as far back as 1792. Full investigation of what compounds were present in the bark started around 1810, and it was not until 1820 that researchers

Cinchona tree from whose bark quinine is obtained.
(Photo courtesy of L. Keith Wade)

Joseph Pelletier and Joseph Caventou managed to extract and purify quinine. The Paris Institute of Science awarded these French chemists a sum of ten thousand francs for their valuable work.

Among the almost thirty alkaloids found in cinchona bark, quinine was quickly identified as the active ingredient. Its structure was not fully determined until well into the twentieth century, so early attempts to synthesize the compound had little chance of success. One of these was the effort of the young English chemist William Perkin (whom we met in Chapter 9) to combine two molecules of allyltoluidine with three oxygen atoms to form quinine and water.

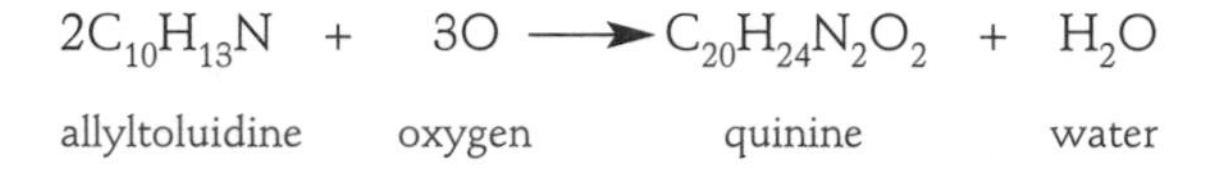

Working in 1856 on the basis that the formula of allyltoluidine ($C_{10}H_{13}N$) was almost half that of quinine ($C_{20}H_{24}N_2O_2$), his experiment was doomed to fail. We now know that the structure of allyltoluidine and the more complicated structure of quinine are as follows:

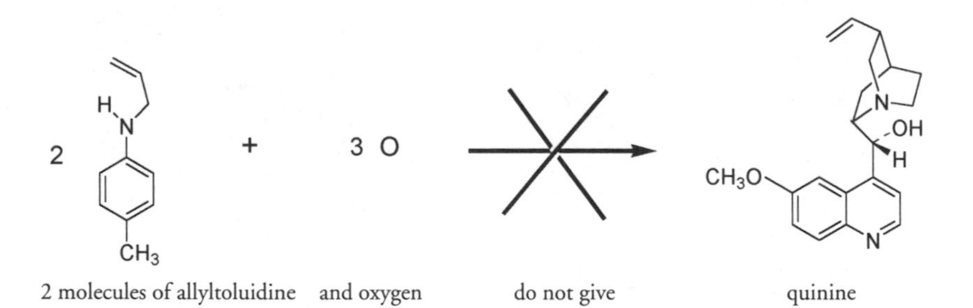

2 molecules of allyltoluidine and oxygen do not give quinine

While Perkin failed to make quinine, his work was extremely fruitful for making mauve—and money—for the dye industry and for the development of the science of organic chemistry.

As the Industrial Revolution brought prosperity to Britain and other parts of Europe during the nineteenth century, capital became available to tackle the problem of unhealthy, marshy farmland. Extensive drainage schemes turned bogs and fens into more productive farms, meaning that less stagnant water was available for breeding mosquitoes, and the incidence of malaria decreased in regions where it had been most prevalent. But the demand for quinine did not decrease. On the contrary, as European colonization increased in Africa and Asia, there was more demand for protection against malaria. The British habit of taking quinine as a prophylactic precaution against malaria developed into the evening "gin and tonic"—the gin being considered necessary to make the bitter-tasting quinine in the tonic water palatable. The British Empire depended on supplies of quinine, as many of its most valuable colonies—in India, Malaya, Africa, and the Caribbean—were in regions of the world where malaria was endemic. The Dutch, French, Spanish, Portuguese, Germans, and Belgians also colonized malarial areas. Worldwide demand for quinine was enormous.

With no synthetic route to quinine in sight, a different solution was sought—and found: the cultivation of cinchona species from the Amazon in other countries. Profit from the sale of cinchona bark was so great that the governments of Bolivia, Ecuador, Peru, and Colombia, in order to maintain their monopoly over the quinine trade, prohibited the export of living cinchona plants or seeds. In 1853, the Dutchman Justus Hasskarl, director of a botanical garden on the island of Java in the Dutch East Indies, managed to smuggle a bag of seeds from *Cinchona calisaya* out of South America. They were successfully grown in Java, but unfortunately for Hasskarl and the Dutch, this species of the cinchona tree had a relatively low quinine content. The British had a similar experience with smuggled seeds from *Cinchona pubescens,* which they planted in India and Ceylon. The trees grew, but the bark had less than the 3 percent quinine content needed for cost-effective production.

In 1861, Charles Ledger, an Australian who had spent a number of years as a *quina* bark trader, managed to persuade a Bolivian Indian to sell him seeds of a species of cinchona tree that supposedly had a very high quinine content. The British government was not interested in buying Ledger's seeds; their experience with cultivation of cinchona had probably led them to decide that it was not economically viable. But the Dutch government purchased a pound of seeds of the species, which became known as *Cinchona ledgeriana,* for about twenty dollars. Although the British had made the smart choice nearly two hundred years previously in ceding the isoeugenol molecule of the nutmeg trade to the Dutch in exchange for the island of Manhattan, it was the Dutch who made the correct call this time. Their twenty-dollar purchase has been called the best investment in history, as the quinine levels in *Cinchona ledgeriana* bark were found to be as high as 13 percent.

The *C. ledgeriana* seeds were planted in Java and carefully cultivated. As the trees matured and their quinine-rich bark was harvested, the export of native bark from South America declined. It was a scenario repeated fifteen years later, when smuggled seeds from another South American tree, *Hevea brasiliensis,* signaled the demise of indigenous rubber production (see Chapter 8).

By 1930 over 95 percent of the world's quinine came from plantations on Java. These cinchona estates were hugely profitable for the Dutch. The quinine molecule, or perhaps more correctly the monopoly on the cultivation of the quinine molecule, almost tipped the scales of World War II. In 1940 Germany invaded the Netherlands and confiscated the complete European stock of quinine from the Amsterdam premises of the "kina bureau." The 1942 Japanese conquest of Java further imperiled the supply of this essential antimalarial. American botanists, led by Raymond Fosberg of the Smithsonian Institution, were sent to the eastern side of the Andes to secure a supply of *quina* bark from trees still growing naturally in the area. Although they did manage to procure a number of tons of bark, they never found any specimens of the highly productive *Cinchona ledgeriana* with which the Dutch had had such astounding success. Quinine was essential to protect Allied troops in the tropics, so once again its synthesis—or that of a similar molecule with antimalarial properties—became extremely important.

Quinine is a derivative of the quinoline molecule. During the 1930s a few synthetic derivatives of quinoline had been created and had proven successful at treating acute malaria. Extensive research on antimalarial drugs during World War II resulted in a 4-aminoquinoline derivative, now known as *chloroquine,* originally made by German chemists before the war, as the best synthetic choice.

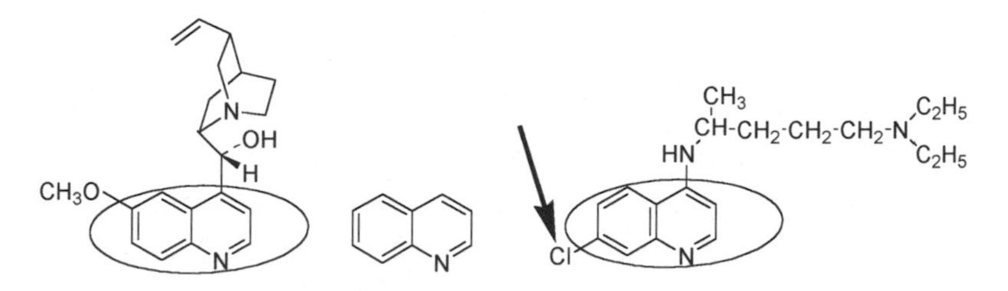

Both quinine (left) and chloroquine (right) incorporate (circled) the quinoline structure (center). The chlorine atom in chloroquine is arrowed.

Chloroquine contains a chlorine atom—another example of a chlorocarbon molecule that has been extremely beneficial to humanity. For over forty years chloroquine was a safe and effective antimalarial drug, well tolerated by most people and with little of the toxicity of the other synthetic quinolines. Unfortunately, chloroquine-resistant strains of the malaria parasite have spread rapidly in the past few decades, reducing the effectiveness of chloroquine, and compounds such as fansidar and mefloquine, with their greater toxicity and sometimes alarming side effects, are now being used for malarial protection.

THE SYNTHESIS OF QUININE

The quest to synthesize the actual quinine molecule was supposedly fulfilled in 1944, when Robert Woodward and William Doering of Harvard University converted a simple quinoline derivative into a molecule that previous chemists, in 1918, had allegedly been able to transform into quinine. The total synthesis of quinine was finally presumed complete. But this was not the case. The published report of the earlier work had been so sketchy that it was not possible to ascertain what had really been done and whether the claim of chemical transformation was valid.

Organic natural product chemists have a saying: "The final proof of structure is synthesis." In other words, no matter how much the evidence points to the correctness of a proposed structure, to be absolutely sure it is correct, you have to synthesize the molecule by an independent route. And in 2001, 145 years after Perkin's now-famous attempt to make quinine, Gilbert Stork, professor emeritus at New York's Columbia University, together with a group of coworkers, did just that. They started with a different quinoline derivative, they followed an alternative route, and they carried out each and every step of their synthesis themselves.

As well as being a reasonably complicated structure, quinine, like many other molecules made in nature, presents the particular challenge of determining which way various bonds around certain of the carbon atoms are positioned in space. The quinine structure has an H atom pointing out of the page (indicated by a solid wedge ▬) and an OH directed behind the page (indicated by a dashed line ----) around the carbon atom adjacent to the quinoline ring system.

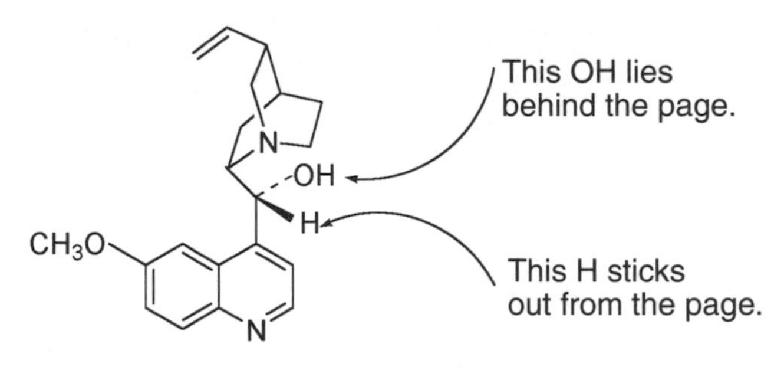

The quinine molecule

An example of the different spatial arrangements of these bonds is shown on page 341 for quinine and a version inverted around the same carbon atom.

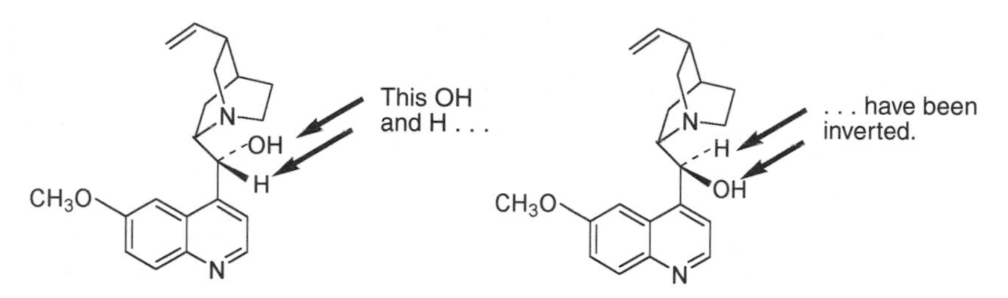

Quinine (left) and the very similar version (right) that would also be synthesized in the laboratory at the same time as quinine

Nature often makes only one of a pair of compounds like this. But when chemists attempt to make the same molecule synthetically, they cannot avoid making an equal mixture of the two. Because they are so similar, separating the two molecules of the pair is tricky and time-consuming. There are three other carbon atom positions in the quinine molecule where both the natural and the inverted versions are unavoidably made during a laboratory synthesis, so these painstaking operations must be repeated four times in all. It was a challenge that Stork and his group overcame—and there is no evidence that the problem was even fully appreciated in 1918.

Quinine continues to be harvested from plantations in Indonesia, India, Zaire, and other African countries, with lesser amounts coming from natural sources in Peru, Bolivia, and Ecuador. Its main uses today are in quinine water, tonic water, and other bitter drinks and in the production of quinidine, a heart medication. Quinine is still thought to provide some measure of protection against malaria in chloroquine-resistant regions.

MAN'S SOLUTION TO MALARIA

While people were searching for ways to harvest more quinine or make it synthetically, physicians were still trying to understand what caused malaria. In 1880 a doctor in the French army in Algeria, Charles-Louis-

Alphonse Laveran, made a discovery that ultimately opened the way for a new molecular approach to the fight against this disease. Laveran, using a microscope to check slides of blood samples, found that the blood of patients with malaria contained cells that we now know are a stage of the malarial protozoa *Plasmodium.* Laveran's findings, initially dismissed by the medical establishment, were confirmed over the next few years with the identification of *P. vivax* and *P. malariae,* and later *P falciparum.* By 1891 it was possible to identify the specific malaria parasite by staining the *Plasmodium* cell with various dyes.

Although it had been hypothesized that mosquitoes were somehow involved in the transmission of malaria, it was not until 1897 that Ronald Ross, a young Englishman who was born in India and was serving as a physician in the Indian Medical Service, identified another life stage of *Plasmodium* in gut tissue of the anopheles mosquito. Thus the complex association between parasite, insect, and man was recognized. It was then realized that the parasite was vulnerable to attack at various points in its life cycle.

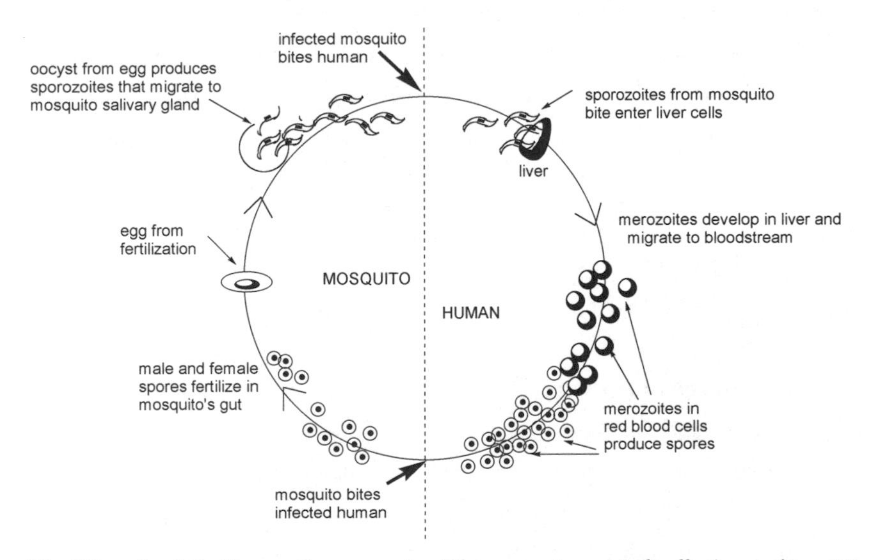

The life cycle of the Plasmodium *parasite. The merozoites periodically (every 48 or 72 hours) break out of their host's red blood cells, causing a fever to spike.*

There are several possible ways to break this disease cycle, such as killing the merozoite stage of the parasite in the liver and blood. Another obvious line of attack is the disease "vector," the mosquito itself. This could involve preventing mosquito bites, killing adult mosquitoes, or preventing them from breeding. It is not always easy to avoid mosquito bites; in places where the cost of reasonable housing is beyond the means of most of the population, window screens are just not feasible. Nor is it practical to drain all stagnant or slow-moving waters to prevent mosquitoes from breeding. Some control of the mosquito population is possible by spreading a thin film of oil over the surface of water, so mosquito larvae in the water cannot breathe. Against the anopheles mosquito itself, however, the best line of attack is powerful insecticides.

Initially the most important of these was the chlorinated molecule DDT, which acts by interfering with a nerve control process unique to insects. For this reason, DDT—at the levels used as an insecticide—is not toxic to other animals, even while it is lethal to insects. The estimated fatal dose for a human is thirty grams. This is a considerable amount; there are no reported human deaths from DDT.

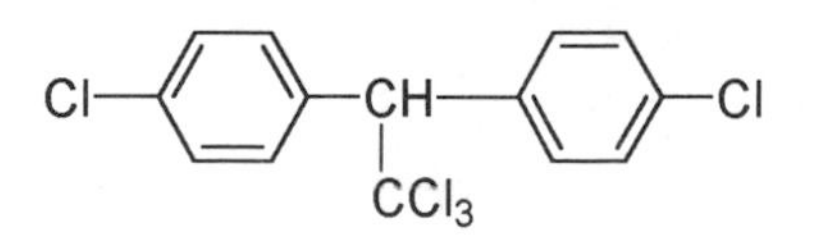

The DDT molecule

Thanks to a variety of factors—improved public health systems, better housing, fewer people living in rural areas, widespread drainage of stagnant water, and almost universal access to antimalarial drugs—the incidence of malaria had, by the early years of the twentieth century, greatly decreased in western Europe and North America. DDT was the final step necessary to eliminate the parasite in developed countries. In 1955 the World Health Organization (WHO) began a massive campaign using DDT to eliminate malaria from the rest of the world.

When DDT spraying started, about 1.8 billion people lived in malarial areas. By 1969 malaria had been eradicated for nearly 40 percent of these people. In some countries the results were phenomenal: in 1947 Greece had approximately two million cases of malaria, while in 1972 the number was seven. If any one molecule can be said to be responsible for the increase in economic prosperity of Greece during the last quarter of the twentieth century, it surely must be DDT. Before DDT spraying began in India, in 1953, there were an estimated 75 million cases a year; by 1968 there were only 300,000. Similar results were reported from countries all around the world. It was no wonder that DDT was considered a miracle molecule. By 1975 the WHO had declared Europe to be malaria free.

As it was such a long-lasting insecticide, treatment every six months—or even yearly where the disease was seasonal—was sufficient to give protection against the disease. DDT was sprayed on the inside walls of houses where the female mosquito clung, waiting for nighttime to seek her meal of blood. DDT stayed in place where it was sprayed, and it was thought that very little would ever make its way into the food chain. It was an inexpensive molecule to produce, and it seemed at the time to have little toxicity for other forms of animal life. Only later did the devastating effect of DDT bioaccumulation become obvious. We have also since realized how overuse of chemical insecticides can upset ecological balance, causing more serious pest problems.

Although the WHO's crusade against malaria had initially looked promising, the global eradication of the parasite proved more difficult than expected for a number of reasons, including the development of resistance to DDT by the mosquito, human population increase, ecological changes that reduced the number of species preying on mosquitoes or their larvae, war, natural disasters, decline of public health services, and the increase of *Plasmodium* resistance to antimalarial molecules. By the early 1970s the WHO had abandoned its dream of complete eradication of malaria and concentrated its efforts on control.

If molecules can be said to go in and out of fashion, then in the developed world DDT is definitely unfashionable—even the name seems to have an ominous ring. Although it is now outlawed in many countries, this insecticide is estimated to have saved fifty million human lives. The threat of death from malaria has largely gone from developed countries—a direct and huge benefit from a much-maligned molecule—but for millions who still live in malarial regions of the world it remains.

HEMOGLOBIN–NATURE'S PROTECTION

In many of these places few people can afford the insecticide molecules that control anopheles mosquitoes or the synthetic quinine substitutes that provide protection for tourists from the West. But nature has bestowed quite a different form of defense against malaria in these regions. As many as 25 percent of sub-Saharan Africans carry a genetic trait for the painful and debilitating disease known as sickle-cell anemia. When both parents are carriers of this trait, a child has a one-in-four chance of having the disease, a one-in-two chance of being a carrier, and a one-in-four chance of neither having the disease nor being a carrier.

Normal red blood cells are round and flexible, allowing them to squeeze through small blood vessels in the body. But in sickle-cell anemia patients, approximately half of the red blood cells become rigid and take on an elongated crescent or sickle shape. These stiffer sickled red blood cells have difficulty squeezing through narrow blood capillaries and can cause blockages in tiny blood vessels, leaving the cells of muscle tissue and vital organs without blood and oxygen. This leads to a sickling "crisis" causing severe pain and sometimes permanently damaging affected organs and tissue. The body destroys abnormal sickle-shaped cells at a faster rate than normal, resulting in an overall reduction in red blood cells—the source of the anemia.

Until recently sickle-cell anemia was usually fatal in childhood; cardiac problems, renal failure, liver failure, infection, and strokes took a toll at an early age. Present-day treatments—but not cures—can allow patients to live longer and healthier lives. Carriers of sickle-cell anemia can be affected by sickling, though usually not enough to compromise blood circulation.

For carriers of the sickle cell trait who live in malarial areas, the disease offers a valuable compensation: a significant degree of immunity to malaria. The definite correlation between incidence of malaria and high carrier frequency of sickle-cell anemia is explained by the evolutionary advantage of being a carrier. Those who inherited the sickle-cell trait from both parents would usually die from it in childhood. Those who did not inherit the trait from either parent were much more likely to succumb, often in childhood, to malaria. Those inheriting the sickle cell gene from only one parent had some immunity to the malarial parasite and survived to a reproductive age. Thus the inherited disorder of sickle-cell anemia not only continued in a population, it increased over generations. Where malaria did not exist, there was no benefit from being a carrier, and the trait would not have persisted in the inhabitants. The absence of an abnormal hemoglobin that provides malarial immunity in the American Indian population is considered crucial evidence that the American continents were malaria-free before the arrival of Columbus.

The red color of red blood cells is due to the presence of molecules of hemoglobin—the function of which is to transport oxygen around the body. One extremely small change in the chemical structure of hemoglobin is responsible for the life-threatening condition of sickle-cell anemia. Hemoglobin is a protein; like silk, it is a polymer comprising amino acid units, but unlike silk, whose chains of variably arranged amino acids may contain thousands of units, hemoglobin's precisely ordered amino acids are arranged in two sets of two identical strands. The four strands are coiled together around four iron-containing entities—the site where oxygen atoms attach. Patients with sickle-cell anemia have just a single amino acid unit difference on one of the sets of strands. On what is called the β strand, the sixth amino acid is valine instead of the glutamic acid present in normal hemoglobin.

Glutamic acid *Valine*

Valine differs from glutamic acid only in the structure of the side chain (outlined).

The β strand consists of 146 amino acids; the α strand has 141 amino acids. So the overall variation in amino acids is only one out of 287—about a third of 1 percent difference in amino acids. Yet the result for the person inheriting the sickle-cell trait from both parents is devastating. If we say that the side group is only about a third of the structure of the amino acid, then the percent difference in actual chemical structure becomes even less—a change in only around a tenth of 1 percent of the molecular structure.

This alteration in protein structure explains the symptoms of sickle-cell anemia. The side group of glutamic acid has COOH as part of its structure, while the side group of valine doesn't. Without this COOH on the sixth amino acid residue of the β strand, the deoxygenated form of sickle-cell anemia hemoglobin is much less soluble; it precipitates inside red blood cells, accounting for their changed shape and loss of flexibility. The solubility of the oxygenated form of sickle-cell anemia is little affected. Thus there is more sickling when there is more deoxygenated hemoglobin.

Once sickled cells start to block capillaries, local tissues become deficient in oxygen, oxygenated hemoglobin is converted to the deoxygenated form, and even more sickling occurs—a vicious cycle that rapidly leads to a crisis. This is why carriers of the sickle-cell trait are also susceptible to sickling: though only about 1 percent of their red blood cells are normally in a sickled state, 50 percent of their hemoglobin molecules have the potential to become sickled. This may happen at low oxygen pressures in unpressurized planes or after exercise at high altitudes; both are conditions in which the deoxygenated form of hemoglobin can build up in the body.

More than 150 different variations in the chemical structure of human hemoglobin have been found, and although some of them are lethal or cause problems, many others are apparently benign. Partial resistance to malaria is thought to be conferred on carriers of the hemoglobin variations that produce other forms of anemia, such as alpha thalassemia, endemic among people of Southeast Asian heritage, and beta thalassemia, most common in those of Mediterranean ancestry, such as Greeks and Italians, and also found in those from the Middle East, India, Pakistan, and parts of Africa. It is probable that as many as five in every thousand humans have some sort of variation in the structure of their hemoglobin, and most will never know.

It is not just the difference in the side group structure between glutamic acid and valine that causes the debilitating problems of sickle-cell anemia; it is also the position in which this occurs in the β strand. We do not know if the same change at a different position would have a similar effect on hemoglobin solubility and red blood cell shape. Nor do we know exactly why this change confers immunity to malaria. Obviously something about a red blood cell containing hemoglobin with valine at position six hinders the life cycle of the *Plasmodium* parasite.

The three molecules at the center of the ongoing struggle against malaria are very different chemically, but each one has had a major influence over events of the past. The alkaloids of the cinchona bark, throughout their long history of benefit to man, brought little economic advantage to the indigenous people of the eastern slopes of the Andes where the *quina* trees grew. Outsiders profited from the quinine molecule, exploiting a unique natural resource of a less developed country for their own advantage. European colonization of much of the world was made possible by the antimalarial properties of quinine, which like many another natural product has provided a molecular model for chemists attempting to reproduce or enhance its effects by making alterations to the original chemical structure.

Although the quinine molecule, in the nineteenth century, allowed the growth of the British Empire and the expansion of other European colonies, it was the success of the DDT molecule as an insecticide that finally eradicated malaria from Europe and North America in the twentieth century. DDT is a synthetic organic molecule that has no natural analogue. There is always a risk when such molecules are manufactured—we have no way of knowing for sure which will be beneficial and which may have harmful effects. Yet how many of us would be prepared to give up entirely the whole spectrum of novel molecules, the products of chemists' innovation that enhance our lives: the antibiotics and antiseptics, the plastics and polymers, the fabrics and flavors, the anesthetics and additives, the colors and coolants?

The repercussions of the small molecular change that produced sickled hemoglobin were felt on three continents. Resistance to malaria was a crucial factor in the rapid growth of the African slave trade in the seventeenth century. The vast majority of slaves imported to the New World came from the region of Africa where malaria was endemic and where the sickle-cell anemia gene is common. Slave traders and slave owners quickly exploited the evolutionary advantage of valine replacing glutamic acid at position six on the hemoglobin molecule. Of course they did not know the chemical reason for the immunity of African slaves to malaria. All they knew was that slaves from Africa could generally survive the fevers in the tropical climates suitable for sugar and cotton cultivation, whereas native Americans, brought from other parts of the New World to labor in the plantations, would rapidly succumb to diseases. This small molecular switch doomed generations of Africans to slavery.

The slave trade would not have flourished as it did had the slaves and their descendants fallen victim to malaria. The profits from the great sugar plantations of the New World would not have been available for economic growth in Europe. There might not have been any great sugar plantations. Cotton would not have developed as a major crop in the southern United States, the Industrial Revolution in Britain might have been delayed or taken a very different direction, and there might not have been civil war in the United States. The events of the past half millennium would have been very different but for this tiny change in the chemical structures of hemoglobin.

Quinine, DDT, and hemoglobin—these three very different structures are united historically by their connections to one of our world's greatest killers. They also typify the molecules discussed in previous chapters. Quinine is a naturally occurring plant product, as are many compounds that have had far-reaching effects on the development of civilization. Hemoglobin too is a natural product, but of animal origin. As well, hemoglobin belongs to the group of molecules classified as polymers, and again polymers of all types have been instrumental in major changes throughout history. And DDT illustrates the dilemmas often associated with man-made compounds. How different our world would be—for better or for worse—without synthetic substances produced through the ingenuity of those who create new molecules.

EPILOGUE

HISTORICAL EVENTS almost always have more than one cause, so it would be far too simplistic to attribute the events mentioned in this book solely to chemical structures. But neither is it an exaggeration to say that chemical structures have played an essential and often unrecognized role in the development of civilization. When a chemist determines the structure of a different natural product or synthesizes a new compound, the effect of a small chemical change—a double bond moved here, an oxygen atom substituted there, an alteration to a side group—often seems of little consequence. It is only with hindsight that we recognize the momentous effect that very small chemical changes can have.

Initially, the chemical structures shown in these chapters may have appeared foreign and perplexing to you. Hopefully we've now removed some of the mystery from such diagrams, and you can see how atoms that make up the molecules of chemical compounds conform to well-defined rules. Yet within the boundaries of these rules there are seemingly endless possibilities for different structures.

The compounds we selected as having interesting and important stories fall into two main groups. The first includes molecules from natural sources—valuable molecules sought after by man. Desire for these molecules governed many aspects of early history. Over the past century and a half the second group of molecules became more important. These are compounds made in laboratories or factories—some of them, like indigo, absolutely identical to molecules from a natural product, and others, like aspirin, variations of the structure of the natural product. Sometimes, like CFCs, they are totally new molecules with no analogs in nature.

To these groups, we can now add a third classification: molecules that may have a tremendous but unpredictable effect on our civilization in the future. These are molecules produced by nature but at the direction and intervention of man. Genetic engineering (or biotechnology, or whatever term is used for the artificial process by which new genetic material is inserted into an organism) results in the production of molecules where they did not previously exist. "Golden rice," for example, is a strain of rice genetically engineered to produce β-carotene, the yellow-orange coloring matter abundant in carrots and other yellow fruits and vegetables and also present in dark green leafy vegetables.

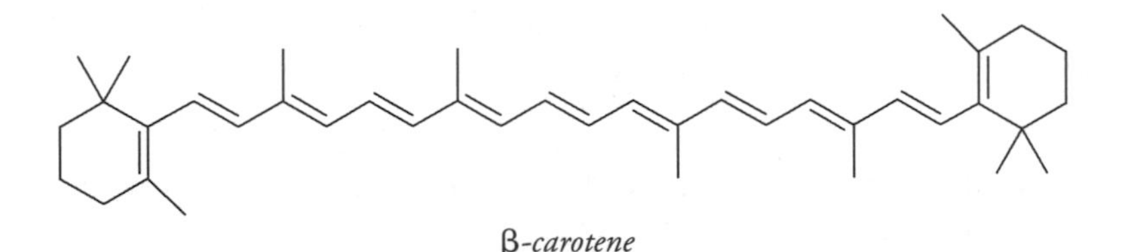

β-carotene

Our bodies need β-carotene to make vitamin A, essential to human nutrition. The diet of millions of people around the world, but particularly in Asia, where rice is the staple crop, is low in β-carotene. Vitamin A deficiency brings diseases that can cause blindness and even death. Rice grains contain virtually no β-carotene, and for parts of the rice-eating world where this molecule is not generally obtained from other sources, the addition of β-carotene to golden rice brings the promise of better health.

But there are downsides to such genetic engineering. Even though the β-carotene molecule itself is found naturally in many plants, critics of biotechnology question whether it will be safe to insert this molecule into places where it does not normally occur. Could such molecules react adversely with other compounds already present? Is there a possibility that they could become allergens for some people? What are the long-term effects of tampering with nature? As well as the many chemical and biological questions, other issues have been raised concerning genetic engineering, such as the profit motive that drives much of this research, the likely loss of crop diversity, and the globalization of agriculture. For all these reasons and uncertainties we need to act cautiously despite what may seem to be obvious advantages in forcing nature to produce molecules where and how we want them. Just as with molecules like PCBs and DDT, chemical compounds can be both a blessing and a curse, and we don't always know which is which at the time of invention. It may be that human manipulation of the complex chemicals that control life will eventually play an important part in developing better crops, in reducing the use of pesticides, and in eradicating diseases. Or it may be that such manipulation will lead to totally unexpected problems that could—in a worst-case scenario—threaten life itself.

In the future, if people look back on our civilization, what will they identify as the molecules that most influenced the twenty-first century? Will it be natural herbicide molecules added to genetically modified crops that inadvertently eliminate hundreds of other plant species? Will it be pharmaceutical molecules that improve our bodily health and our mental well-being? Will it be new varieties of illegal drugs with links to terrorism and organized crime? Will it be toxic molecules that further pollute our environment? Will it be molecules that provide a pathway to new or more efficient sources of energy? Will it be overuse of antibiotics, resulting in the development of resistant "superbugs"?

Columbus could not have foreseen the results of his search for piperine, Magellan was unaware of the long-term effects of his quest for isoeugenol, and Schönbein would surely have been astonished that the nitrocellulose he made from his wife's apron was the start of great industries as diverse as explosives and textiles. Perkin could not have anticipated that his small experiment would eventually lead not only to a huge synthetic dye trade but also to the development of antibiotics and pharmaceuticals. Marker, Nobel, Chardonnet, Carothers, Lister, Baekeland, Goodyear, Hofmann, Leblanc, the Solvay brothers, Harrison, Midgley, and all the others whose stories we have told had little idea of the historical importance of their discoveries. So we are perhaps in good company if we hesitate to try to predict whether today there already exists an unsuspected molecule that will eventually have such a profound and unanticipated effect on life as we know it that our descendants will say, "This changed our world."

ACKNOWLEDGMENTS

THIS BOOK COULD not have been written without the enthusiastic support of our families, friends, and colleagues. We would like to thank everyone; we appreciated each suggestion and comment, even if we did not use them all.

Professor Con Cambie of the University of Auckland, New Zealand, could not have expected to spend time in his retirement checking structural diagrams and chemical formulas. We are grateful for his willingness to do so, for his eagle eye, and for his wholehearted endorsement of the project. Any errors remaining are ours.

We would also like to thank our agent, Jane Dystel, of Jane Dystel Literary Management, who saw the possibilities in our interest in the relationship between chemical structures and historical events.

Wendy Hubbert, our editor at Tarcher/Putnam, claims she has learned a lot (about chemistry) through the editing process, but we think we learned a lot more from Wendy. It was her insistence on narrative, through lines and transitions, that made this a book. We knew the connections were there; Wendy—in never allowing a loose end—encouraged us to tie them all together.

Lastly we acknowledge the curiosity and ingenuity of those chemists who came before us. Without their efforts we would never have experienced the understanding and the fascination that is the joy of chemistry.

SELECTED BIBLIOGRAPHY

Allen, Charlotte. "The Scholars and the Goddess." *Atlantic Monthly.* January 2001.

Arlin, Marian. *The Science of Nutrition.* New York: Macmillan, 1977.

Asbell, Bernard. *The Pill: A Biography of the Drug That Changed the World.* New York: Random House, 1995.

Aspin, Chris. *The Cotton Industry.* Series 63. Aylesbury: Shire Publications, 1995.

Atkins, P. W. *Molecules.* Scientific American Library series, no. 21. New York: Scientific American Library, 1987.

Balick, Michael J., and Paul Alan Cox. *Plants, People, and Culture: The Science of Ethnobotany.* Scientific American Library series, no. 60. New York: Scientific American Library, 1997.

Ball, Philip. "What a Tonic." *Chemistry in Britain* (October 2001): 26–29.

Bangs, Richard, and Christian Kallen. *Islands of Fire, Islands of Spice: Exploring the Wild Places of Indonesia.* San Francisco: Sierra Club Books, 1988.

Brown, G. I. *The Big Bang: A History of Explosives.* Gloucestershire: Sutton Publications, 1998.

Brown, Kathryn. "Scary Spice." *New Scientist* (December 23–30, 2000): 53.

Brown, William H., and Christopher S. Foote. *Organic Chemistry.* Orlando, Fla.: Harcourt Brace, 1998.

Bruce, Ginny. *Indonesia: A Travel Survival Kit.* Australia: Lonely Planet Publications, 1986.

Bruice, Paula Yurkanis. *Organic Chemistry.* Englewood Cliffs, N.J.: Prentice-Hall, 1998.

Cagin, S., and P. Day. *Between Earth and Sky: How CFCs Changed Our World and Endangered the Ozone Layer.* New York: Pantheon Books, 1993.

Campbell, Neil A. *Biology.* Menlo Park, Calif.: Benjamin/Cummings, 1987.

Carey, Francis A. *Organic Chemistry.* New York: McGraw-Hill, 2000.

Caton, Donald. *What a Blessing She Had Chloroform: The Medical and Social Responses to the Pain of Childbirth from 1800 to the Present.* New Haven: Yale University Press, 1999.

Chang, Raymond. *Chemistry.* New York: McGraw-Hill, 1998.

Chester, Ellen. *Woman of Valor: Margaret Sanger and the Birth Control Movement in America.* New York: Simon and Schuster, 1992.

Clow, A., and N. L. Clow. *The Chemical Revolution: A Contribution to Social Technology.* London: Batchworth Press, 1952.

Collier, Richard. *The River That God Forgot: The Story of the Amazon Rubber Boom.* New York: E. P. Dutton, 1968.

Coon, Nelson. *The Dictionary of Useful Plants.* Emmaus, Pa.: Rodale Press, 1974.

Cooper, R. C., and R. C. Cambie. *New Zealand's Economic Native Plants.* Auckland: Oxford University Press, 1991.

Davidson, Basil. *Black Mother: The Years of the African Slave Trade.* Boston: Little, Brown, 1961.

Davis, Lee N. *The Corporate Alchemists: The Power and Problems of the Chemical Industry.* London: Temple-Smith, 1984.

Davis, M. B., J. Austin, and D. A. Partridge. *Vitamin C: Its Chemistry and Biochemistry.* London: Royal Society of Chemistry, 1991.

De Bono, Edward, ed. *Eureka: An Illustrated History of Inventions from the Wheel to the Computer.* New York: Holt, Rinehart, and Winston, 1974.

Delderfield, R. F. *The Retreat from Moscow.* London: Hodder and Stoughton, 1967.

Djerassi, C. *The Pill, Pygmy Chimps and Degas' Horse: The Autobiography of Carl Djerassi.* New York: Harper and Row, 1972.

DuPuy, R. E., and T. N. DuPuy. *The Encyclopedia of Military History from 3500 B.C. to the Present.* Rev. ed. New York: Harper and Row, 1977.

Ege, Seyhan. *Organic Chemistry: Structure and Reactivity.* Lexington, Mass.: D. C. Heath, 1994.

Ellis, Perry. "Overview of Sweeteners." *Journal of Chemical Education* 72, no. 8 (August 1995): 671–75.

Emsley, John. *Molecules at an Exhibition: Portraits of Intriguing Materials in Everyday Life.* New York: Oxford University Press, 1998.

Fairholt, F. W. *Tobacco: Its History and Associations.* Detroit: Singing Tree Press, 1968.

Feltwell, John. *The Story of Silk.* New York: St. Martin's Press, 1990.

Fenichell, S. *Plastic: The Making of a Synthetic Century.* New York: HarperCollins, 1996.

Fessenden, Ralph J., and Joan S. Fessenden. *Organic Chemistry.* Monterey, Calif.: Brooks/Cole, 1986.

Fieser, Louis F., and Mary Fieser. *Advanced Organic Chemistry.* New York: Reinhold, 1961.

Finniston, M., ed. *Oxford Illustrated Encyclopedia of Invention and Technology.* Oxford: Oxford University Press, 1992.

Fisher, Carolyn. "Spices of Life." *Chemistry in Britain* (January 2002).

Fox, Marye Anne, and James K. Whitesell. *Organic Chemistry.* Sudbury: Jones and Bartlett, 1997.

Frankforter, A. Daniel. *The Medieval Millennium: An Introduction.* Englewood Cliffs, N.J.: Prentice-Hall, 1998.

Garfield, Simon. *Mauve: How One Man Invented a Colour That Changed the World.* London: Faber and Faber, 2000.

Gilbert, Richard. *Caffeine, the Most Popular Stimulant: Encyclopedia of Psychoactive Drugs.* London: Burke, 1988.

Goodman, Sandra. *Vitamin C: The Master Nutrient.* New Canaan, Conn.: Keats, 1991.

Gottfried, Robert S. *The Black Death: Natural and Human Disaster in Medieval Europe.* New York: Macmillan, 1983.

Harris, Nathaniel. *History of Ancient Greece.* London: Hamlyn, 2000.

Heiser, Charles B., Jr. *The Fascinating World of the Nightshades: Tobacco, Mandrake, Potato, Tomato, Pepper, Eggplant, etc.* New York: Dover, 1987.

Herold, J. Christopher. *The Horizon Book of the Age of Napoleon.* New York: Bonanza Books, 1983.

Hildebrand, J. H., and R. E. Powell. *Reference Book of Inorganic Chemistry.* New York: Macmillan, 1957.

Hill, Frances. *A Delusion of Satan: The Full Story of the Salem Witch Trials.* London: Hamish Hamilton, 1995.

Hough, Richard. *Captain James Cook: A Biography.* New York: W. W. Norton, 1994.

Huntford, Roland. *Scott and Amundsen (The Last Place on Earth).* London: Hodder and Stoughton, 1979.

Inglis, Brian. *The Opium Wars.* New York: Hodder and Stoughton, 1976.

Jones, Maitland, Jr. *Organic Chemistry.* New York: W. W. Norton, 1997.

Kauffman, George B. "Historically Significant Coordination Compounds. 1. Alizarin dye." *Chem 13 News* (May 1988).

Kauffman, George B., and Raymond B. Seymour. "Elastomers. 1. Natural Rubber." *Journal of Chemical Education* 67, no. 5 (May 1990): 422–25.

Kaufman, Peter B. *Natural Products from Plants.* Boca Raton, Fla.: CRC Press, 1999.

Kolander, Cheryl. *A Silk Worker's Notebook.* Colo.: Interweave Press, 1985.

Kotz, John C., and Paul Treichel, Jr. *Chemistry and Chemical Reactivity.* Orlando, Fla.: Harcourt Brace College, 1999.

Kurlansky, Mark. *Salt: A World History.* Toronto: Alfred A. Knopf Canada, 2002.

Lanman, Jonathan T. *Glimpses of History from Old Maps: A Collector's View.* Tring, Eng.: Map Collector, 1989.

Latimer, Dean, and Jeff Goldberg. *Flowers in the Blood: The Story of Opium.* New York: Franklin Watts, 1981.

Lehninger, Albert L. *Biochemistry: The Molecular Basis of Cell Structure and Function.* New York: Worth, 1975.

Lewis, Richard J. *Hazardous Chemicals Desk Reference.* New York: Van Nostrand Reinhold, 1993.

Loudon, G. Marc. *Organic Chemistry.* Menlo Park, Calif.: Benjamin/Cummings, 1988.

MacDonald, Gayle. "Mauve with the Times." *Toronto Globe and Mail,* April 28, 2001.

Magner, Lois N. *A History of Life Sciences.* New York: Marcel Dekker, 1979.

Manchester, William. *A World Lit Only by Fire: The Medieval Mind and the Renaissance: Portrait of an Age.* Boston: Little, Brown, 1992.

Mann, John. *Murder, Magic and Medicine.* Oxford: Oxford University Press, 1992.

McGee, Harold. *On Food and Cooking: The Science and Lore of the Kitchen.* New York: Charles Scribner's Sons, 1984.

McKenna, Terence. *Food of the Gods.* New York: Bantam Books, 1992.

McLaren, Angus. *A History of Conception from Antiquity to the Present Day.* Oxford: Basil Blackwell, 1990.

McMurry, John. *Organic Chemistry.* Monterey, Calif.: Brooks/Cole, 1984.

Meth-Cohn, Otto, and Anthony S. Travis. "The Mauveine Mystery." *Chemistry in Britain* (July 1995): 547–49.

Miekle, Jeffrey L. *American Plastic: A Cultural History.* New Brunswick, N.J.: Rutgers University Press, 1995.

Milton, Giles. *Nathaniel's Nutmeg.* New York: Farrar, Straus and Giroux, 1999.

Mintz, Sidney W. *Sweetness and Power: The Place of Sugar in Modern History.* New York: Viking Penguin, 1985.

Multhauf, R. P. *Neptune's Gift: A History of Common Salt.* Baltimore, Md.: Johns Hopkins University Press, 1978.

Nikiforuk, Andrew. *The Fourth Horseman: A Short History of Epidemics, Plagues, Famine and Other Scourges.* Toronto: Penguin Books Canada, 1992.

Noller, Carl R. *Chemistry of Organic Compounds.* Philadelphia: W. B. Saunders, 1966.

Orton, James M., and Otto W. Neuhaus. *Human Biochemistry.* St. Louis: C. V. Mosby, 1975.

Pakenham, Thomas. *The Scramble for Africa: 1876–1912.* London: Weidenfeld and Nicolson, 1991.

Pauling, Linus. *Vitamin C, the Common Cold and the Flu.* San Francisco: W. H. Freeman, 1976.

Pendergrast, Mark. *Uncommon Grounds: The History of Coffee and How It Transformed the World.* New York: Basic Books, 1999.

Peterson, William. *Population.* New York: Macmillan, 1975.

Radel, Stanley R., and Marjorie H. Navidi. *Chemistry.* St. Paul, Minn.: West, 1990.

Rayner-Canham, G., P. Fisher, P. Le Couteur, and R. Raap. *Chemistry: A Second Course.* Reading, Mass.: Addison-Wesley, 1989.

Robbins, Russell Hope. *The Encyclopedia of Witchcraft and Demonology.* New York: Crown, 1959.

Roberts, J. M. *The Pelican History of the World.* Middlesex: Penguin Books, 1980.

Rodd, E. H. *Chemistry of Carbon Compounds.* 5 vols. Amsterdam: Elsevier, 1960.

Rosenblum, Mort. *Olives: The Life and Lore of a Noble Fruit.* New York: North Point Press, 1996.

Rudgley, Richard. *Essential Substances: A Cultural History of Intoxicants in Society.* New York: Kodansha International, 1994.

Russell, C. A., ed. *Chemistry, Society and the Environment: A New History of the British Chemical Industry.* Cambridge: Royal Society of Chemistry.

Savage, Candace. *Witch: The Wild Ride from Wicked to Wicca.* Vancouver, B.C.: Douglas and McIntyre, 2000.

Schivelbusch, Wolfgang. *Tastes of Paradise: A Social History of Spices, Stimulants, and Intoxicants.* Translated by David Jacobson. New York: Random House, 1980.

Schmidt, Julius. Rev. and ed. by Neil Campbell. *Organic Chemistry.* London: Oliver and Boyd, 1955.

Seymour, R. B., ed. *History of Polymer Science and Technology.* New York: Marcel Dekker, 1982.

Snyder, Carl H. *The Extraordinary Chemistry of Ordinary Things.* New York: John Wiley and Sons, 1992.

Sohlman, Ragnar, and Henrik Schuck. *Nobel, Dynamite and Peace.* New York: Cosmopolitan, 1929.

Solomons, Graham, and Craig Fryhle. *Organic Chemistry.* New York: John Wiley and Sons, 2000.

Stamp, L. Dudley. *The Geography of Life and Death.* Ithaca, N.Y.: Cornell University Press, 1964.

Stine, W. R. *Chemistry for the Consumer.* Boston: Allyn and Bacon, 1979.

Strecher, Paul G. *The Merck Index: An Encyclopedia of Chemicals and Drugs.* Rahway, N.J.: Merck, 1968.

Streitwieser, Andrew, Jr., and Clayton H. Heathcock. *Introduction to Organic Chemistry.* New York: Macmillan, 1981.

Styer, Lubert. *Biochemistry.* San Francisco: W. H. Freeman, 1988.

Summers, Montague. *The History of Witchcraft and Demonology.* Castle Books, 1992.

Tannahill, Reay. *Food in History.* New York: Stein and Day, 1973.

Thomlinson, Ralph. *Population Dynamics: Causes and Consequences of World Demographic Changes.* New York: Random House, 1976.

Time-Life Books, ed. *Witches and Witchcraft: Mysteries of the Unknown.* Virginia: Time-Life Books, 1990.

Travis, A. S. *The Rainbow Makers: The Origins of the Synthetic Dyestuffs Industry in Western Europe.* London and Toronto: Associated University Presses, 1993.

Visser, Margaret. *Much Depends on Dinner: The Extraordinary History and Mythology, Allure and Obsessions. Perils and Taboos of an Ordinary Meal.* Toronto: McClelland and Stewart, 1986.

Vollhardt, Peter C., and Neil E. Schore. *Organic Chemistry: Structure and Function.* New York: W. H. Freeman, 1999.

Watts, Geoff. "Twelve Scurvy Men." *New Scientist* (February 24, 2001): 46–47.

Watts, Sheldon. *Epidemics and History: Disease, Power and Imperialism.* Wiltshire: Redwood Books, 1997.

Webb, Michael. *Alfred Nobel: Inventor of Dynamite.* Mississauga, Canada: Copp Clark Pitman, 1991.

Weinburg, B. A., and B. K. Bealer. *The World of Caffeine: The Science and Culture of the World's Most Popular Drug.* New York: Routledge, 2001.

Wright, James W. *Ecocide and Population.* New York: St. Martin's Press, 1972.

Wright, Lawrence. *Clean and Decent: The Fascinating History of the Bathroom and the Water Closet.* Cornwall: T.J. Press (Padstow), 1984.

INDEX

Index

ABOUT THE AUTHORS

PENNY CAMERON LE COUTEUR, PH.D., teaches chemistry at Capilano College in British Columbia, Canada, where she has been a professor for over thirty years. Winner of a Polysar Award for Outstanding Chemistry Teaching in Canadian Colleges, she was formerly the head of Capilano's chemistry department and was chair of pure and applied sciences. She has written chemistry distance-education courses, co-authored a chemistry textbook, and served as a project adviser in chemistry for universities in eastern Indonesia. She was born in Auckland, New Zealand, and now lives in North Vancouver.

JAY BURRESON, PH.D., has worked as an industrial chemist for a number of years, and has held a National Institutes of Health special fellowship for postdoctoral research into marine natural products at the University of Hawaii. He is presently general manager of a high-tech company and lives near Corvallis, Oregon.